LECTURES ON
ALGEBRAIC
TOPOLOGY

LECTURES ON ALGEBRAIC TOPOLOGY

HAYNES MILLER

Massachusetts Institute of Technology, USA

NEW JERSEY · LONDON · SINGAPORE · BEIJING · SHANGHAI · HONG KONG · TAIPEI · CHENNAI · TOKYO

Published by

World Scientific Publishing Co. Pte. Ltd.

5 Toh Tuck Link, Singapore 596224

USA office: 27 Warren Street, Suite 401-402, Hackensack, NJ 07601

UK office: 57 Shelton Street, Covent Garden, London WC2H 9HE

Library of Congress Cataloging-in-Publication Data

Names: Miller, Haynes R., 1948–   author.

Title: Lectures on algebraic topology / Haynes Miller, Massachusetts Institute of Technology, USA.

Description: New Jersey : World Scientific Publishing Co., [2022] |
   Includes bibliographical references and index.

Identifiers: LCCN 2021038221 | ISBN 9789811231247 (hardcover) |
   ISBN 9789811232855 (paperback) | ISBN 9789811231254 (ebook) |
   ISBN 9789811231261 (ebook other)

Subjects: LCSH: Algebraic topology.

Classification: LCC QA612 .M5215 2022 | DDC 514/.2--dc23

LC record available at https://lccn.loc.gov/2021038221

British Library Cataloguing-in-Publication Data

A catalogue record for this book is available from the British Library.

The graphic on the cover comes from mathcurve.com.

For any available supplementary material, please visit
https://www.worldscientific.com/worldscibooks/10.1142/12132#t=suppl

Desk Editor: Liu Yumeng

Printed in Singapore

To Juli

Preface

Algebraic topology is a fundamental and unifying discipline. It was the birthplace of many ideas pervading mathematics today, and its methods are ever more widely utilized.

These notes record lectures in a year-long graduate course at MIT, as presented in 2016–2017. The second semester was given again in the spring of 2020. My goal was to give a pretty standard classical approach to this subject, but with an eye to more recent perspectives. I wanted to introduce students to the basic language of category theory, homological algebra, and simplicial sets, so useful throughout mathematics and finding their first real manifestations in algebraic topology. On the other hand I barely touched on some important subjects. I did not talk about simplicial complexes at all, nor about the Lefschetz fixed point theorem. I gave only a brief summary of the theory of covering spaces and the fundamental group, which are regarded as prerequisites for this course.

In the first part, I especially wanted to give an honest account of the machinery — relative cap product and Čech cohomology — needed in the proof of Poincaré duality. The present document contains a bit more detail on these matters than was presented in the course itself. The 2020 course was disrupted by the COVID-19 pandemic, and the entire simplicial development of classifying spaces, Lectures 57–59, were consequently omitted. The pace picked up speed as the course went along, and we ended with a cursory treatment of Thom's work on cobordism. In this second half, I probably didn't cover quite as much in the lectures as is written in this text.

This is a volume of lecture notes, not a textbook! (There are good ones: [10, 15, 24, 36, 50, 69, 72] for example.) I have been inspired by the admirable examples set by the authors of [14] and [59]. I have opted for variety rather than completeness. Most lectures conclude with a series of exercises, most of which were actually assigned as part of the course. They vary widely in difficulty.

I was lucky enough to have in the audience a student, Sanath Devalapurkar, who spontaneously decided to liveTEX the entire course. This resulted in a remarkably accurate record of what happened in the classroom — right down to random alarms ringing and embarrassing jokes and mistakes on the blackboard. Sanath's TEX forms the basis of these notes, and I am grateful to him for making them available. The attractive drawings in the first half were provided by another student, Xianglong Ni, who also carefully proofread the manuscript. Chapters 4–8 reflect the 2020 class and so depart more from the original notes.

In addition to Sanath and Xianglong, I am delighted to thank the generations of students who have kept me on track and honest over several decades of teaching this subject. I owe a particular debt to Phil Hirschhorn, Stefan Jackowski, Calder Morton-Ferguson, Timothy Ngotiaoco, and Manuel Rivera, each of whom pointed out errors in the text and suggested corrections. Of course many inaccuracies are guaranteed to remain, a reality for which I apologize.

I am grateful to mathcurve.com for the beautiful image of Boy's surface that graces the cover of this book. See Exercise 71.11 for more on this immersion.

Newton, MA
December, 2020

Contents

Chapter 1

Singular homology

1 Introduction: singular simplices and chains

This is a course on algebraic topology. The objects of study are of course topological spaces, and the machinery we develop in this course is designed to be applicable to a general space. But we are really mainly interested in geometrically important spaces. Here are some examples.

- The most basic example is n-*dimensional Euclidean space*, $\mathbb{R}^n$.
- The n-*sphere* $S^n = \{x \in \mathbb{R}^{n+1} : |x| = 1\}$, topologized as a subspace of $\mathbb{R}^{n+1}$.
- Identifying antipodal points in S^n gives *real projective space* $\mathbb{RP}^n = S^n/(x \sim -x)$, i.e. the space of lines through the origin in $\mathbb{R}^{n+1}$.
- Call an ordered collection of k orthonormal vectors in a real inner product space an *orthonormal k-frame*. The space of orthonormal k-frames in $\mathbb{R}^n$, topologized as a subspace of $(S^{n-1})^k$, forms the *Stiefel manifold* $V_k(\mathbb{R}^n)$. For example, $V_1(\mathbb{R}^n) = S^{n-1}$.
- The *Grassmannian* $\mathrm{Gr}_k(\mathbb{R}^n)$ is the space of k-dimensional linear subspaces of $\mathbb{R}^n$. Forming the span gives us a surjection $V_k(\mathbb{R}^n) \to \mathrm{Gr}_k(\mathbb{R}^n)$, and the Grassmannian is given the quotient topology. For example, $\mathrm{Gr}_1(\mathbb{R}^n) = \mathbb{RP}^{n-1}$.

All these examples are *manifolds*; that is, they are Hausdorff spaces locally homeomorphic to Euclidean space. Aside from $\mathbb{R}^n$ itself, they are also compact. Such spaces exhibit a hidden symmetry, which is the culmination of the first half of this course: Poincaré duality.

1

Simplices and chains

As the name suggests, the central aim of algebraic topology is the usage of algebraic tools to study topological spaces. A common technique is to probe topological spaces via maps to them from simpler spaces. In different ways, this approach gives rise to singular homology and homotopy groups. We now detail the former; the latter takes the stage in the second half.

Definition 1.1. For $n \geq 0$, the *standard n-simplex* Δ^n is the convex hull of the standard basis $\{e_0, \ldots, e_n\}$ in $\mathbb{R}^{n+1}$:

$$\Delta^n = \left\{ \sum t_i e_i : \sum t_i = 1, t_i \geq 0 \right\} \subseteq \mathbb{R}^{n+1}.$$

Each e_i is a *vertex* of the simplex (plural "vertices"). The t_i are called *barycentric coordinates*.

The word "simplex" comes from the Latin, and should suggest "simple" in the sense of "not compound." In mathematics its plural is always "simplices." There is a well-developed theory of *simplicial complexes*, appropriately organized unions of simplices, which however we will not develop in these lectures. Here the word "complex" is used, as it is in "complex number," not to denote complexity but rather "compound" (of real and imaginary parts, in the case of numbers).

The standard simplices are related by face inclusions $d^i \colon \Delta^{n-1} \to \Delta^n$ for $0 \leq i \leq n$, where d^i is the affine map that sends vertices to vertices, in order, and omits the vertex e_i.

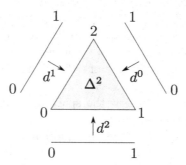

Definition 1.2. Let X be any topological space. A *singular n-simplex* in X is a continuous map $\sigma : \Delta^n \to X$. We will often drop the adjective "singular." Denote by $\mathrm{Sin}_n(X)$ the set of all n-simplices in X.

This seems like a rather bold construction to make, as $\mathrm{Sin}_n(X)$ is huge. But be patient! For the moment, notice the peculiar use of the word "singular." It derives from the notion that the image of the map σ might have cusps or kinks or other kinds of "singularities" — another specialized mathematical term, indicating that these points are unusual and special.

For $0 \le i \le n$, precomposition by the face inclusion d^i produces a map $d_i \colon \mathrm{Sin}_n(X) \to \mathrm{Sin}_{n-1}(X)$ sending σ to $\sigma \circ d^i$. This is the "ith face" of σ. This allows us to make sense of the "boundary" of a simplex.

For example, if σ is a 1-simplex that forms a closed loop, then $d_1\sigma = d_0\sigma$. We would like to re-express this equality as a statement that the boundary "vanishes" — we would like to write "$d_0\sigma - d_1\sigma = 0$." Here we don't mean to subtract one point of X from another! Rather we mean to form a "formal difference." To accommodate such formal sums and differences, we will enlarge $\mathrm{Sin}_n(X)$ still further by forming the free abelian group it generates.

Definition 1.3. The abelian group $S_n(X)$ of *singular n-chains* in X is the free abelian group generated by n-simplices,

$$S_n(X) = \mathbb{Z}\mathrm{Sin}_n(X) \,.$$

So an n-chain is a finite linear combination of n-simplices,

$$\sum_{i=1}^{k} a_i\sigma_i \,, \quad a_i \in \mathbb{Z}, \quad \sigma_i \in \mathrm{Sin}_n(X) \,.$$

If $n < 0$, $\mathrm{Sin}_n(X)$ is declared to be empty, so $S_n(X) = 0$.

We can now define the *boundary operator*

$$d \colon \mathrm{Sin}_n(X) \to S_{n-1}(X) \,,$$

by

$$d\sigma = \sum_{i=0}^{n} (-1)^i d_i\sigma \,.$$

This extends to a homomorphism $d \colon S_n(X) \to S_{n-1}(X)$ by additivity.

We use this homomorphism to obtain something more tractable than the entirety of $S_n(X)$. First we restrict our attention to chains with vanishing boundary.

Definition 1.4. An *n-cycle* in X is an n-chain c with $dc = 0$. An n-chain is a *boundary* if it is in the image of $d \colon S_{n+1}(X) \to S_n(X)$. Notation:

$$Z_n(X) = \ker(d \colon S_n(X) \to S_{n-1}(X)) \,,$$

$$B_n(X) = \mathrm{im}(d \colon S_{n+1}(X) \to S_n(X)) \,.$$

For example, a 1-simplex is a cycle if its ends coincide. More generally, a sum of 1-simplices is a cycle if the right endpoints match up with the left endpoints. Geometrically, you get a collection of loops, or "cycles." This is the origin of the term "cycle."

Every 0-chain is a cycle, since $S_{-1}(X) = 0$.

Singular homology

It turns out that there's a cheap way to produce cycles:

Theorem 1.5. *Any boundary is a cycle; that is, $d^2 = 0$.*

We'll leave the verification of this important result as a homework problem. What we have found, then, is that the singular chains form a "chain complex," as in the following definition.

Definition 1.6. A *graded abelian group* is a sequence of abelian groups, indexed by the integers. A *chain complex* is a graded abelian group $\{A_n\}$ together with homomorphisms $d : A_n \to A_{n-1}$ with the property that $d^2 = 0$.

We have just defined the *singular chain complex* $S_*(X)$ of a space X.

The chains that are cycles by virtue of being boundaries are the "cheap" ones. If we quotient by them, what's left is the "interesting cycles," captured in the following definition.

Definition 1.7. The *nth singular homology group* of X is:

$$H_n(X) = \frac{Z_n(X)}{B_n(X)} = \frac{\ker(d : S_n(X) \to S_{n-1}(X))}{\operatorname{im}(d : S_{n+1}(X) \to S_n(X))}.$$

We use the same language for any chain complex: it has cycles, boundaries, and homology groups. The homology forms a graded abelian group. Two cycles that differ by a boundary are said to be *homologous*. (The word "homology" arose first in biology to indicate a shared evolutionary origin.)

Both $Z_n(X)$ and $B_n(X)$ are free abelian groups because they are subgroups of the free abelian group $S_n(X)$, but the quotient $H_n(X)$ isn't necessarily free. While $Z_n(X)$ and $B_n(X)$ are uncountably generated, $H_n(X)$ turns out to be finitely generated for the spaces we are interested in! If T is the torus, for example, then we will see that $H_1(T) \cong \mathbb{Z} \oplus \mathbb{Z}$, with generators given by the 1-cycles illustrated below.

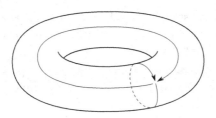

We will learn to compute the homology groups of a wide variety of spaces. The n-sphere for example has the following homology groups:

$$H_q(S^n) = \begin{cases} \mathbb{Z} & \text{if } q = n > 0 \\ \mathbb{Z} & \text{if } q = 0, n > 0 \\ \mathbb{Z} \oplus \mathbb{Z} & \text{if } q = n = 0 \\ 0 & \text{otherwise}. \end{cases}$$

There is an interesting n-cycle, that, roughly speaking, covers every point of the sphere exactly once. Any q-cycle with q different from 0 and n is a boundary, so it doesn't contribute to the homology.

Exercises

Exercise 1.8. (a) Let $[n]$ denote the totally ordered set $\{0, 1, \ldots, n\}$. Let $\phi : [m] \to [n]$ be an order preserving function (so that if $i \leq j$ then $\phi(i) \leq \phi(j)$). Identifying the elements of $[n]$ with the vertices of the standard simplex Δ^n, ϕ extends to an affine map $\Delta^m \to \Delta^n$ that we also denote by ϕ. Give a formula for this map in terms of barycentric coordinates: If we write $\phi(s_0, \ldots, s_m) = (t_0, \ldots, t_n)$, what is t_j as a function of $(s_0, \ldots, s_m)$?

(b) Write $d^j : [n-1] \to [n]$ for the order preserving injection that omits j as a value. Show that an order preserving injection $\phi : [n-k] \to [n]$ is uniquely a composition of the form $d^{j_k} d^{j_{k-1}} \cdots d^{j_1}$, with $0 \leq j_1 < j_2 < \cdots < j_k \leq n$. Do this by describing the integers $j_1, \ldots, j_k$ directly in terms of ϕ, and then verify the straightening rule

$$d^i d^j = d^{j+1} d^i \quad \text{for} \quad i \leq j.$$

(c) Show that any order preserving map $\phi : [m] \to [n]$ factors uniquely as the composition of an order preserving surjection followed by an order preserving injection.

(d) Write $s^i : [m+1] \to [m]$ for the order-preserving surjection that repeats the value i. Show that any order-preserving surjection $\phi : [m] \to [n]$ has a unique expression $s^{i_1} s^{i_2} \cdots s^{i_k}$ with $n \geq i_1 \geq i_2 \geq \cdots i_k \geq 0$. Do this by describing the numbers $i_1, \ldots, i_k$, directly in terms of ϕ, and finding a straightening rule of the form $s^i s^j = \cdots$ for $i < j$.

(e) Finally, implement your assertion that any order preserving map factors as a surjection followed by an injection by establishing a straightening rule of the form $s^i d^j = \cdots$.

Recall the notation $\mathrm{Sin}_n(X)$ for the set of continuous maps from Δ^n to the space X. The affine extension $\phi : \Delta^m \to \Delta^n$ of an order-preserving map $\phi : [m] \to [m]$ induces a map $\phi^* : \mathrm{Sin}_n(X) \to \mathrm{Sin}_m(X)$. In particular, write

$$d_i = (d^i)^* \qquad s_j = (s^j)^*.$$

The d_i's are *face maps*, the s_i's are *degeneracies*.

(f) Write down the identities satisfied by these operators, resulting from the identities you found relating the d^i's and s^j's.

A *simplicial set* is a sequence of sets $K_0, K_1, \ldots$, with maps $d_i : K_n \to K_{n-1}$, $0 \leq i \leq n$, and $s_i : K_n \to K_{n+1}$, $0 \leq i \leq n$, satisfying these identities. The elements of K_n are the "n-simplices" of K; when $n = 0$ they are the "vertices" of K. For example, we have the *singular simplicial set* $\mathrm{Sin}_*(X)$ of a space X.

(g) Use the relations among the d_i's to prove that

$$d^2 = 0 : S_n(X) \to S_{n-2}(X).$$

2 Homology

In the last lecture we introduced the standard n-simplex $\Delta^n \subseteq \mathbb{R}^{n+1}$. Singular simplices in a space X are maps $\sigma : \Delta^n \to X$ and constitute the set $\mathrm{Sin}_n(X)$. For example, $\mathrm{Sin}_0(X)$ consists of points of X. We also described the face inclusions $d^i : \Delta^{n-1} \to \Delta^n$, and the induced "face maps"

$$d_i : \mathrm{Sin}_n(X) \to \mathrm{Sin}_{n-1}(X), \quad 0 \leq i \leq n,$$

given by precomposing with face inclusions: $d_i \sigma = \sigma \circ d^i$. For homework you established some quadratic relations satisfied by these maps. A collection of sets $K_n, n \geq 0$, together with maps $d_i : K_n \to K_{n-1}$ related to each other in this way, is a *semi-simplicial set*. So we have assigned to any space X a semi-simplicial set $S_*(X)$. (You actually get a *simplicial* set; but, while the "degeneracies" will ultimately play an important role, they do not enter into the definition of singular homology. Simplicial sets were originally called "complete semi-simplicial complexes"; "semi-simplicial" because they weren't necessarily simplicial complexes, and "complete" because they included degeneracies. Current usage recycles the "semi-" to mean that only the face maps are used, not the degeneracies.)

To the semi-simplicial set $\{\mathrm{Sin}_n(X), d_i\}$ we then applied the free abelian group functor, obtaining a semi-simplicial abelian group. Forming alternating sums of the d_is, we constructed a boundary map d which makes $S_*(X)$ a *chain complex* — that is, $d^2 = 0$. We capture this process in a diagram:

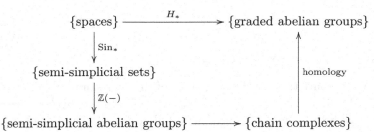

Example 2.1. Suppose we have $\sigma\colon \Delta^1 \to X$. Define $\phi\colon \Delta^1 \to \Delta^1$ by sending $(t, 1-t)$ to $(1-t, t)$. Precomposing σ with ϕ gives another singular simplex $\overline{\sigma}$ which reverses the orientation of σ. It is *not* true that $\overline{\sigma} = -\sigma$ in $S_1(X)$.

However, we claim that $\overline{\sigma} \equiv -\sigma \bmod B_1(X)$. This means that there is a 2-chain in X whose boundary is $\overline{\sigma} + \sigma$. If $d_0\sigma = d_1\sigma$, so that $\sigma \in Z_1(X)$, then $\overline{\sigma}$ and $-\sigma$ are homologous cycles, so that $[\overline{\sigma}] = -[\sigma]$ in $H_1(X)$.

To construct an appropriate "homology" — a 2-chain τ with the property that $d\tau = \overline{\sigma} + \sigma$ — consider the projection map $\pi : \Delta^2 \to \Delta^1$ that is the affine extension of the map sending e_0 and e_2 to e_0 and e_1 to e_1. (Incidentally this is not a degeneracy since it is not order-preserving.)

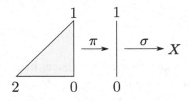

We'll compute $d(\sigma \circ \pi)$. Some of the terms will be constant singular simplices. Let's write

$$c_x^n : \Delta^n \to X$$

for the constant map with value $x \in X$. Then

$$d(\sigma \circ \pi) = \sigma \pi d^0 - \sigma \pi d^1 + \sigma \pi d^2 = \overline{\sigma} - c^1_{\sigma(0)} + \sigma \, .$$

The constant simplex $c^1_{\sigma(0)}$ is an "error term," and we wish to eliminate it. To achieve this we can use the constant 2-simplex $c^2_{\sigma(0)}$ at $\sigma(0)$; its boundary is

$$c^1_{\sigma(0)} - c^1_{\sigma(0)} + c^1_{\sigma(0)} = c^1_{\sigma(0)} \, .$$

So

$$\overline{\sigma} + \sigma = d(\sigma \circ \pi + c^2_{\sigma(0)}) \, ,$$

and $\overline{\sigma} \equiv -\sigma \bmod B_1(X)$ as claimed.

Let's compute the singular homology of the very simplest spaces, $\varnothing$ and $*$. For the first, $\mathrm{Sin}_n(\varnothing) = \varnothing$, so $S_n(\varnothing) = 0$. Hence $S_*(\varnothing)$ is the zero chain complex. This means that $Z_*(\varnothing) = B_*(\varnothing) = 0$. The homology in all dimensions is therefore 0.

For $*$, we have $\mathrm{Sin}_n(*) = \{c^n_*\}$ for all $n \geq 0$. Consequently $S_n(*) = \mathbb{Z}$ for $n \geq 0$ and 0 for $n < 0$. Suppose $n > 0$. For each i, $d_i c^n_* = c^{n-1}_*$, so the boundary maps $d \colon S_n(*) \to S_{n-1}(*)$ in the chain complex depend on the parity of n as follows:

$$d(c^n_*) = \sum_{i=0}^{n}(-1)^i c^{n-1}_* = \begin{cases} c^{n-1}_* & \text{for } n \text{ even, and} \\ 0 & \text{for } n \text{ odd.} \end{cases}$$

This means that our chain complex is:

$$0 \leftarrow \mathbb{Z} \xleftarrow{0} \mathbb{Z} \xleftarrow{1} \mathbb{Z} \xleftarrow{0} \mathbb{Z} \xleftarrow{1} \cdots \, .$$

The boundaries coincide with the cycles except in dimension zero, where $B_0(*) = 0$ while $Z_0(*) = \mathbb{Z}$. Therefore $H_0(*) = \mathbb{Z}$ and $H_i(*) = 0$ for $i \neq 0$.

Induced maps

We've defined homology groups for each space, but haven't yet considered what happens to maps between spaces. A continuous map $f \colon X \to Y$ induces a map $f_* \colon \mathrm{Sin}_n(X) \to \mathrm{Sin}_n(Y)$ by composition:

$$f_* : \sigma \mapsto f \circ \sigma \, .$$

We claim that f_* is a map of semi-simplicial sets; that is, it commutes with face maps: $f_* \circ d_i = d_i \circ f_*$. The four maps involved in this equality form the arrows in the diagram

$$\begin{array}{ccc} \text{Sin}_n(X) & \xrightarrow{f_*} & \text{Sin}_n(Y) \\ \downarrow{\scriptstyle d_i} & & \downarrow{\scriptstyle d_i} \\ \text{Sin}_{n-1}(X) & \xrightarrow{f_*} & \text{Sin}_{n-1}(Y) \end{array}$$

which also displays their sources and targets. A diagram like this is said to "commute" or to "be commutative" if any two directed paths with the same source and target are equal. So we want to see that this diagram is commutative.

Well, $d_i f_* \sigma = (f_* \sigma) \circ d^i = f \circ \sigma \circ d^i$, while $f_*(d_i \sigma) = f_*(\sigma \circ d^i) = f \circ \sigma \circ d^i$. The diagram remains commutative when we pass to the free abelian groups of chains.

If C_* and D_* are chain complexes, a *chain map* $f \colon C_* \to D_*$ is a collection of homomorphisms $f_n \colon C_n \to D_n$ such that the following diagram commutes for every n:

$$\begin{array}{ccc} C_n & \xrightarrow{f_n} & D_n \\ \downarrow{\scriptstyle d_C} & & \downarrow{\scriptstyle d_D} \\ C_{n-1} & \xrightarrow{f_{n-1}} & D_{n-1} \end{array}$$

For example, we just showed that if $f \colon X \to Y$ is a continuous map then $f_* \colon S_*(X) \to S_*(Y)$ is a chain map.

A chain map induces a map in homology, $f_* \colon H_n(C) \to H_n(D)$. The method of proof is a "diagram chase" and it will be the first of many. We check that we get a map $Z_n(C) \to Z_n(D)$. Let $c \in Z_n(C)$, so that $d_C c = 0$. Then $d_D f_n(c) = f_{n-1} d_C c = f_{n-1}(0) = 0$, because f is a chain map. This means that $f_n(c)$ is also an n-cycle, i.e., f gives a map $Z_n(C) \to Z_n(D)$.

Similarly, f_* sends $B_n(C)$ to $B_n(D)$: Let $c \in B_n(C)$, so that there exists $c' \in C_{n+1}$ such that $d_C c' = c$. Then $f_n(c) = f_n d_C c' = d_D f_{n+1}(c')$. Thus $f_n(c)$ is the boundary of $f_{n+1}(c')$, and f gives a map $B_n(C) \to B_n(D)$.

We have another commutative diagram! —

$$\begin{array}{ccc} B_n(C) & \xrightarrow{f_*} & B_n(D) \\ \uparrow & & \uparrow \\ Z_n(C) & \xrightarrow{f_*} & Z_n(D)\,. \end{array}$$

Forming the quotients gives us a map on homology: $f_* \colon H_n(X) \to H_n(Y)$.

Exercises

Exercise 2.2. Write down a singular 2-cycle representing the "fundamental class" of the torus $T^2 = S^1 \times S^1$. We will give a precise definition of the fundamental class of a manifold later, but for now let's just say that this cycle should be made up of singular 2-simplices that together cover all but a small (e.g. nowhere dense) subset of T^2 exactly once.

Exercise 2.3. Construct an isomorphism

$$H_n(X) \oplus H_n(Y) \to H_n(X \amalg Y).$$

3   Categories, functors, and natural transformations

From spaces and continuous maps, we constructed graded abelian groups and homomorphisms. We now recast this kind of construction in the more general language of category theory. This is a very general framework for discussing relationships between mathematical structures. It was formalized by Samuel Eilenberg and Saunders Mac Lane in 1945. Both did much to establish the foundations of algebraic topology. Mac Lane (1909–2005) founded a school of topology at the University of Chicago. Born in Poland, "Sammy" Eilenberg (1913–1998) worked at Columbia University and in addition to his work with Mac Lane he collaborated with Norman Steenrod to write the founding document [18] in modern algebraic topology and, with Henri Cartan, an equally definitive book [13] on homological algebra.

Our discussion of category theory will be interspersed throughout the text, introducing new concepts as they are needed. Here we begin by introducing the basic definitions.

Definition 3.1. A *category* $\mathcal{C}$ consists of the following data.

- a class ob($\mathcal{C}$) of *objects*;
- for every pair of objects X and Y, a set of *morphisms* $\mathcal{C}(X, Y)$;
- for every object X an *identity morphism* $1_X \in \mathcal{C}(X, X)$; and
- for every triple of objects X, Y, Z, a *composition* map $\mathcal{C}(Y, Z) \times \mathcal{C}(X, Y) \to \mathcal{C}(X, Z)$, written $(g, f) \mapsto g \circ f$.

These data are required to satisfy the following two properties:

- For $\in \mathcal{C}(X, Y)$, $1_Y \circ f = f$ and $f \circ 1_X = f$.
- Composition is associative: $(h \circ g) \circ f = h \circ (g \circ f)$.

Note that we allow the collection of objects to be a class. This enables us to talk about a "category of all sets" for example. But we require each $\mathcal{C}(X,Y)$ to be set, and not merely a class. Some interesting categories have a *set* of objects; they are called *small categories*.

We will often write $X \in \mathcal{C}$ to mean that X is an object of $\mathcal{C}$, and $f\colon X \to Y$ to mean $f \in \mathcal{C}(X,Y)$.

Definition 3.2. If $X, Y \in \mathcal{C}$, then $f\colon X \to Y$ is an *isomorphism* if there exists $g\colon Y \to X$ with $f \circ g = 1_Y$ and $g \circ f = 1_X$. We may write

$$f : X \xrightarrow{\cong} Y$$

to indicate that f is an isomorphism.

It's easy to see that g is unique if it exists; it's the "inverse" of f.

Example 3.3. Many common mathematical structures can be arranged in categories.

- Sets and functions between them form a category **Set**.
- Abelian groups and homomorphisms form a category **Ab**.
- Topological spaces and continuous maps form a category **Top**.
- Chain complexes and chain maps form a category ch**Ab**.
- A monoid is the same as a category with one object, where the elements of the monoid are the morphisms in the category. It's a small category.
- The totally ordered sets $[n] = \{0, \ldots, n\}$ for $n \geq 0$ together with weakly order-preserving maps between them form the *simplex category* $\mathbf{\Delta}$, another small category. It contains as a subcategory the *semi-simplex category* $\mathbf{\Delta}_{inj}$ with the same objects but only injective order-preserving maps.
- A partially ordered set or "poset" forms a category in which there is a morphism from x to y iff $x \leq y$. A small category is a poset exactly when (1) there is at most one morphism between any two objects, and (2) the only isomorphisms are identities. This is to be distinguished from the category of posets and order-preserving maps between them, which is "large."

A *subcategory* of a category $\mathcal{C}$ consists of a collection of objects and morphisms such that the structure maps of $\mathcal{C}$ restrict to give a new category. A subcategory $\mathcal{D} \subseteq \mathcal{C}$ is *full* if whenever $X, Y \in \mathcal{D}$, $\mathcal{D}(X,Y) = \mathcal{C}(X,Y)$. For example, finite sets form a full subcategory of **Set**.

Functors

Categories may be related to each other by rules describing effect on both objects and morphisms.

Definition 3.4. Let $\mathcal{C}, \mathcal{D}$ be categories. A *functor* $F \colon \mathcal{C} \to \mathcal{D}$ consists of the data of

- an assignment $F : \mathrm{ob}(\mathcal{C}) \to \mathrm{ob}(\mathcal{D})$, and
- for all $X, Y \in \mathrm{ob}(\mathcal{C})$, a function $F : \mathcal{C}(X, Y) \to \mathcal{D}(F(X), F(Y))$.

These data are required to satisfy the following two properties:

- For all $X \in \mathrm{ob}(\mathcal{C})$, $F(1_X) = 1_{F(X)} \in \mathcal{D}(F(X), F(X))$, and
- For all composable pairs of morphisms f, g in $\mathcal{C}$, $F(g \circ f) = F(g) \circ F(f)$.

We have defined quite a few functors already:

$$\mathbb{Z} : \mathbf{Set} \to \mathbf{Ab}, \quad (-)_n : \mathrm{ch}\mathbf{Ab} \to \mathbf{Ab},$$

$$\mathrm{Sin}_n : \mathbf{Top} \to \mathbf{Set}, \quad S_n : \mathbf{Top} \to \mathbf{Ab}, \quad H_n : \mathbf{Top} \to \mathbf{Ab},$$

for example. The map $F(f)$ induced by f is often denotes simply f_*, since the name of the functor tends to be already present. So a map of spaces $f : X \to Y$ induces a homomorphism $f_* : H_n(X) \to H_n(Y)$.

We also have defined, for each X, a homomorphism $d : S_n(X) \to S_{n-1}(X)$. This is a "morphism between functors." This property is captured by another definition.

Definition 3.5. Let $F, G \colon \mathcal{C} \to \mathcal{D}$ be two functors. A *natural transformation* or *natural map* $\theta \colon F \to G$ consists of maps $\theta_X \colon F(X) \to G(X)$ for all $X \in \mathcal{C}$ such that for all $f \colon X \to Y$ the following diagram commutes.

$$
\begin{array}{ccc}
F(X) & \xrightarrow{\ \theta_X\ } & G(X) \\
{\scriptstyle F(f)} \downarrow & & \downarrow {\scriptstyle G(f)} \\
F(Y) & \xrightarrow{\ \theta_Y\ } & G(Y)
\end{array}
$$

So for example the boundary map $d \colon S_n \to S_{n-1}$ is a natural transformation of functors $\mathbf{Top} \to \mathbf{Ab}$.

Example 3.6. Suppose that $\mathcal{C}$ and $\mathcal{D}$ are two categories, and assume that $\mathcal{C}$ is small. We may then form the *functor category* $\mathrm{Fun}(\mathcal{C}, \mathcal{D})$. Its objects are the functors from $\mathcal{C}$ to $\mathcal{D}$, and given two functors F, G, $\mathrm{Fun}(\mathcal{C}, \mathcal{D})(F, G)$

is the set of natural transformations from F to G. We let the reader define the rest of the structure of this category, and check the axioms. We assumed that $\mathcal{C}$ is small in order to guarantee that there is no more than a set of natural transformations between functors.

For example, let G be a group (or a monoid) viewed as a one-object category. An object $F \in \mathrm{Fun}(G, \mathbf{Ab})$ is simply a group action of G on $F(*) = A$, i.e. a representation of G in abelian groups. Given another $F' \in \mathrm{Fun}(G, \mathbf{Ab})$ with $F'(*) = A'$, a natural transformation from F to F' is precisely a G-equivariant homomorphism $A \to A'$.

Exercises

Exercise 3.7. Write $\pi_0(X)$ for the set of path-components of a space X. Construct an isomorphism

$$\mathbb{Z}\pi_0(X) \to H_0(X) .$$

Exercise 3.8. Say what it means to assert that the isomorphisms you constructed in Exercise 2.3 and in Exercise 3.7 are *natural*, and make sure that they are.

4   Categorical language

Let Vect_k be the category of vector spaces over a fixed field k, and k-linear transformations between them. Given a vector space V, you can consider the dual $V^* = \mathrm{Hom}(V, k)$. Does this give us a functor? If you have a linear transformation $f : V \to W$, you get a map $f^* : W^* \to V^*$ by sending $\varphi : W \to k$ to $\varphi \circ f : V \to k$. This is like a functor, but the induced map goes the wrong way. This operation does preserve composition and identities, in an appropriate sense. This is an example of a *contravariant functor*.

I'll leave it to you to spell out the definition, but notice that there is a universal example of a contravariant functor out of a category $\mathcal{C}$: $\mathcal{C} \to \mathcal{C}^{op}$, where $\mathcal{C}^{op}$ has the same objects as $\mathcal{C}$, but $\mathcal{C}^{op}(X, Y)$ is declared to be the set $\mathcal{C}(Y, X)$. The identity morphisms remain the same. To describe the composition in $\mathcal{C}^{op}$, I'll write f^{op} for $f \in \mathcal{C}(Y, X)$ regarded as an element of $\mathcal{C}^{op}(X, Y)$; then $f^{op} \circ g^{op} = (g \circ f)^{op}$.

Then a contravariant functor from $\mathcal{C}$ to $\mathcal{D}$ is the same thing as a ("*covariant*") functor from $\mathcal{C}^{op}$ to $\mathcal{D}$.

Let $\mathcal{C}$ be a category, and fix $Y \in \mathcal{C}$. Define a functor $\mathcal{C}^{op} \to \mathbf{Set}$ by sending X to $\mathcal{C}(X, Y)$, and a map $f : W \to X$ to the map $\mathcal{C}(X, Y) \to$

$\mathcal{C}(W, Y)$ sending $\varphi : X \to Y$ to $\varphi \circ f$. This is called the functor *represented by Y*. It is very important to note that $\mathcal{C}(-, Y)$ is contravariant, while, on the other hand, for any fixed X, $\mathcal{C}(X, -)$ is a covariant functor (and is said to be "corepresented" by X). $\mathcal{C}(-, -)$ is a "bifunctor," contravariant in the first variable and covariant in the second.

Example 4.1. Recall from Example 3.3 that the simplex category $\mathbf{\Delta}$ has objects the totally ordered sets $[n] = \{0, 1, \ldots, n\}$, $n \geq 0$, with order preserving maps as morphisms. The "standard simplex" gives us a functor $\Delta : \mathbf{\Delta} \to \mathbf{Top}$. Now fix a space X, and consider

$$[n] \mapsto \mathbf{Top}(\Delta^n, X) \,.$$

This gives us a contravariant functor $\mathbf{\Delta} \to \mathbf{Set}$, or a covariant functor $\mathbf{\Delta}^{op} \to \mathbf{Set}$. This functor carries in it all the face and degeneracy maps we discussed earlier, and their compositions. Let us make a definition.

Definition 4.2. Let $\mathcal{C}$ be any category. A *simplicial object* in $\mathcal{C}$ is a functor $K : \mathbf{\Delta}^{op} \to \mathcal{C}$. Simplicial objects in $\mathcal{C}$ form a category with natural transformations as morphisms. Similarly, *semi-simplicial object* in $\mathcal{C}$ is a functor $\mathbf{\Delta}_{inj}^{op} \to \mathcal{C}$.

So the singular functor Sin_* gives a functor from spaces to simplicial sets (and so, by restriction, to semi-simplicial sets).

I want to interject one more bit of categorical language that will often be useful to us.

Definition 4.3. A morphism $f : X \to Y$ in a category $\mathcal{C}$ is a *split epimorphism* ("split epi" for short) if there exists $g : Y \to X$ (called a *section* or a *splitting*) such that the composite $Y \xrightarrow{g} X \xrightarrow{f} Y$ is the identity.

Example 4.4. In the category of sets, a map $f : X \to Y$ is a split epimorphism exactly when, for every element of Y there exists some element of X whose image in Y is the original element. So f is surjective. Is every surjective map a split epimorphism? This is equivalent to the axiom of choice! because a section of f is precisely a choice of $x \in f^{-1}(y)$ for every $y \in Y$.

Every categorical definition is accompanied by a "dual" definition.

Definition 4.5. A map $g : Y \to X$ is a *split monomorphism* ("split mono" for short) if there is $f : X \to Y$ such that $f \circ g = 1_Y$.

Example 4.6. Again let $C = $ **Set**. Any split monomorphism is an injection: If $y, y' \in Y$, and $g(y) = g(y')$, we want to show that $y = y'$. Apply f, to get $y = f(g(y)) = f(g(y')) = y'$. But the injection $\varnothing \to Y$ is a split monomorphism only if $Y = \varnothing$. So there's an asymmetry in the category of sets.

Lemma 4.7. *A map is an isomorphism if and only if it is both a split epimorphism and a split monomorphism.*

Proof. Easy! $\qquad\qquad\square$

Example 4.8. Suppose $C = $ **Ab**, and you have a split epi $f : A \to B$. Let $g : B \to A$ be a section. We also have the inclusion $i : \ker f \to A$, and hence a map

$$[\,g \quad i\,] : B \oplus \ker f \to A\,.$$

I leave it to you to check that this map is an isomorphism, and to formulate a dual statement.

The importance of these definitions is this: Functors will not in general respect "monomorphisms" or "epimorphisms," but:

Lemma 4.9. *Any functor sends split epis to split epis and split monos to split monos.*

Proof. Apply the functor to the diagram establishing f as a split epi or mono. $\qquad\qquad\square$

Exercises

Exercise 4.10. Here are a couple more "categorical" definitions, giving you some practice with the idea of constructions being defined by universal mapping properties.

Let C be a category, A a set, and $a \mapsto X_a$ an assignment of an object of C to each element of A. A *product* of these objects is an object Y together with maps $\mathrm{pr}_a : Y \to X_a$ with the following property. For any object Z and any family of maps $f_a : Z \to X_a$, there is a unique map $Z \to Y$ such that $f_a = \mathrm{pr}_a \circ f$ for all $a \in A$. A *coproduct* of these objects is an object Y together with maps $\mathrm{in}_a : X_a \to Y$ with the following property. For any object Z and any family of maps $f_a : X_a \to Z$, there is a unique map $Y \to Z$ such that $f_a = f \circ \mathrm{in}_a$ for all $a \in A$.

(a) Describe constructions of the product and coproduct (if they exist) in the following categories: sets, pointed sets, spaces, abelian groups. (A *pointed set* is a pair $(S, *)$ where S is a set and $* \in S$.)

(b) What should be meant by the product when $A = \varnothing$? How about the coproduct? What are these objects in the four categories mentioned in (a)? Give an example of a category in which neither one of these constructions exists.

(c) Show that if $(Y, \{\mathrm{pr}_a\})$ and $(Y', \{\mathrm{pr}'_a\})$ are both products of a family $\{X_a : a \in A\}$, then there is a unique map $f : Y \to Y'$ such that $\mathrm{pr}'_a \circ f = \mathrm{pr}_a$ for all $a \in A$, and that this map is an isomorphism.

(d) Endow the reals $\mathbb{R}$ with its natural partial order, and consider a map $A \to \mathbb{R}$. Under what conditions does the product of these objects exist, and if it does what is it? Same question for the coproduct.

5   Homotopy, star-shaped regions

We've computed the homology of a point. Let's now compare the homology of a general space X to this example. There's always a unique map $X \to *$: $*$ is a *terminal object* in **Top**. We have an induced map

$$H_n(X) \to H_n(*) = \begin{cases} \mathbb{Z} & n = 0 \\ 0 & \text{otherwise}. \end{cases}$$

This map may be described on the chain level: A 0-cycle is a formal linear combination $c = \sum a_i x_i$ of points of X. Define $\varepsilon : S_0(X) \to \mathbb{Z}$ by sending c to $\sum a_i \in \mathbb{Z}$. We can use this map to form the *augmented singular complex* $\widetilde{S}_*(X)$ by defining $\widetilde{S}_n(X) = S_n(X)$ for $n \neq -1$ and $\widetilde{S}_{-1} = \mathbb{Z}$, and using ε for $d : \widetilde{S}_0(X) \to \widetilde{S}_{-1}(X)$. Its homology will be called the *augmented (singular) homology* of X, $\widetilde{H}_*(X)$.

The surjection $\widetilde{S}_*(X) \to S_*(X)$ induces an isomorphism in positive dimensions. If X is nonempty,

$$\widetilde{H}_*(X) = \ker(\varepsilon : H_*(X) \to H_*(*)) \,.$$

In fact any choice of point in X — a "basepoint" — provides a splitting of $\varepsilon : S_*(X) \to \mathbb{Z}$, and an isomorphism

$$H_*(X) \cong \widetilde{H}_*(X) \oplus \mathbb{Z} \,.$$

But if X is empty, we find

$$\widetilde{H}_q(\varnothing) = \begin{cases} \mathbb{Z} & \text{for } q = -1 \\ 0 & \text{for } q \neq -1 \,. \end{cases}$$

This convention isn't universally accepted, but I find it useful.

What other spaces besides a point have trivial homology? More generally we can ask:

Question 5.1. When do two maps $X \to Y$ induce the same map in homology?

For example, when do $1_X : X \to X$ and $X \to * \to X$ induce the same map in homology? If they do, then $\varepsilon : H_*(X) \to \mathbb{Z}$ is an isomorphism.

The key idea is that homology is a discrete invariant, so it should be unchanged by deformation. Here's the definition that makes "deformation" precise.

Definition 5.2. Let $f_0, f_1 : X \to Y$ be two maps. A *homotopy* from f_0 to f_1 is a map $h : X \times I \to Y$ (continuous, of course) such that $h(x, 0) = f_0(x)$ and $f(x, 1) = f_1(x)$. We say that f_0 and f_1 are *homotopic*, and that h is a *homotopy* between them. This relation is denoted by $f_0 \simeq f_1$.

Homotopy is an equivalence relation on maps from X to Y. Transitivity follows from the gluing lemma of point set topology. We denote by $[X, Y]$ the set of *homotopy classes* of maps from X to Y. A key result about homology is this.

Theorem 5.3 (Homotopy invariance of homology). *If $f_0 \simeq f_1$, then $H_*(f_0) = H_*(f_1)$: homology cannot distinguish between homotopic maps.*

Suppose I have two maps $f_0, f_1 : X \to Y$ with a homotopy $h : f_0 \simeq f_1$, and a map $g : Y \to Z$. Composing h with g gives a homotopy between $g \circ f_0$ and $g \circ f_1$. Precomposing also works: If $g : W \to X$ is a map, then $h \circ (g \times 1) : f_0 \simeq f_1 : X \to Y$. These facts let us compose homotopy classes: we can complete the diagram of categories and functors:

$$
\begin{array}{ccc}
\mathbf{Top}(Y, Z) \times \mathbf{Top}(X, Y) & \longrightarrow & \mathbf{Top}(X, Z) \\
\downarrow & & \downarrow \\
[Y, Z] \times [X, Y] & \dashrightarrow & [X, Z]
\end{array}
$$

Definition 5.4. The *homotopy category* (of topological spaces) Ho($\mathbf{Top}$) has the same objects as $\mathbf{Top}$, but

$$
\mathrm{Ho}(\mathbf{Top})(X, Y) = [X, Y] = \mathbf{Top}(X, Y)/\simeq .
$$

We may restate Theorem 5.3 as follows: For each n, the homology functor H_n : **Top** $\to$ **Ab** factors as **Top** $\to$ Ho(**Top**) $\to$ **Ab**; it is a "homotopy functor."

We will start to work on a proof of this theorem in a minute, and complete it in the next lecture, but let's stop now and think about some consequences.

Definition 5.5. A map $f : X \to Y$ is a *homotopy equivalence* if $[f] \in [X, Y]$ is an isomorphism in Ho(**Top**). In other words, there is a map $g : Y \to X$ such that $f \circ g \simeq 1_Y$ and $g \circ f \simeq 1_X$.

Such a map g is a *homotopy inverse* for f; it is well-defined only up to homotopy.

Most topological properties are not preserved by homotopy equivalences. For example, compactness is not a homotopy-invariant property: Consider the inclusion $i : S^{n-1} \subseteq \mathbb{R}^n - \{0\}$. A homotopy inverse $p : \mathbb{R}^n - \{0\} \to S^{n-1}$ can be obtained by dividing a (nonzero!) vector by its length. Clearly $p \circ i = 1_{S^{n-1}}$. We have to find a homotopy $i \circ p \simeq 1_{\mathbb{R}^n - \{0\}}$. This is a map $(\mathbb{R}^n - \{0\}) \times I \to \mathbb{R}^n - \{0\}$, and we can use $(v, t) \mapsto tv + (1 - t)\frac{v}{||v||}$.

On the other hand:

Corollary 5.6. *Homotopy equivalences induce isomorphisms in homology.*

Proof. If f has homotopy inverse g, then f_* has inverse g_*. $\qquad\square$

Definition 5.7. A space X is *contractible* if the map $X \to *$ is a homotopy equivalence.

Corollary 5.8. *Let X be a contractible space. The augmentation ε : $H_*(X) \to \mathbb{Z}$ is an isomorphism.*

Homotopy equivalences in general may be somewhat hard to visualize. A particularly simple and important class of homotopy equivalences is given by the following definition.

Definition 5.9. An inclusion $A \hookrightarrow X$ is a *deformation retract* provided that there is a map $h : X \times I \to X$ such that $h(x, 0) = x$ and $h(x, 1) \in A$ for all $x \in X$ and $h(a, t) = a$ for all $a \in A$ and $t \in I$.

For example, S^{n-1} is a deformation retract of $\mathbb{R}^n - \{0\}$.

Chain homotopy

We now set about constructing a proof of homotopy invariance of homology. The first step is to understand the analogue of homotopy on the level of chain complexes.

Definition 5.10. Let C_*, D_* be chain complexes, and $f_0, f_1 : C_* \to D_*$ be chain maps. A *chain homotopy* $h : f_0 \simeq f_1$ is a collection of homomorphisms $h : C_n \to D_{n+1}$ such that $dh + hd = f_1 - f_0$.

This definition takes some getting used to. Here's a picture (not a commutative diagram).

$$\cdots \longrightarrow C_{n+1} \xrightarrow{\ d\ } C_n \xrightarrow{\ d\ } C_{n-1} \longrightarrow \cdots$$

$$\cdots \longrightarrow D_{n+1} \xrightarrow{\ d\ } D_n \xrightarrow{\ d\ } D_{n-1} \longrightarrow \cdots$$

Lemma 5.11. *If $f_0, f_1 : C_* \to D_*$ are chain homotopic, then $f_{0*} = f_{1*} :$ $H_*(C) \to H_*(D)$.*

Proof. We want to show that for every $c \in Z_n(C_*)$, the difference $f_1 c - f_0 c$ is a boundary. Well,

$$f_1 c - f_0 c = (f_1 - f_0)c = (dh + hd)c = dhc + hdc = dhc. \qquad \square$$

So homotopy invariance of homology will follow from:

Proposition 5.12. *Let $f_0, f_1 : X \to Y$ be homotopic. Then $f_{0*}, f_{1*} :$ $S_*(X) \to S_*(Y)$ are chain homotopic.*

To prove this we will begin with a special case.

Definition 5.13. A subset $X \subseteq \mathbb{R}^n$ is *star-shaped* with respect to $b \in X$ if for every $x \in X$ the interval

$$\{tb + (1-t)x : t \in [0,1]\}$$

lies in X.

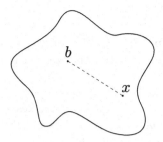

Any nonempty convex region is star shaped with respect to any of its points. Any star-shaped region X is contractible: A homotopy inverse to $X \to *$ is given by sending $*$ to b. One composite is the identity. A homotopy from the other composite to the identity 1_X is given by $(x, t) \mapsto tb + (1 - t)x$.

Now that we have the notion of chain homotopy, we can hope for something similar to happen on the chain level. A chain map $f : C_* \to D_*$ is a *chain homotopy equivalence* if there is a chain map $g : D_* \to C_*$ such that the composites fg and gf are chain homotopic to the respective identity maps. Here is the result we might hope for:

Proposition 5.14. *If X is a star-shaped region then $\varepsilon : S_*(X) \to \mathbb{Z}$ is a chain homotopy equivalence.*

Proof. We have maps $S_*(X) \xrightarrow{\varepsilon} \mathbb{Z} \xrightarrow{\eta} S_*(X)$ where $\eta(1) = c_b^0$. Clearly $\varepsilon\eta = 1$, and the claim is that $\eta\varepsilon \simeq 1 : S_*(X) \to S_*(X)$. The chain map $\eta\varepsilon$ concentrates everything at the point b: $\eta\varepsilon\sigma = c_b^0$ if $\sigma \in \mathrm{Sin}_0(X)$ and $\eta\varepsilon\sigma = 0$ if $\sigma \in \mathrm{Sin}_n(X)$ for $n > 0$. Our chain homotopy $h : S_q(X) \to S_{q+1}(X)$ will actually send simplices to simplices. For $\sigma \in \mathrm{Sin}_q(X)$, define the chain homotopy evaluated on σ by means of the following "cone construction": $h(\sigma) = b * \sigma$, where

$$(b * \sigma)(t_0, \ldots, t_{q+1}) = t_0 b + (1 - t_0)\sigma\left(\frac{(t_1, \ldots, t_{q+1})}{1 - t_0}\right).$$

Explanation: The denominator $1 - t_0$ makes the entries sum to 1, as they must if we are to apply σ to this vector. When $t_0 = 1$, this isn't defined, but it doesn't matter since we are multiplying by $1 - t_0$. So $(b * \sigma)(1, 0, \ldots, 0) = b$; this is the vertex of the cone.

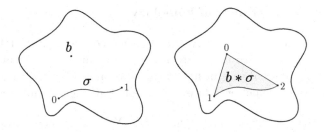

Setting $t_0 = 0$, we find

$$d_0(b * \sigma) = \sigma.$$

Setting $t_i = 0$ for $i > 0$, we find

$$d_i(b * \sigma) = h d_{i-1} \sigma.$$

Using the formula for the boundary operator, we find

$$d(b * \sigma) = \sigma - b * d\sigma$$

... *unless* $q = 0$, when

$$d(b * \sigma) = \sigma - c_b^0.$$

This can be assembled into the equation $d(b * \sigma) + b * (d\sigma) = \sigma - \eta\varepsilon\sigma$, or

$$dh + hd = 1 - \eta\varepsilon,$$

which is what we wanted. $\qquad\square$

Exercises

Exercise 5.15. (a) Let A_* be a chain complex. It is *acyclic* if $H_*(A_*) = 0$, and *contractible* if it is chain-homotopy-equivalent to the trivial chain complex. Prove that a chain complex is contractible if and only if it is acyclic *and* for every n the inclusion $Z_n A \hookrightarrow A_n$ is a split monomorphism of abelian groups.
(b) Give an example of an acyclic chain complex that is not contractible.

Exercise 5.16. Propose a construction of the product and the coproduct of two spaces in the homotopy category, and check that your proposal serves the purpose.

6   Homotopy invariance of homology

We now know that the homology of a star-shaped region is trivial: in such a space, every cycle with augmentation 0 is a boundary. We will use that fact, which is a special case of homotopy invariance of homology, to prove the general result, which we state in somewhat stronger form:

Theorem 6.1. *A homotopy* $h : f_0 \simeq f_1 : X \to Y$ *determines a natural chain homotopy* $f_{0*} \simeq f_{1*} : S_*(X) \to S_*(Y)$.

The proof uses naturality (a lot). For a start, notice that if $k : g_0 \simeq g_1 : C_* \to D_*$ is a chain homotopy, and $j : D_* \to E_*$ is another chain map, then the composites $j \circ k_n : C_n \to E_{n+1}$ give a chain homotopy $j \circ g_0 \simeq j \circ g_1$. So if we can produce a chain homotopy k between the chain maps induced by the two inclusions $i_0, i_1 : X \to X \times I$, we can get a chain homotopy between $f_{0*} = h_* \circ i_{0*}$ and $f_{1*} = h_* \circ i_{1*}$ in the form $h_* \circ k$.

So now we want to produce a natural chain homotopy, with components $k_n : S_n(X) \to S_{n+1}(X \times I)$. The unit interval hosts a natural 1-simplex given by an identification $\Delta^1 \to I$, and we should imagine k as being given by "multiplying" by that 1-chain. This "multiplication" is a special case of a chain map

$$\times : S_*(X) \times S_*(Y) \to S_*(X \times Y) \,,$$

defined for any two spaces X and Y, with lots of good properties. It will ultimately be used to compute the homology of a product of two spaces in terms of the homology groups of the factors.

Here's the general result.

Theorem 6.2. *There exists a map* $\times : S_p(X) \times S_q(Y) \to S_{p+q}(X \times Y)$, *the* cross product, *that (with $a, a' \in S_p(X)$ and $b, b' \in S_q(Y)$) is:*

- *Natural, in the sense that if $f : X \to X'$ and $g : Y \to Y'$ then $f_*(a) \times g_*(b) = (f \times g)_*(a \times b)$.*
- *Bilinear, in the sense that $(a+a') \times b = (a \times b) + (a' \times b)$, and $a \times (b+b') = a \times b + a \times b'$.*
- *Satisfies the Leibniz rule, i.e., $d(a \times b) = (da) \times b + (-1)^p a \times db$.*
- *Normalized, in the following sense. Let $x \in X$ and $y \in Y$. Write $j_x : Y \to X \times Y$ for $y \mapsto (x, y)$, and write $i_y : X \to X \times Y$ for $x \mapsto (x, y)$. Then $c_x^0 \times b = (j_x)_* b \in S_q(X \times Y)$ and $a \times c_y^0 = (i_y)_* a \in S_p(X \times Y)$.*

The Leibniz rule contains the first occurrence of the "topologist's sign rule"; we'll see these signs appearing often. Watch for when it appears in our proof.

Proof. We're going to use induction on $p+q$; the normalization axiom gives us the cases $p + q = 0, 1$. Let's assume that we've constructed the cross-product in total dimension $p+q-1$. We want to define $\sigma \times \tau$ for $\sigma \in S_p(X)$ and $\tau \in S_q(Y)$.

Naturality allows us to focus on universal examples. There is a universal example of a singular p-simplex! — namely the identity map $\iota_p : \Delta^p \to \Delta^p$. It's universal in the sense any p-simplex $\sigma : \Delta^p \to X$ can be written as $\sigma_*(\iota_p)$ where $\sigma_* : \mathrm{Sin}_p(\Delta^p) \to \mathrm{Sin}_p(X)$ is the map induced by σ. To define $\sigma \times \tau$ in general, then, it suffices to define $\iota_p \times \iota_q \in S_{p+q}(\Delta^p \times \Delta^q)$; we can (and must, to achieve naturality) then take $\sigma \times \tau = (\sigma \times \tau)_*(\iota_p \times \iota_q)$. This is not circular! The second occurrence of "$\sigma \times \tau$" means the map $\Delta^p \times \Delta^q \to X \times Y$, not the cross-product of σ and τ.

Our long list of axioms is useful in the induction. For one thing, if $p = 0$ or $q = 0$, normalization gets us started. So now assume that both p and q are positive. We want the cross-product to satisfy the Leibniz rule:

$$d(\iota_p \times \iota_q) = (d\iota_p) \times \iota_q + (-1)^p \iota_p \times d\iota_q \in S_{p+q-1}(\Delta^p \times \Delta^q) .$$

Since $d^2 = 0$, a necessary condition for $\iota_p \times \iota_q$ to exist is that $d((d\iota_p) \times \iota_q + (-1)^p \iota_p \times d\iota_q) = 0$. Let's compute what this is, using the Leibniz rule in dimension $p + q - 1$ where we have it by the inductive assumption:

$$d((d\iota_p) \times \iota_q + (-1)^p \iota_p \times (d\iota_q)) =$$

$$(d^2\iota_p) \times \iota_q + (-1)^{p-1}(d\iota_p) \times (d\iota_q) + (-1)^p(d\iota_p) \times (d\iota_q) + \iota_p \times (d^2\iota_q) = 0$$

because $d^2 = 0$. Note that this calculation would not have worked without the sign!

The subspace $\Delta^p \times \Delta^q \subseteq \mathbb{R}^{p+1} \times \mathbb{R}^{q+1}$ is convex and nonempty, and hence star-shaped. Therefore we know that $H_{p+q-1}(\Delta^p \times \Delta^q) = 0$ (remember, $p + q > 1$), which means that every cycle is a boundary. In other words, our necessary condition is also sufficient! So, choose any element with the right boundary and declare it to be $\iota_p \times \iota_q$.

The induction is now complete provided we can check that this choice satisfies naturality, bilinearity, and the Leibniz rule. I leave this as a relaxing exercise for the listener. □

The essential point here is that the space supporting the universal pair of singular simplices — $\Delta^p \times \Delta^q$ — has trivial homology. Naturality transports the result in that case to the general situation.

The cross-product that this procedure constructs is not unique; it depends on a choice of the chain $\iota_p \times \iota_q$ for each pair p, q with $p + q > 1$. The cone construction in the proof that star-shaped regions have vanishing homology does provide us with a specific choice. But this specific formula isn't that useful, and it turns out that any two choices lead to naturally chain homotopy equivalent cross products.

Completion of the proof of homotopy invariance, Theorem 6.1. To define our chain homotopy $h_X : S_n(X) \to S_{n+1}(X \times I)$, pick any 1-simplex $\iota : \Delta^1 \to I$ such that $d_0\iota = c_1^0$ and $d_1\iota = c_0^0$, and define

$$h_X\sigma = (-1)^n \sigma \times \iota.$$

Let's compute:

$$dh_X\sigma = (-1)^n d(\sigma \times \iota) = (-1)^n (d\sigma) \times \iota + \sigma \times (d\iota).$$

But $d\iota = c_1^0 - c_0^0 \in S_0(I)$, which means that we can continue (remembering that $|\partial\sigma| = n - 1$):

$$\cdots = -h_X d\sigma + (\sigma \times c_1^0 - \sigma \times c_0^0) = -h_X d\sigma + (\iota_{1*}\sigma - \iota_{0*}\sigma),$$

using the normalization axiom of the cross-product. This is the result. □

Exercises

Exercise 6.3. Complete the proof of Theorem 6.2 by checking the axioms at the inductive step.

7   Homology cross product

In the last lecture we proved homotopy invariance of homology using the construction of a chain level bilinear cross-product

$$\times : S_p(X) \times S_q(Y) \to S_{p+q}(X \times Y)$$

that satisfied the Leibniz formula

$$d(a \times b) = (da) \times b + (-1)^p a \times (db).$$

What else does this map give us?

Let's abstract a little bit. Suppose we have three chain complexes A_*, B_*, and C_*, and suppose we have maps $\times : A_p \times B_q \to C_{p+q}$ that satisfy bilinearity and the Leibniz formula. What does this induce in homology?

Lemma 7.1. *These data determine a bilinear map* $\times : H_p(A) \times H_q(B) \to H_{p+q}(C)$.

Proof. Let $a \in Z_p(A)$ and $b \in Z_q(B)$. We want to define $[a] \times [b] \in H_{p+q}(C)$, and we hope that the obvious guess $[a] \times [b] = [a \times b]$ actually works. For a start, $a \times b$ is a cycle: By Leibniz, $d(a \times b) = da \times b + (-1)^p a \times db$, which vanishes because a and b are cycles.

Now we need to check that the homology class of $a \times b$ depends only on the homology classes we started with. So pick other cycles a' and b' in the same homology classes. We want $[a \times b] = [a' \times b']$. In other words, we need to show that $a \times b$ differs from $a' \times b'$ by a boundary. We can write $a' = a + d\bar{a}$ and $b' = b + d\bar{b}$, and compute, using bilinearity:

$$a' \times b' = (a + d\bar{a}) \times (b + d\bar{b}) = a \times b + a \times d\bar{b} + (d\bar{a}) \times b + (d\bar{a}) \times (d\bar{b}).$$

We need to deal with the last three terms here. Since $da = 0$,

$$d(a \times \bar{b}) = (-1)^p a \times (d\bar{b}).$$

Since $d\bar{b} = 0$,

$$d(\bar{a} \times b) = (d\bar{a}) \times b.$$

And since $d^2\bar{b} = 0$,

$$d(\bar{a} \times d\bar{b}) = (d\bar{a}) \times (d\bar{b}).$$

This means that $a' \times b'$ and $a \times b$ differ by

$$d\left((-1)^p(a \times \bar{b}) + \bar{a} \times b + \bar{a} \times d\bar{b}\right),$$

and so are homologous.

The last step is to check bilinearity, which is left to the listener. $\square$

This gives the following result.

Theorem 7.2. *There is a map*

$$\times : H_p(X) \times H_q(Y) \to H_{p+q}(X \times Y)$$

that is natural, bilinear, and normalized.

We will see (Theorem 25.13) that this map is also *uniquely defined* by these conditions, unlike the chain-level cross product.

I just want to mention an explicit choice of $\iota_p \times \iota_q$. This is called the Eilenberg-Zilber or shuffle chain, though it was introduced by Eilenberg and Mac Lane. You're highly encouraged to think about this yourself. It comes from a triangulation of the prism.

The simplices in this triangulation are indexed by injections

$$\omega : [p+q] \to [p] \times [q]$$

such that each coordinate, $[p+q] \to [p]$ and $[p+q] \to [q]$, is order-preserving. Injectivity forces $\omega(0) = (0,0)$ and $\omega(p+q) = (p,q)$. Each such map determines an affine map $\Delta^{p+q} \to \Delta^p \times \Delta^q$ of the same name. These will be the singular simplices making up $\iota_p \times \iota_q$. To specify the coefficients, think of ω as a staircase in the rectangle $[0,p] \times [0,q]$. Let $A(\omega)$ denote the area under that staircase. Then the Eilenberg-Zilber chain is given by

$$\iota_p \times \iota_q = \sum (-1)^{A(\omega)} \omega \,.$$

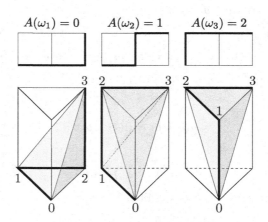

This description appears in a paper [19] by Eilenberg and Moore. It's very pretty, and it provides an explicit chain map

$$\zeta_{X,Y} : S_*(X) \times S_*(Y) \to S_*(X \times Y)$$

that satisfies many good properties on the nose and not just up to chain homotopy. For example, it's *associative* —

$$S_*(X) \times S_*(Y) \times S_*(Z) \xrightarrow{\zeta_{X,Y} \times 1} S_*(X \times Y) \times S_*(Z)$$

$$\Big\downarrow {\scriptstyle 1 \times \zeta_{Y,Z}} \qquad\qquad\qquad\qquad \Big\downarrow {\scriptstyle \zeta_{X \times Y, Z}}$$

$$S_*(X) \times S_*(Y \times Z) \xrightarrow{\zeta_{X, Y \times Z}} S_*(X \times Y \times Z)$$

commutes, and *commutative* —

$$S_*(X) \times S_*(Y) \xrightarrow{\zeta_{X,Y}} S_*(X \times Y)$$

$$\Big\downarrow {\scriptstyle \tau} \qquad\qquad\qquad\qquad \Big\downarrow {\scriptstyle S_*(T)}$$

$$S_*(Y) \times S_*(X) \xrightarrow{\zeta_{Y,X}} S_*(Y \times X)$$

commutes, where on spaces $T(x,y) = (y,x)$, and on chain complexes $\tau(a,b) = (-1)^{pq}(b,a)$ when a has degree p and b has degree q.

We will see that these properties hold up to chain homotopy for any choice of chain-level cross-product.

Exercises

Exercise 7.3. Let S and T be sets and A an abelian group. Establish a bijection between the set of maps of sets from $S \times T$ to A and the set of bilinear maps $\mathbb{Z}S \times \mathbb{Z}T \to A$.

Exercise 7.4. For positive integers m,n, let $\mathbb{Z}/m$, $\mathbb{Z}/n$ denote the cyclic groups of order m,n. Construct a surjective bilinear map $\mu : \mathbb{Z}/m \times \mathbb{Z}/n \to \mathbb{Z}/\gcd\{m,n\}$. Show that any bilinear map $\mathbb{Z}/m \times \mathbb{Z}/n \to A$ factors uniquely as $f \circ \mu$ where $f : \mathbb{Z}/\gcd\{m,n\} \to A$ is a homomorphism.

8 Relative homology

An ultimate goal of algebraic topology is to find means to compute the set of homotopy classes of maps from one space to another. This is important because many geometrical problems can be rephrased as such a computation. It's a lot more modest than wanting to characterize, somehow, all continuous maps from X to Y; but the very fact that it still contains a great deal of interesting information means that it is still a very challenging problem.

Homology is in a certain sense the best "additive" approximation to this problem; and its additivity makes it much more computable. To justify this, we want to describe the sense in which homology is "additive." Here are two related aspects of this claim.

(1) If $A \subseteq X$ is a subspace, then $H_*(X)$ a combination of $H_*(A)$ and $H_*(X - A)$.
(2) The homology $H_*(A \cup B)$ is like $H_*(A) + H_*(B) - H_*(A \cap B)$.

The first hope is captured by the long exact sequence of a pair, the second by the Mayer-Vietoris Theorem. Both facts show that homology behaves like a measure. The precise statement of both facts uses the machinery of exact sequences. I'll use the following language.

Definition 8.1. A *sequence* of abelian groups is a diagram of abelian groups of the form

$$\cdots \to C_{n+1} \xrightarrow{f_n} C_n \xrightarrow{f_{n-1}} C_{n-1} \to \cdots$$

(which may terminate on the left or on the right at some finite stage) in which all composites are zero; that is, $\operatorname{im} f_n \subseteq \ker f_{n-1}$ for all n. It is *exact* at C_n provided that this inequality is an equality.

Thus a chain complex is a sequence that is unbounded in both directions. A sequence is exact at C_n if and only if $H_n(C_*) = 0$. So homology measures the failure of exactness.

Example 8.2. The sequence $0 \to A \xrightarrow{i} B$ is exact if and only if i is injective, and $B \xrightarrow{p} C \to 0$ is exact if and only if p is surjective.

Exactness was a key concept in the development of algebraic topology, and "exact" is a great word for the concept. A foundational treatment [18] of algebraic topology was published by Sammy Eilenberg and Norman Steenrod in 1952. The story goes that in the galleys for the book they left a blank space whenever the word representing this concept was used, and filled it in at the last minute.

Definition 8.3. A *short exact sequence* is an exact sequence of the form

$$0 \to A \xrightarrow{i} B \xrightarrow{p} C \to 0 \,.$$

Any sequence of the form $A \to B \to C$ expands to a commutative diagram

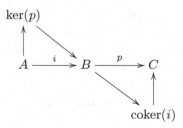

It is exact at B if and only if $A \to \ker p$ is surjective, or, equivalently, $\operatorname{coker}(i) \to C$ is injective. It is short exact if furthermore i is injective and p is surjective.

We will compare the homology groups of a space X with those of a subspace A. Let's formalize this a bit. Along with the category **Top** of spaces, we have the category **Top₂** of *pairs* of spaces. An object of **Top₂** is a space X together with a subspace A. A map $(X, A) \to (Y, B)$ is a continuous map $X \to Y$ that sends A into B.

There are four obvious functors relating **Top** and **Top₂**:

$$X \mapsto (X, \varnothing), \quad X \mapsto (X, X),$$

$$(X, A) \mapsto X, \quad (X, A) \mapsto A.$$

We'll define *relative homology groups* of a pair (X, A). First, just divide $S_n(X)$ by the subgroup $S_n(A)$:

$$S_n(X, A) = S_n(X)/S_n(A).$$

This is the group of *relative n-chains* of the pair (X, A).

Do the relative chains form themselves into a chain complex?

Lemma 8.4. *Let A_* be a subcomplex of the chain complex B_*. There is a unique structure of chain complex on the quotient graded abelian group C_*, with entries $C_n = B_n/A_n$, such that $B_* \to C_*$ is a chain map.*

Proof. To define $d : C_n \to C_{n-1}$, represent a class in C_n by $b \in B_n$, and hope that $[db] \in B_{n-1}/A_{n-1}$ is well defined. If we replace b by $b + a$ for $a \in A_n$, we find

$$d(b + a) = db + da \equiv db \mod A_{n-1},$$

so our hope is justified. Then $d^2[b] = [d^2 b] = 0$, so we do get a chain complex. □

Definition 8.5. The *relative singular chain complex* of the pair (X, A) is

$$S_*(X, A) = \frac{S_*(X)}{S_*(A)}.$$

This is a functor from pairs of spaces to chain complexes. Of course

$$S_*(X, \varnothing) = S_*(X), \quad S_*(X, X) = 0.$$

Definition 8.6. The *relative singular homology* of the pair (X, A) is the homology of the relative singular chain complex:

$$H_n(X, A) = H_n(S_*(X, A)).$$

Relative homology was introduced by Solomon Lefschetz (1884–1972). Russian, educated in France as a chemical engineer, he lost the use of his hands in an accident and then received a PhD in mathematics at Clark University in Worcester. He became the backbone of the great development of algebraic topology at Princeton University. A fragment of his academic family tree appears at the end of this unit.

One of the nice features of the absolute chain group $S_n(X)$ is that it is free as an abelian group. This is also the case for its quotient $S_n(X, A)$, since the map $S_n(A) \to S_n(X)$ takes basis elements to basis elements. $S_n(X, A)$ is freely generated by the cosets of the singular n-simplices in X that do not lie entirely in A.

Example 8.7. Consider Δ^n, relative to its boundary,

$$\partial \Delta^n := \bigcup_i \operatorname{im} d_i \cong S^{n-1}.$$

We have the identity map $\iota_n : \Delta^n \to \Delta^n$, the universal n-simplex, in $\operatorname{Sin}_n(\Delta^n) \subseteq S_n(\Delta^n)$. It is not a cycle; its boundary $d\iota_n \in S_{n-1}(\Delta^n)$ is the alternating sum of the faces of the n-simplex. Each of these singular simplices lies in $\partial \Delta^n$, so $d\iota_n \in S_{n-1}(\partial \Delta^n)$, and $[\iota_n] \in S_n(\Delta_n, \partial \Delta_n)$ *is a relative* cycle. We will see that the relative homology $H_n(\Delta^n, \partial \Delta^n)$ is infinite cyclic, with generator the class of $[\iota_n]$.

Exercises

Exercise 8.8. Let $0 \to A \xrightarrow{i} B \xrightarrow{p} C \to 0$ be a short exact sequence. Establish bijections among the following three sets.

(i) The set of homomorphisms $\sigma : C \to B$ such that $p\sigma = 1_C$.

(ii) The set of homomorphisms $\pi : B \to A$ such that $\pi i = 1_A$.

(iii) The set of homomorphisms $\alpha : A \oplus C \to B$ such that $\alpha(a, 0) = ia$ for all $a \in A$ and $p\alpha(a, c) = c$ for all $(a, c) \in A \oplus C$.

Show that any homomorphism as in (iii) is an isomorphism.

Any one of these is a *splitting* of the short exact sequence, and the sequence is then said to be *split*.

Exercise 8.9. Construct a "semi-relative cross product," natural in X and the pair (Y, B):

$$\times : H_p(X) \times H_q(Y, B) \to H_{p+q}(X \times Y, X \times B)$$

that agrees with the cross product we constructed in Lecture 7 if $B = \varnothing$

and that makes

$$
\begin{array}{ccc}
H_p(X) \times H_q(B) & \xrightarrow{\ \times\ } & H_{p+q}(X \times B) \\
\downarrow{\scriptstyle 1 \times \partial} & & \downarrow{\scriptstyle \partial} \\
H_p(X) \times H_{q-1}(Y, B) & \xrightarrow{\ \times\ } & H_{p+q-1}(X \times Y, X \times B)
\end{array}
$$

commute, at least up to sign.

9 Homology long exact sequence

A pair of spaces (X, A) gives rise to a short exact sequence of chain complexes:

$$0 \to S_*(A) \to S_*(X) \to S_*(X, A) \to 0.$$

In homology, this will relate $H_*(A)$, $H_*(X)$, and $H_*(X, A)$.

To investigate this, let's suppose we have a general short exact sequence of chain complexes,

$$0 \to A_* \xrightarrow{f} B_* \xrightarrow{g} C_* \to 0,$$

and study what happens in homology. Clearly the composite $H_*(A) \to H_*(B) \to H_*(C)$ is trivial. Is this sequence exact? Let $[b] \in H_n(B)$ be such that $g_*([b]) = 0$. It's determined by some $b \in B_n$ such that $d(b) = 0$. Since $g_*([b]) = 0$, there is some $\bar{c} \in C_{n+1}$ such that $d\bar{c} = g(b)$. Now g is surjective, so there is some $\bar{b} \in B_{n+1}$ such that $g(\bar{b}) = \bar{c}$. Then we can consider $d\bar{b} \in B_n$, and $g(d\bar{b}) = d\bar{c} \in C_n$. We have a new representative for $[b]$, namely $b - d\bar{b}$. This maps to zero in C_n, so by exactness there is some $a \in A_n$ such that $f(a) = b - d\bar{b}$. Is a a cycle? Well, $f(da) = df(a) = d(b - d\bar{b}) = db - d^2\bar{b} = db$, but we assumed that $db = 0$, so $f(da) = 0$. This means that da is zero because f is an injection. Therefore a is a cycle. Does $[a] \in H_n(A)$ do the job? Well, $f([a]) = [b - d\bar{b}] = [b]$. This proves exactness of $H_n(A) \to H_n(B) \to H_n(C)$ at $H_n(B)$.

On the other hand, $H_*(A) \to H_*(B)$ may fail to be injective, and $H_*(B) \to H_*(C)$ may fail to be surjective. Instead:

Theorem 9.1 (The homology long exact sequence). *Let* $0 \to A_* \xrightarrow{f} B_* \xrightarrow{g} C_* \to 0$ *be a short exact sequence of chain complexes. Then there is a* *natural homomorphism* $\partial : H_n(C) \to H_{n-1}(A)$ *such that the sequence*

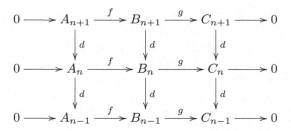

$$\cdots \xrightarrow{\;g_*\;} H_{n+1}(C)$$

$$H_n(A) \xleftarrow{f_*} H_n(B) \xrightarrow{\;g_*\;} H_n(C)$$

$$H_{n-1}(A) \xleftarrow{f_*} \cdots$$

is exact.

Proof. We'll construct the "boundary map" ∂, and leave the rest as an exercise. Here's an expanded part of the short exact sequence:

$$
\begin{array}{ccccccccc}
0 & \longrightarrow & A_{n+1} & \xrightarrow{f} & B_{n+1} & \xrightarrow{g} & C_{n+1} & \longrightarrow & 0 \\
& & \downarrow{\scriptstyle d} & & \downarrow{\scriptstyle d} & & \downarrow{\scriptstyle d} & & \\
0 & \longrightarrow & A_n & \xrightarrow{f} & B_n & \xrightarrow{g} & C_n & \longrightarrow & 0 \\
& & \downarrow{\scriptstyle d} & & \downarrow{\scriptstyle d} & & \downarrow{\scriptstyle d} & & \\
0 & \longrightarrow & A_{n-1} & \xrightarrow{f} & B_{n-1} & \xrightarrow{g} & C_{n-1} & \longrightarrow & 0
\end{array}
$$

Let $c \in C_n$ be a cycle: $dc = 0$. The map g is surjective, so pick a $b \in B_n$ such that $g(b) = c$, and consider $db \in B_{n-1}$. Now $g(db) = dg(b) = dc = 0$, so by exactness there is some $a \in A_{n-1}$ such that $f(a) = db$. Actually there's a unique such a because f is injective. We need to check that a is a cycle. What is da? Well, $d^2 b = 0$, so da maps to 0 under f. But because f is injective, $da = 0$, i.e., a is a cycle. This means we can try to define $\partial[c]$ as $[a]$.

To make sure that this is well-defined, let's check that the homology class $[a]$ doesn't depend on the b that we chose. Pick some other b' such that $g(b') = c$. Then there is $a' \in A_{n-1}$ such that $f(a') = db'$. We want $a - a'$ to be a boundary, so that $[a] = [a']$: We need to find $\bar{a} \in A_n$ such that $d\bar{a} = a - a'$. Well, $g(b - b') = 0$, so by exactness there is $\bar{a} \in A_n$ such that $f(\bar{a}) = b - b'$. Then $df(\bar{a}) = f(d\bar{a}) = d(b - b') = f(a - a')$.

I leave the rest of what needs checking to the listener. $\square$

Corollary 9.2 (The Snake Lemma). *Suppose that we have a map of short exact sequences,*

$$0 \longrightarrow A \longrightarrow B \longrightarrow C \longrightarrow 0$$
$$\downarrow i \qquad \downarrow j \qquad \downarrow k$$
$$0 \longrightarrow A' \longrightarrow B' \longrightarrow C' \longrightarrow 0 .$$

There is a naturally associated 6-term exact sequence

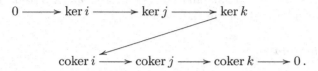

Proof. Regard each column as a chain complex, and apply Theorem 9.1. □

Example 9.3. A pair of spaces (X, A) gives rise to a natural long exact sequence in homology:

$$\cdots \longrightarrow H_{n+1}(X, A)$$
$$\overset{\partial}{\swarrow}$$
$$H_n(A) \longleftarrow H_n(X) \longrightarrow H_n(X, A)$$
$$\overset{\partial}{\swarrow}$$
$$H_{n-1}(A) \longrightarrow \cdots$$

Example 9.4. Let's think again about the pair (D^n, S^{n-1}). The result is cleaner in augmented homology (as in Lecture 5). Any map induces an isomorphism in $\widetilde{S}_{-1}$, so to a pair (X, A) we can associate a short exact sequence

$$0 \to \widetilde{S}_*(A) \to \widetilde{S}_*(X) \to S_*(X, A) \to 0$$

and hence a long exact sequence

$$\cdots \longrightarrow H_{n+1}(X, A)$$
$$\overset{\partial}{\swarrow}$$
$$\widetilde{H}_n(A) \longleftarrow \widetilde{H}_n(X) \longrightarrow H_n(X, A)$$
$$\overset{\partial}{\swarrow}$$
$$\widetilde{H}_{n-1}(A) \longrightarrow \cdots$$

In the example (D^n, S^{n-1}), $\widetilde{H}_*(D^n) = 0$ and so

$$\partial : H_q(D^n, S^{n-1}) \xrightarrow{\cong} \widetilde{H}_{q-1}(S^{n-1})$$

for all n and q. This even works when $n = 0$; remember that $S^{-1} = \varnothing$ and $\widetilde{H}_{-1}(\varnothing) = \mathbb{Z}$. This is why I like this convention.

To go any further in this analysis, we'll need another tool, known as "excision," coming right up in the next lecture.

The five lemma

The homology long exact sequence is often used in conjunction with an elementary fact about a map between exact sequences known as the *five lemma*. Suppose you have two exact sequences of abelian groups and a map between them — a "ladder":

$$
\begin{array}{ccccccccc}
A_4 & \xrightarrow{d} & A_3 & \xrightarrow{d} & A_2 & \xrightarrow{d} & A_1 & \xrightarrow{d} & A_0 \\
\downarrow{\scriptstyle f_4} & & \downarrow{\scriptstyle f_3} & & \downarrow{\scriptstyle f_2} & & \downarrow{\scriptstyle f_1} & & \downarrow{\scriptstyle f_0} \\
B_4 & \xrightarrow{d} & B_3 & \xrightarrow{d} & B_2 & \xrightarrow{d} & B_1 & \xrightarrow{d} & B_0
\end{array}
$$

When can we guarantee that the middle map f_2 is an isomorphism? We're going to "diagram chase" again. Just follow your nose, making assumptions as necessary.

Surjectivity: Let $b_2 \in B_2$. We want to show that there is something in A_2 mapping to b_2. We can consider $db_2 \in B_1$. Let's assume that f_1 is surjective. Then there's $a_1 \in A_1$ such that $f_1(a_1) = db_2$. What is da_1? Well, $f_0(da_1) = df_1(a_1) = ddb = 0$. So assume that f_0 is injective. Then da_1 is zero, so by exactness of the top sequence there is some $a_2 \in A_2$ such that $da_2 = a_1$. What is $f_2(a_2)$? To answer this, begin by asking: What is $df_2(a_2)$? By commutativity, $df_2(a_2) = f_1(da_2) = f_1(a_1) = db_2$. Let's consider $b_2 - f_2(a_2)$. This maps to zero under d. So by exactness there is $b_3 \in B_3$ such that $db_3 = b_2 - f_2(a_2)$. If we assume that f_3 is surjective, then there is $a_3 \in A_3$ such that $f_3(a_3) = b_3$. But now $da_3 \in A_2$, and $f_2(da_3) = df_3(a_3) = b_2 - f_2(a_2)$. This means that $b_2 = f(a_2 + da_3)$, verifying surjectivity of f_2.

This proves the first half of the following important fact. The second half is "dual" to the first.

Proposition 9.5 (Five lemma). *In the map of exact sequences above,*

- *If f_0 is injective and f_1 and f_3 are surjective, then f_2 is surjective.*

- *If f_4 is surjective and f_3 and f_1 are injective, then f_2 is injective.*

Very commonly one knows that f_0, f_1, f_3, and f_4 are all isomorphisms, and concludes that f_2 is also an isomorphism. For example:

Corollary 9.6. *Let*

$$
\begin{array}{ccccccccc}
0 & \longrightarrow & A'_* & \longrightarrow & B'_* & \longrightarrow & C'_* & \longrightarrow & 0 \\
& & \downarrow f & & \downarrow g & & \downarrow h & & \\
0 & \longrightarrow & A_* & \longrightarrow & B_* & \longrightarrow & C_* & \longrightarrow & 0
\end{array}
$$

be a map of short exact sequences of chain complexes. If two of the three maps induced in homology by f, g, and h are isomorphisms, then so is the third.

Here's an application.

Proposition 9.7. *Let $(A, X) \to (B, Y)$ be a map of pairs, and assume that two of $A \to B$, $X \to Y$, and $(X, A) \to (Y, B)$ induce isomorphisms in homology. Then the third one does as well.*

Proof. Just apply the five lemma to the map between the two homology long exact sequences. □

Exercises

Exercise 9.8. Suppose that

$$
\begin{array}{ccccccccc}
\cdots & \longrightarrow & A_n & \longrightarrow & B_n & \longrightarrow & C_n & \longrightarrow & A_{n-1} & \longrightarrow & \cdots \\
& & \downarrow & & \downarrow & & \cong\downarrow & & \downarrow & & \\
\cdots & \longrightarrow & A'_n & \longrightarrow & B'_n & \longrightarrow & C'_n & \longrightarrow & A'_{n-1} & \longrightarrow & \cdots
\end{array}
$$

is a "ladder": a map of long exact sequences. So both rows are exact and each square commutes. Suppose also that every third vertical map is an isomorphism, as indicated. Prove that these data determine a long exact sequence

$$
\cdots \to A_n \to A'_n \oplus B_n \to B'_n \to A_{n-1} \to \cdots .
$$

Exercise 9.9 ("3 × 3 lemma"). Let

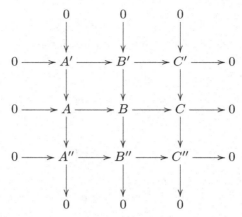

be a commutative diagram of abelian groups. Assume that all three columns are exact, that all but one of the rows is exact, and that the compositions in the remaining row are trivial. Prove that the remaining row is also exact. (Hint: view each row as a chain complex ....)

Exercise 9.10. (Another 3 × 3 puzzle.) Suppose all the rows and columns in the commutative diagram in Exercise 9.9 are exact. Construct from this a natural exact sequence

$$0 \to A' \to A \oplus B' \to B \to C'' \to 0.$$

It may be easiest to construct two short exact sequences and then splice them together.

    What is the dual statement?

Exercise 9.11. (Long exact homology sequence of a triple.) Let (C, B, A) be a "triple," so C is a space, B is a subspace of C, and A is a subspace of B. Show that there are natural transformations $\partial : H_n(C, B) \to H_{n-1}(B, A)$ such that

$$\cdots \to H_n(B, A) \xrightarrow{i_*} H_n(C, A) \xrightarrow{j_*} H_n(C, B) \xrightarrow{\partial} H_{n-1}(B, A) \to \cdots$$

is exact, where $i : (B, A) \to (C, A)$ and $j : (C, A) \to (C, B)$ are the inclusions of pairs.

10 Excision and applications

We have found two general properties of singular homology: homotopy invariance and the long exact sequence of a pair. We also claimed that

$H_*(X, A)$ "depends only on $X - A$." You have to be careful about this. The following definition gives conditions that will capture the sense in which the relative homology of a pair (X, A) depends only on the complement of A in X.

Definition 10.1. A triple (X, A, U), where $U \subseteq A \subseteq X$, is *excisive* if $\overline{U} \subseteq \text{Int}(A)$. The inclusion $(X - U, A - U) \subseteq (X, A)$ is then called an *excision*.

Theorem 10.2. *An excision induces an isomorphism in homology,*

$$H_*(X - U, A - U) \xrightarrow{\cong} H_*(X, A).$$

So you can cut out closed bits of the interior of A without changing the relative homology. The proof will take us a couple of days. Before we give applications, let me pose a different way to interpret the motto "$H_*(X, A)$ depends only on $X - A$." Collapsing the subspace A to a point gives us a map of pairs

$$(X, A) \to (X/A, *).$$

When does this map induce an isomorphism in homology? Excision has the following consequence.

Corollary 10.3. *Let (X, A) be a pair of spaces, and assume that there is a subspace B of X such that (1) $\overline{A} \subseteq \text{Int} B$ and (2) $A \to B$ is a deformation retract. Then*

$$H_*(X, A) \to H_*(X/A, *)$$

is an isomorphism.

Proof. The diagram of pairs

$$
\begin{array}{ccccc}
(X, A) & \xrightarrow{\ i\ } & (X, B) & \xleftarrow{\ j\ } & (X - A, B - A) \\
\downarrow & & \downarrow & & \downarrow{\scriptstyle k} \\
(X/A, *) & \xrightarrow{\ \overline{i}\ } & (X/A, B/A) & \xleftarrow{\ \overline{j}\ } & (X/A - *, B/A - *)
\end{array}
$$

commutes. We want the left vertical to be a homology isomorphism, and will show that the rest of the perimeter consists of homology isomorphisms. The map k is a homeomorphism of pairs, and j is an excision by assumption (1). The map i induces an isomorphism in homology by assumption (2), the long exact sequences, and the five-lemma. Since I is a compact Hausdorff space, the map $B \times I \to (B/A) \times I$ is again a quotient map (see e.g.

[51, pp. 186 and 289]), so the deformation $B \times I \to B$, which restricts to the constant deformation on A, descends to show that $* \to B/A$ is a deformation retract. So the map $\bar{\imath}$ is also a homology isomorphism. Finally, $\bar{*} \subseteq \mathrm{Int}(B/A)$ in X/A, by definition of the quotient topology, so $\bar{\jmath}$ induces an isomorphism by excision. $\square$

Definition 10.4. A *pointed space* is a pair $(X, *)$ in which the subspace is a singleton. The category of pointed spaces is denoted by **Top**$_*$. The *reduced homology* $\overline{H}_*(X)$ of a pointed space is the relative homology $H_*(X, *)$.

So Corollary 10.3 expresses rather general relative homology groups as reduced homology of the quotient space. Note that there is a canonical isomorphism

$$\overline{H}_*(X) \xrightarrow{\cong} \widetilde{H}_*(X)$$

with the augmented homology of the underlying unpointed space. This isomorphism notwithstanding, the distinction is useful. Reduced homology is a functor of *pointed* spaces; augmented homology is a functor of *unpointed* spaces (which might even be empty).

The homology groups of spheres

Now what are some consequences? For a start, we'll finally get around to computing the homology of the sphere. It happens simultaneously with a computation of $H_*(D^n, S^{n-1})$. To describe generators, for each $n \geq 0$ pick a homeomorphism

$$(\Delta^n, \partial\Delta^n) \to (D^n, S^{n-1}),$$

and write

$$\iota_n \in S_n(D^n, S^{n-1})$$

for the corresponding relative n-chain.

Proposition 10.5. *Let $n > 0$ and let $* \in S^{n-1}$ be any point. Then:*

$$H_q(S^n) = \begin{cases} \mathbb{Z} = \langle [\partial \iota_{n+1}] \rangle & \text{if} \quad q = n > 0 \\ \mathbb{Z} = \langle [c_*^0] \rangle & \text{if} \quad q = 0, n > 0 \\ \mathbb{Z} \oplus \mathbb{Z} = \langle [c_*^0], [\partial \iota_1] \rangle & \text{if} \quad q = n = 0 \\ 0 & \text{otherwise} \end{cases}$$

and

$$H_q(D^n, S^{n-1}) = \begin{cases} \mathbb{Z} = \langle [\iota_n] \rangle & \text{if } q = n \\ 0 & \text{otherwise}. \end{cases}$$

Proof. The division into cases for $H_q(S^n)$ can be eased by employing augmented homology. We already know that for $n \geq 0$

$$\partial : H_q(D^n, S^{n-1}) \to \widetilde{H}_{q-1}(S^{n-1})$$

is an isomorphism, so what remains is to check that

$$\widetilde{H}_q(S^{n-1}) = \begin{cases} \mathbb{Z} & \text{if } q = n - 1 \\ 0 & \text{if } q \neq n - 1. \end{cases}$$

This follows by an induction, using the pair of isomorphisms

$$\widetilde{H}_{q-1}(S^{n-1}) \xleftarrow{\cong} H_q(D^n, S^{n-1}) \xrightarrow{\cong} H_q(D^n/S^{n-1}, *),$$

since $D^n/S^{n-1} \cong S^n$. The right hand arrow is an isomorphism since S^{n-1} is a deformation retract of a neighborhood in D^n. □

Applications

Why should you care about this complicated homology calculation?

Corollary 10.6. *If $m \neq n$, then S^m and S^n are not homotopy equivalent.*

Proof. Their homology groups are not isomorphic. □

Corollary 10.7. *If $m \neq n$, then $\mathbb{R}^m$ and $\mathbb{R}^n$ are not homeomorphic.*

Proof. If m or n is zero, this is clear, so let $m, n > 0$. Assume we have a homeomorphism $f : \mathbb{R}^m \to \mathbb{R}^n$. This restricts to a homeomorphism $\mathbb{R}^m - \{0\} \to \mathbb{R}^n - \{f(0)\}$. But these spaces are homotopy equivalent to spheres of different dimension. □

Corollary 10.8 (Brouwer fixed-point theorem). *If $f : D^n \to D^n$ is continuous, then there is some point $x \in D^n$ such that $f(x) = x$.*

Proof. Suppose not. Then you can draw a ray from $f(x)$ through x. It meets the boundary of D^n at a point $g(x) \in S^{n-1}$. Check that $g : D^n \to S^{n-1}$ is continuous. If x is on the boundary, then $x = g(x)$, so g provides a factorization of the identity map on S^{n-1} through D^n. This is inconsistent with our computation because the identity map induces the identity map on $\widetilde{H}_{n-1}(S^{n-1}) \cong \mathbb{Z}$, while $\widetilde{H}_{n-1}(D^n) = 0$. □

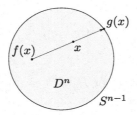

Our computation of the homology of a sphere also implies that there are many homotopy classes of self-maps of S^n, for any $n \geq 1$. We will distinguish them by means of the *degree*: A map $f : S^n \to S^n$ induces an endomorphism of the infinite cyclic group $H_n(S^n)$. Any endomorphism of an infinite cyclic group is given by multiplication by an integer. This integer is well defined (independent of a choice of generator), and any integer occurs. Thus $\mathrm{End}(\mathbb{Z}) = \mathbb{Z}_\times$, the monoid of integers under multiplication. The homotopy classes of self-maps of S^n also form a monoid, under composition, and:

Theorem 10.9. *Let $n \geq 1$. The degree map provides us with a surjective monoid homomorphism*

$$\deg : [S^n, S^n] \to \mathbb{Z}_\times .$$

Proof. Degree is multiplicative by functoriality of homology.

We construct a map of degree k on S^n by induction on n. If $n = 1$, this is just the winding number; an example is given by regarding S^1 as unit complex numbers and sending z to z^k. The proof that this has degree k is an exercise.

Suppose $n > 1$, and that we've constructed a map $f_k : S^{n-1} \to S^{n-1}$ of degree k. Extend it to a map $\overline{f}_k : D^n \to D^n$ by defining $\overline{f}_k(tx) = t f_k(x)$ for $t \in [0,1]$. We may then collapse the sphere to a point and identify the quotient with S^n. This gives us a new map $g_k : S^n \to S^n$ making the diagram below commute.

$$
\begin{array}{ccccc}
H_{n-1}(S^{n-1}) & \xleftarrow{\;\cong\;} & H_n(D^n, S^{n-1}) & \xrightarrow{\;\cong\;} & H_n(S^n) \\
\Big\downarrow{\scriptstyle f_{k*}} & & \Big\downarrow & & \Big\downarrow{\scriptstyle g_{k*}} \\
H_{n-1}(S^{n-1}) & \xleftarrow{\;\cong\;} & H_n(D^n, S^{n-1}) & \xrightarrow{\;\cong\;} & H_n(S^n)
\end{array}
$$

The horizontal maps are isomorphisms, so $\deg g_k = k$ as well. $\qquad\square$

We will see in Lecture 65 that this map is in fact an isomorphism.

Exercises

Exercise 10.10. This exercise generalizes our computation of the homology of spheres, and introduces several important constructions.

The *cone* on a space X is the quotient space $CX = X \times I / X \times \{0\}$, where I is the unit interval $[0, 1]$. The cone is a pointed space, with basepoint $*$ given by the "cone point," i.e. the image of $X \times \{0\}$. (By convention, the cone on the empty space $\varnothing$ is a single point, the cone point.) Regard X as the subspace of CX of all points of the form $(x, 1)$.

Define the *suspension* of a space X to be $SX = CX/X$. Make SX a pointed space by declaring the image of $X \subseteq CX$ to be the basepoint in SX. (By convention, the quotient $W/\varnothing$ is the disjoint union of W with a single point, which is declared to be the basepoint. So $S\varnothing = */\varnothing$ is the discrete two-point space, with the new point as basepoint.)

The quotient map induces a map of pairs $f : (CX, X) \to (SX, *)$.

(a) Show that CX is contractible.

For any $a, b \in I$ with $a \leq b$, let $C_a^b X$ denote the image of $X \times [a, b]$ in CX. Thus $C_0^1 X = CX$, $C_0^0 X = *$, and $C_1^1 X = X$.

Let $p : CX \to CX$ send (x, t) to $(x, 3t)$ for $t \leq 1/3$ and to $(x, 1)$ if $t \geq 1/3$.

(b) Show that p defines a homotopy equivalence of pairs $(C_0^{2/3} X, C_{1/3}^{2/3} X) \to (CX, X)$.

(c) Show that the evident map $e : (C_0^{2/3} X, C_{1/3}^{2/3} X) \to (SX, C_{1/3}^1 X/X)$ is an excision.

(d) Show that p defines a homotopy equivalence of pairs $(SX, C_{1/3}^1 X/X) \to (SX, *)$.

(e) Conclude from the commutativity of

$$
\begin{array}{ccc}
(C_0^{2/3} X, C_{1/3}^{2/3} X) & \xrightarrow{\ e\ } & (SX, C_{1/3}^1 X/X) \\
\downarrow & & \downarrow \\
(CX, X) & \xrightarrow{\ f\ } & (SX, *)
\end{array}
$$

that f induces an isomorphism in homology.

(f) Show that there is a natural isomorphism between augmented and reduced homology groups, $\widetilde{H}_{n-1}(X) \to \overline{H}_n(SX)$, for any n.

Exercise 10.11. (a) Verify the claim that the map $z \mapsto z^d$, sending the unit circle in the complex numbers to itself, has degree d.

(b) Regard S^{n-1} as the unit sphere in $\mathbb{R}^n$. Let L be a line through the origin in $\mathbb{R}^n$, and $L^\perp$ its orthogonal complement. Let ρ_L be the linear map given by -1 on L and $+1$ on $L^\perp$. What is $\deg(\rho_L|_{S^{n-1}})$?

(c) What is the degree of the "antipodal map," $\alpha : S^{n-1} \to S^{n-1}$ sending x to $-x$?

(d) The tangent space to a point x on the sphere S^{n-1} can be regarded as the subspace of $\mathbb{R}^n$ of vectors perpendicular to x. A "vector field" on S^{n-1} is thus a continuous function $v : S^{n-1} \to \mathbb{R}^n$ such that $v(x) \perp x$ for all $x \in S^{n-1}$. Show that if n is odd then every vector field vanishes at some point on the sphere. (When $n - 1 = 2$, this is the "hairy ball theorem," proved by Poincaré in 1885.) On the other hand, construct a nowhere vanishing vector field on S^{n-1} for any even n.

11 Eilenberg-Steenrod axioms and the locality principle

Before we proceed to prove the excision theorem, let's review the properties of singular homology as we have developed them. They are captured by a set of axioms, due to Sammy Eilenberg and Norman Steenrod [18]. (Steenrod (1910–1971) was a highly influential topologist, a student and later colleague of Lefschetz's at Princeton.)

Definition 11.1. A *homology theory* (on **Top**) is:

- a sequence of functors $h_n : \mathbf{Top_2} \to \mathbf{Ab}$ for all $n \in \mathbb{Z}$ and
- a sequence of natural transformations $\partial : h_n(X, A) \to h_{n-1}(A, \varnothing)$

such that:

- (Homotopy invariance) If $f_0, f_1 : (X, A) \to (Y, B)$ are homotopic, then $f_{0*} = f_{1*} : h_n(X, A) \to h_n(Y, B)$.
- (Excision) Excisions induce isomorphisms.
- (Long exact sequence) For any pair (X, A), the sequence

$$\cdots \to h_{q+1}(X, A) \xrightarrow{\partial} h_q(A) \to h_q(X) \to h_q(X, A) \xrightarrow{\partial} \cdots$$

  is exact, where we have written $h_q(X)$ for $h_q(X, \varnothing)$.
- (Dimension axiom) The group $h_n(*)$ is nonzero only for $n = 0$.

We add the following "Milnor axiom" [43] to our definition. To state it, let I be a set and suppose that for each $i \in I$ we have a space X_i. We can form their disjoint union or *coproduct* $\coprod X_i$. The inclusion maps

$X_i \to \coprod X_i$ induce maps $h_n(X_i) \to h_n(\coprod X_i)$, and these in turn induce a map from the direct sum, or coproduct, in **Ab**:

$$\alpha : \bigoplus_{i \in I} h_n(X_i) \to h_n \left(\coprod_{i \in I} X_i \right).$$

Then:

- (Milnor axiom) The map α is an isomorphism for all n.

Singular homology satisfies these, with $h_0(*) = \mathbb{Z}$. We will soon add "coefficients" to homology, producing a homology theory whose value on a point is any prescribed abelian group. Eilenberg and Steenrod enunciated these axioms with the goal of bringing order to the plethora of variants of singular homology that were appearing at the time. But in later developments, it emerged that the dimension axiom is rather like the parallel postulate in Euclidean geometry: It's "obvious," but, as it turns out, the remaining axioms accommodate extremely interesting alternatives, in which $h_n(*)$ is nonzero for many — often infinitely many — values of n, often both positive and negative. With the dimension axiom in place, one has "ordinary homology."

Locality

Excision is a statement that homology is "localizable." To make this precise, we need some definitions.

Definition 11.2. Let X be a topological space. A family $\mathcal{A}$ of subsets of X is a *cover* if X is the union of the interiors of elements of $\mathcal{A}$.

Definition 11.3. Let $\mathcal{A}$ be a cover of X. An n-simplex σ is $\mathcal{A}$-*small* if there is $A \in \mathcal{A}$ such that the image of σ is entirely in A.

Notice that if $\sigma : \Delta^n \to X$ is $\mathcal{A}$-small, then so is $d_i\sigma$; in fact, for any simplicial operator ϕ, $\phi^*\sigma$ is again $\mathcal{A}$-small. Let's denote by $\mathrm{Sin}_*^{\mathcal{A}}(X)$ the graded set of $\mathcal{A}$-small simplices. This is a sub-simplicial set of $\mathrm{Sin}_*(X)$. Applying the free abelian group functor, we get the subcomplex

$$S_*^{\mathcal{A}}(X) \subseteq S_*(X)$$

of $\mathcal{A}$-*small singular chains*. Write $H_*^{\mathcal{A}}(X)$ for its homology.

Theorem 11.4 (The locality principle). *The inclusion $S_*^{\mathcal{A}}(X) \subseteq S_*(X)$ induces an isomorphism in homology, $H_*^{\mathcal{A}}(X) \xrightarrow{\cong} H_*(X)$.*

This will take a little time to prove. Let's see right now how it implies excision.

Suppose $X \supset A \supset U$ is excisive, so that $\overline{U} \subseteq \mathrm{Int}(A)$, or $\mathrm{Int}(X - U) \cup \mathrm{Int}(A) = X$. Thus if we let $B = X - U$, then $\mathcal{A} = \{A, B\}$ is a cover of X. Rewriting in terms of B,

$$(X - U, A - U) = (B, A \cap B),$$

so we aim to show that

$$S_*(B, A \cap B) \to S_*(X, A)$$

induces an isomorphism in homology. We have the following diagram of chain complexes with exact rows:

$$
\begin{array}{ccccccccc}
0 & \longrightarrow & S_*(A) & \longrightarrow & S_*^{\mathcal{A}}(X) & \longrightarrow & S_*^{\mathcal{A}}(X)/S_*(A) & \longrightarrow & 0 \\
 & & \downarrow{=} & & \downarrow & & \downarrow & & \\
0 & \longrightarrow & S_*(A) & \longrightarrow & S_*(X) & \longrightarrow & S_*(X, A) & \longrightarrow & 0 .
\end{array}
$$

The middle vertical induces an isomorphism in homology by the locality principle, so the homology long exact sequences combined with the five-lemma shows that the right hand vertical is also a homology isomorphism. But

$$S_n^{\mathcal{A}}(X) = S_n(A) + S_n(B) \subseteq S_n(X)$$

and a simple result about abelian groups provides an isomorphism

$$\frac{S_n(B)}{S_n(A \cap B)} = \frac{S_n(B)}{S_n(A) \cap S_n(B)} \xrightarrow{\cong} \frac{S_n(A) + S_n(B)}{S_n(A)} = \frac{S_n^{\mathcal{A}}(X)}{S_n(A)},$$

so excision follows.

This case of a cover with two elements leads to another expression of excision, known as the "Mayer-Vietoris sequence." In describing it we will use the following notation for the various inclusion.

$$
\begin{array}{ccc}
A \cap B & \xrightarrow{\ j_1\ } & A \\
\downarrow{j_2} & & \downarrow{i_1} \\
B & \xrightarrow[\ i_2\ ]{} & X
\end{array}
$$

Theorem 11.5 (Mayer-Vietoris). *Assume that $\{A, B\}$ is a cover of X. There are natural maps $\partial : H_n(X) \to H_{n-1}(A \cap B)$ such that the sequence*

$$\cdots \xrightarrow{\beta} H_{n+1}(X)$$

$$H_n(A \cap B) \xleftarrow{\ \ \alpha\ \ } H_n(A) \oplus H_n(B) \xrightarrow{\ \beta\ } H_n(X)$$

$$H_{n-1}(A \cap B) \xrightarrow{\ \ \alpha\ \ } \cdots$$

is exact, where

$$\alpha = \begin{bmatrix} j_{1*} \\ -j_{2*} \end{bmatrix}, \qquad \beta = \begin{bmatrix} i_{1*} & i_{2*} \end{bmatrix}.$$

Proof. This is the homology long exact sequence associated to the short exact sequence of chain complexes

$$0 \to S_*(A \cap B) \xrightarrow{\alpha} S_*(A) \oplus S_*(B) \xrightarrow{\beta} S_*^A(X) \to 0,$$

combined with the locality principle. $\square$

The Mayer-Vietoris theorem follows from excision as well, via the following simple observation. Suppose we have a map of long exact sequences

$$\cdots \longrightarrow C'_{n+1} \xrightarrow{k} A'_n \longrightarrow B'_n \longrightarrow C'_n \longrightarrow \cdots$$
$$\downarrow h \qquad \downarrow f \qquad \downarrow \cong \qquad \downarrow h$$
$$\cdots \longrightarrow C_{n+1} \xrightarrow{k} A_n \longrightarrow B_n \longrightarrow C_n \longrightarrow \cdots$$

in which every third arrow is an isomorphism as indicated. Define a map

$$\partial : A_n \to B_n \xleftarrow{\cong} B'_n \to C'_n.$$

An easy diagram chase shows:

Lemma 11.6. *The sequence*

$$\cdots \longrightarrow C'_{n+1} \xrightarrow{\begin{bmatrix} h \\ -k \end{bmatrix}} C_{n+1} \oplus A'_n \xrightarrow{\begin{bmatrix} k & f \end{bmatrix}} A_n \xrightarrow{\partial} C'_n \longrightarrow \cdots$$

is exact.

To get the Mayer-Vietoris sequence, let $\{A, B\}$ be a cover of X and apply the lemma to the ladder

$$\cdots \to H_n(A \cap B) \to H_n(B) \to H_n(B, A \cap B) \to H_{n-1}(A \cap B) \to \cdots$$
$$\downarrow \qquad \downarrow \qquad \downarrow \cong \qquad \downarrow$$
$$\cdots \longrightarrow H_n(A) \longrightarrow H_n(X) \longrightarrow H_n(X, A) \longrightarrow H_{n-1}(A) \longrightarrow \cdots.$$

Exercises

Exercise 11.7. Verify Lemma 11.6

Exercise 11.8. (a) Use the Mayer-Vietoris sequence to compute the homology groups of the projective plane P, the Klein bottle K, and the torus T. (The projective plane is obtained by sewing a disk onto a Möbius band along their boundaries. The Klein bottle is obtained either by sewing two Möbius bands together, or by sewing the two boundary components of a cylinder together in a funny way. A torus is obtained by sewing the boundary components of a cylinder together in a less funny way. In each case, it's a good idea to give yourself a hem: glue open "collars" together.)
(b) Hopefully you computed that $H_2(T)$ is an infinite cyclic group. Say something sensible about whether the "fundamental class" you constructed in Exercise 2.2 is indeed a generator of that abelian group.

Exercise 11.9. State and prove a version of the Mayer-Vietoris long exact sequence for relative homology.

12   Subdivision

We will begin the proof of the locality principle today, and finish it in the next lecture. The key is a process of subdivision of singular simplices. It will use the "cone construction" $b*$ from Lecture 5. The cone construction dealt with a region X in Euclidean space, star-shaped with respect to $b \in X$, and gave a chain-homotopy between the identity and the "constant map" on $S_*(X)$:

$$db* + b* \, d = 1 - \eta_b \varepsilon$$

where $\varepsilon : S_*(X) \to \mathbb{Z}$ is the augmentation and $\eta_b : \mathbb{Z} \to S_*(X)$ sends 1 to the constant 0-chain c_b^0.

The cone construction can be used to "subdivide" an "affine simplex." An *affine simplex* is the convex hull of a finite set of points in Euclidean space. To make this non-degenerate, assume that the points $v_0, v_1, \ldots, v_n$, have the property that $\{v_1 - v_0, \ldots, v_n - v_0\}$ is linearly independent. The *barycenter* of this simplex is the center of mass of the vertices,

$$b = \frac{1}{n+1} \sum v_i \, .$$

Start with $n = 1$. To subdivide a 1-simplex, just cut it in half. For the 2-simplex, look at the subdivision of each face, and form the cone of them with the barycenter of the 2-simplex. This gives us a decomposition of the 2-simplex into six sub-simplices.

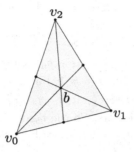

We want to formalize this process, and extend it to singular simplices (using naturality, of course). Define a natural transformation

$$\$: S_n(X) \to S_n(X),$$

the *subdivision operator*, by defining it first on the standard n-simplex, namely by specifying what $\$(\iota_n)$ is where $\iota_n : \Delta^n \to \Delta^n$ is the universal n-simplex, and then extending by naturality:

$$\$(\sigma) = \sigma_* \$(\iota_n).$$

Here's the definition. When $n = 0$, define $\$$ to be the identity; i.e., $\$\iota_0 = \iota_0$. For $n > 0$, define

$$\$\iota_n = b_n * \$d\iota_n,$$

where b_n is the barycenter of Δ^n. This makes a lot of sense if you draw a picture, and it's a very clever definition that captures the geometry we described.

The dollar sign symbol is a little odd, but consider: it derives from the symbol for the Spanish piece of eight, which was meant to be subdivided (so for example two bits is a quarter).

Here's what we'll prove.

Proposition 12.1. $\$$ *is a natural chain map* $S_*(X) \to S_*(X)$ *that is naturally chain-homotopic to the identity.*

Proof. Let's begin by proving that it's a chain map. We'll use induction on n. It's enough to show that $d\$\iota_n = \$d\iota_n$, because then, for any n-simplex σ,

$$d\$\sigma = d\$\sigma_*\iota_n = \sigma_* d\$\iota_n = \sigma_* \$d\iota_n = \$d\sigma_*\iota_n = \$d\sigma.$$

Dimension zero is easy: since $S_{-1} = 0$, $d\$\iota_0$ and $\$d\iota_0$ are both zero and hence equal.

For $n \geq 1$, we want to compute $d\$\iota_n$. Using the fact that $b*$ is a chain homotopy, we compute:

$$d\$\iota_n = d(b_n * \$d\iota_n)$$
$$= (1 - \eta_b\varepsilon - b_n * d)(\$d\iota_n).$$

The map ε here is nonzero when $n = 1$, but

$$\eta_b\varepsilon\$d\iota_1 = \eta_b\varepsilon\$(c_1^0 - c_0^0) = \eta_b\varepsilon(c_1^0 - c_0^0) = 0 ,$$

since ε takes sums of coefficients. So the $\eta_b\varepsilon$ term drops out for any $n \geq 1$. Now we can continue, using the inductive hypothesis:

$$d\$\iota_n = (1 - b_n * d)(\$d\iota_n)$$
$$= \$d\iota_n - b_n * d\$d\iota_n$$
$$= \$d\iota_n - b_n\$d^2\iota_n$$
$$= \$d\iota_n$$

because $d^2 = 0$. So $\$$ is a chain map.

To define the chain homotopy T, we'll just write down a formula and then justify the choice. Making use of naturality, it suffices to define $T\iota_n$. Here it is:

$$T\iota_n = b_n * (\$\iota_n - \iota_n - Td\iota_n) \in S_{n+1}(\Delta^n) .$$

This formula is inductive. The expression $Td\iota_n$ is defined using naturality of T. When $n = 0$ the right hand side is 0, and this starts the induction.

Once again, we're going to check that T is a chain homotopy by induction, and, again, we need to check only on the universal case.

When $n = 0$, it's true that $dT\iota_0 - Td\iota_0 = \$\iota_0 - \iota_0$ since $T\iota_0 = 0$ and $\$\iota_0 = \iota_0$. Now let's assume that $dTc - Tdc = \$c - c$ for every $(n-1)$-chain c. Start by computing $dT\iota_n$:

$$dT\iota_n = d(b_n * (\$\iota_n - \iota_n - Td\iota_n))$$
$$= (1 - b_n * d)(\$\iota_n - \iota_n - Td\iota_n)$$
$$= \$\iota_n - \iota_n - Td\iota_n - b_n * (d\$\iota_n - d\iota_n - dTd\iota_n).$$

All we want now is that $b_n * (d\$\iota_n - d\iota_n - dTd\iota_n) = 0$. We can do this using the inductive hypothesis, because $d\iota_n$ is in dimension $n-1$. Compute:

$$dTd\iota_n = -Td(d\iota_n) + \$d\iota_n - d\iota_n$$
$$= \$d\iota_n - d\iota_n$$
$$= d\$\iota_n - d\iota_n .$$

This means that $d\$\iota_n - d\iota_n - dTd\iota_n = 0$, so T is indeed a chain homotopy.

$\square$

Exercises

Exercise 12.2. The constructions sketched in Exercise 11.8 are examples of the following general procedure. Take two closed surfaces, Σ_1 and Σ_2, cut a disk out from each one, and glue them together along the hem. This is the *connected sum* $\Sigma_1 \# \Sigma_2$. Write T_1 for the torus, and $T_g = T_1 \# T_{g-1}$. Write P_1 for the projective plane, and $P_g = P_1 \# P_{g-1}$. A theorem of Rado (e.g. [35]) asserts that this is a complete list of compact connected 2-manifolds.
(a) What is the Klein bottle, in this notation?
(b) Complete the work from Exercise 11.8: compute the homology groups of these closed surfaces. Show that $H_1(\Sigma_1 \# \Sigma_2) \cong H_1(\Sigma_1) \oplus H_1(\Sigma_2)$.

13 Proof of the locality principle

We have constructed the subdivision operator $\$: S_*(X) \to S_*(X)$, with the idea that it will shrink chains and by iteration eventually render any chain $\mathcal{A}$-small. Does $\$$ succeed in making simplices smaller? Let's look first at the affine case. Recall that the "diameter" of a metric space X is given by

$$\mathrm{diam}(X) = \sup\{d(x,y) : x, y \in X\}.$$

Lemma 13.1. *Let σ be an affine n-simplex. Its diameter is the maximum distance between vertices. Let τ be the image of an n-simplex in $\$\sigma$. It is another affine n-simplex, and* $\mathrm{diam}(\tau) \leq \frac{n}{n+1}\mathrm{diam}(\sigma)$.

Proof. Suppose that the vertices of σ are $v_0, v_1, \ldots, v_n$. The simplex is the convex hull of its verticies, so its diameter is the maximal distance between vertices. Let b be the barycenter of σ, and write the vertices of τ as $w_0 = b, w_1, \ldots, w_n$. We want to estimate $|w_i - w_j|$. First, compute

$$|b - v_i| = \left| \frac{v_0 + \cdots + v_n - (n+1)v_i}{n+1} \right| = \left| \frac{(v_0 - v_i) + \cdots + (v_n - v_i)}{n+1} \right|.$$

One of the terms in the numerator is zero, so we can continue:

$$|b - v_i| \leq \frac{n}{n+1} \max_{i,j} |v_i - v_j| = \frac{n}{n+1}\mathrm{diam}(\sigma).$$

Since $w_i \in \sigma$,

$$|b - w_i| \leq \max_i |b - v_i| \leq \frac{n}{n+1}\mathrm{diam}(\sigma).$$

For the other cases, we use induction: The w_i for $i > 0$ are the vertices of a simplex in $d\sigma$, so

$$|w_i - w_j| \le \mathrm{diam}(\text{simplex in } d\sigma) \le \frac{n-1}{n} \max_k \{\mathrm{diam}(d_k\sigma)\}$$

$$\le \frac{n}{n+1}\mathrm{diam}(\sigma). \qquad \square$$

Now let's transfer this calculation to singular simplices in a space X equipped with a cover $\mathcal{A}$.

Lemma 13.2. *For any singular chain c, some iterate of the subdivision operator sends c to an $\mathcal{A}$-small chain.*

Proof. We may assume that c is a single simplex $\sigma : \Delta^n \to X$, because in general you just take the largest of the iterates of \$ needed to send the simplices in c to a $\mathcal{A}$-small chains. We now encounter again the great virtue of singular homology: We pull $\mathcal{A}$ back to a cover of the standard simplex. Define an open cover of Δ^n by

$$\mathcal{U} = \{\sigma^{-1}(\mathrm{Int}(A)) : A \in \mathcal{A}\}.$$

The space Δ^n is a compact metric space, and so is subject to the Lebesgue covering lemma, which we apply to the open cover $\mathcal{U}$ to finish the proof. $\quad\square$

Lemma 13.3 (Lebesgue covering lemma; e.g. [10, I.9.11]). *Let M be a compact metric space, and let $\mathcal{U}$ be an open cover. Then there is $\epsilon > 0$ such that for all $x \in M$, $B_\epsilon(x) \subseteq U$ for some $U \in \mathcal{U}$.*

So iterating the subdivision operator does produce small chains. The remaining ingredient in the proof of the Locality Principal is now the following observation.

Lemma 13.4. *For any $k \ge 1$, $\$^k \simeq 1 : S_*(X) \to S_*(X)$.*

Proof. We construct T_k such that $dT_k + T_k d = \$^k - 1$. To begin, we take $T_1 = T$, the chain homotopy from Proposition 12.1, since $dT + Td = \$ - 1$. Let's apply \$ to this equation. We get $\$dT + \$Td = \$^2 - \$$. Sum up these two equations to get

$$dT + Td + \$dT + \$Td = \$^2 - 1,$$

which simplifies to

$$d(\$ + 1)T + (\$ + 1)Td = \$^2 - 1$$

since $\$d = d\$$.

So define $T_2 = (\$ + 1)T$. Continuing, you see that we can define

$$T_k = (\$^{k-1} + \$^{k-2} + \cdots + 1)T. \qquad \square$$

We are now in position to prove:

Theorem 13.5 (The locality principle). *Let $\mathcal{A}$ be a cover of a space X. The inclusion $S_*^{\mathcal{A}}(X) \subseteq S_*(X)$ is a quasi-isomorphism; that is, $H_*^{\mathcal{A}}(X) \to H_*(X)$ is an isomorphism.*

Proof. To prove surjectivity let c be an n-cycle in X. We want to find an $\mathcal{A}$-small n-cycle that is homologous to c. There's only one thing to do. Pick k such that $\$^k c$ is $\mathcal{A}$-small. It is a cycle because because $\$^k$ is a chain map. I want to compare this new cycle with c. That's what the chain homotopy T_k is designed for:

$$\$^k c - c = dT_k c + T_k dc = dT_k c$$

since c is a cycle. So $\$^k c$ and c are homologous.

Now for injectivity. Suppose c is a cycle in $S_n^{\mathcal{A}}(X)$ such that $c = db$ for some $b \in S_{n+1}(X)$. We want c to also be a boundary of an $\mathcal{A}$-small chain. Use the chain homotopy T_k again: Suppose that k is such that $\$^k b$ is $\mathcal{A}$-small. Compute:

$$d\$^k b - c = d(\$^k - 1)b = d(dT_k + T_k d)b = dT_k c$$

so

$$c = d\$^k b - dT_k c = d(\$^k b - T_k c).$$

Now $\$^k b$ is $\mathcal{A}$-small, by choice of k. I claim that $T_k c$ is also $\mathcal{A}$-small. It's enough to show that $T_k \sigma$ is $\mathcal{A}$-small if σ is. We know that $\sigma = \sigma_* \iota_n$. Because σ is $\mathcal{A}$-small, we know that $\sigma : \Delta^n \to X$ is the composition $i_* \overline{\sigma}$ where $\overline{\sigma} : \Delta^n \to A$ and $i : A \to X$ is the inclusion of some $A \in \mathcal{A}$. By naturality, then, $T_k \sigma = T_k i_* \overline{\sigma} = i_* T_k \overline{\sigma}$, which is certainly $\mathcal{A}$-small. $\square$

This completes the proof of the Eilenberg-Steenrod axioms for singular homology. In the next chapter, we will develop a variety of practical tools, using these axioms to compute the singular homology of many spaces.

Exercises

Exercise 13.6. Let $\mathcal{A}$ be a cover of a space X. For any simplex in X, let $k(\sigma)$ be the smallest integer such that $\$^k \sigma$ is $\mathcal{A}$-small. Define a map $S_*(X) \to S_*^{\mathcal{A}}(X)$ by sending each simplex σ to $\$^{k(\sigma)}\sigma$. Show that this defines a homotopy inverse of the inclusion map.

Lefschetz progeny

According to the Mathematical Genealogy Project, Solomon Lefschetz has had more than 10,000 academic descendants. Here are just a few, with special attention to MIT faculty (marked with an asterisk).

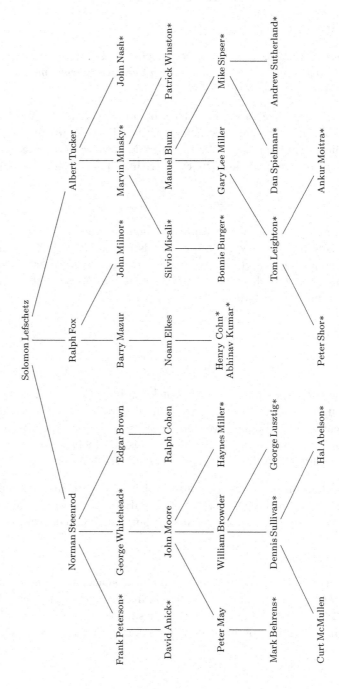

Chapter 2

Computational methods

14 CW complexes I

There are various ways to model geometrically interesting spaces. Manifolds provide one important model, well suited to analysis. Another model is given by simplicial complexes. That theory is very combinatorial, and constructing a simplicial complex model for a given space involves making a lot of choices that are combinatorial rather than topological in character. We won't discuss simplicial complexes further in this course; we refer you to [50] for example for a treatment of them. A more flexible model, one more closely reflecting topological information, is provided by the theory of CW complexes.

In building up a space as a CW complex, we will successively "glue" "cells," dimension by dimension, onto what has been already built. This gluing process is a general construction.

Suppose we have a pair (B, A), and a map $\alpha : A \to X$. Define a space $X \cup_\alpha B$ (or $X \cup_A B$) in the diagram

$$
\begin{array}{ccc}
A & \xrightarrow{\ \alpha\ } & X \\
\uparrow & & \downarrow \\
B & \longrightarrow & X \cup_\alpha B
\end{array}
$$

by

$$ X \cup_\alpha B = X \amalg B / \sim $$

where the equivalence relation on the coproduct is generated by requiring that $a \sim \alpha(a)$ for all $a \in A$. We say that we have "attached B to X along α."

There are two kinds of equivalence classes in $X \cup_\alpha B$: (1) singletons containing elements of $B - A$, and (2) $\{x\} \amalg \alpha^{-1}(x)$ for $x \in X$. The space $X \cup_\alpha B$ is characterized by a universal property: any solid-arrow commutative diagram

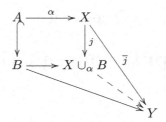

can be uniquely filled in. It's a "push-out."

Example 14.1. With $X = *$, $* \cup_A B = B/A$.

Example 14.2. With $A = \varnothing$, $X \cup_\varnothing B$ is the coproduct $X \amalg B$.

Example 14.3. With both,

$$B/\varnothing = * \cup_\varnothing B = * \amalg B.$$

For example, $\varnothing/\varnothing = *$. This is creation from nothing! We won't get into the religious ramifications.

Example 14.4 (Attaching a cell). A basic collection of pairs of spaces is given by the disks relative to their boundaries: (D^n, S^{n-1}). (Recall that $S^{-1} = \varnothing$.) In this context, D^n is called an "n-cell," and a map $\alpha : S^{n-1} \to X$ allows us to attach an n-cell to X, to form

$$\begin{array}{ccc} S^{n-1} & \xrightarrow{\ \alpha\ } & X \\ \downarrow & & \downarrow \\ D^n & \longrightarrow & X \cup_\alpha D^n. \end{array}$$

You might want to generalize this a little bit, and attach a bunch of n-cells all at once:

$$\begin{array}{ccc} \coprod_{i \in I_n} S_i^{n-1} & \xrightarrow{\ \alpha\ } & X \\ \downarrow & & \downarrow \\ \coprod_{i \in I_n} D_i^n & \longrightarrow & X \cup_\alpha \coprod_{i \in I_n} D_i^n. \end{array}$$

What are some examples? When $n = 0$, $(D^0, S^{-1}) = (*, \varnothing)$, so you are just adding a discrete set to X:

$$X \cup_\alpha \coprod_{i \in I_0} D^0 = X \amalg I_0.$$

More interesting: Let's attach two 1-cells to a point:

$$S^0 \amalg S^0 \xrightarrow{\ \alpha\ } *$$

$$D^1 \amalg D^1 \longrightarrow * \cup_\alpha (D^1 \amalg D^1).$$

Again there's just one choice for α, and $*\cup_\alpha (D^1 \amalg D^1)$ is a figure 8, because you start with two 1-disks and identify the four boundary points together. Let me write $S^1 \vee S^1$ for this space. We can go on and attach a single 2-cell to manufacture a torus. Think of the figure 8 as the perimeter of a square with opposite sides identified.

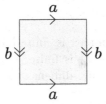

The inside of the square is a 2-cell, attached to the perimeter by a map I'll denote by $aba^{-1}b^{-1}$:

$$S^1 \xrightarrow{\ aba^{-1}b^{-1}\ } S^1 \vee S^1$$

$$D^2 \longrightarrow (S^1 \vee S^1) \cup_{aba^{-1}b^{-1}} D^2 = T^2 .$$

This example illuminates the following definition.

Definition 14.5. A *CW complex* is a space X equipped with a sequence of subspaces

$$\varnothing = \mathrm{Sk}_{-1}X \subseteq \mathrm{Sk}_0X \subseteq \mathrm{Sk}_1X \subseteq \cdots \subseteq X$$

such that

- X is the union of the $\mathrm{Sk}_n X$'s, and
- for all n, there is a pushout diagram like this:

$$\coprod_{i \in I_n} S_i^{n-1} \xrightarrow{\ \alpha_n\ } \mathrm{Sk}_{n-1}X$$

$$\coprod_{i \in I_n} D_i^n \xrightarrow{\ \beta_n\ } \mathrm{Sk}_n X .$$

The subspace $\mathrm{Sk}_n X$ is the *n-skeleton* of X. Sometimes it's convenient to use the alternate notation X_n for the *n*-skeleton. The first condition is intended topologically, so that a subset of X is open if and only if its intersection with each $\mathrm{Sk}_n X$ is open; or, equivalently, a map $f : X \to Y$ is continuous if and only if its restriction to each $\mathrm{Sk}_n X$ is continuous. The maps α_n are the *attaching maps* and the maps β_n are *characteristic maps*.

The sequence of skeleta determines the CW structure. A *cell structure* consists of a choice of attaching and characteristic maps. Generally a CW complex has many cell structures; it's a bit like choosing a basis in a vector space.

Example 14.6. We just constructed the torus as a CW complex with $\mathrm{Sk}_0 T^2 = *$, $\mathrm{Sk}_1 T^2 = S^1 \vee S^1$, and $\mathrm{Sk}_2 T^2 = T^2$.

Definition 14.7. A CW complex is *finite-dimensional* if $\mathrm{Sk}_n X = X$ for some n; of *finite type* if each I_n is finite, i.e., finitely many cell in each dimension; and *finite* if it is finite-dimensional and of finite type.

The *dimension* of a CW complex is the largest n for which there are *n*-cells. This is not obviously a topological invariant, but, have no fear, it turns out that it is.

In "CW," the "C" is for cell, and the "W" is for weak, because of the topology on a CW complex. This definition is due to J. H. C. Whitehead (1904–1960, Oxford University, one of the founding fathers of both homotopy theory and geometric topology).

Here are a couple of important facts about them.

Theorem 14.8. *(1) Any CW complex is Hausdorff, and it is compact if and only if it is finite.*
(2) Any compact smooth manifold admits a CW structure.

Proof. See [10] Prop. IV.8.1, and [24] Prop. A.3.   □

A *cell* of a CW complex is the image of a characteristic map. The *interior* of D^n is $D^n \backslash S^{n-1}$ (so for example the interior of D^0 is just D^0), and a *cell interior* is the image of the interior of a disk under a characteristic map. Neglecting the topology, a CW complex is the disjoint union of its cell interiors. A CW complex is *locally finite* if each cell meets only finitely many other cells. An infinite bouquet of circles is thus not locally finite.

Theorem 14.9. *A CW complex is locally compact if and only if it is locally finite. A connected CW complex is metrizable if and only if it is locally finite.*

Proof. See [20], Prop. 1.5.17. □

Exercises

Exercise 14.10. Provide the Euclidean space $\mathbb{R}^n$ with the structure of a CW complex.

Exercise 14.11. Provide each compact connected surface with the structure of a CW complex with just a single 0-cell and a single 2-cell.

15   CW complexes II

There are a few more general things to say about CW complexes, and some important examples.

Definition 15.1. Let X be a CW complex with a cell structure $\{\beta_i : D_i^n \to X_n : i \in I_n, n \geq 0\}$. A *subcomplex* of X (with this cell structure) is a subspace $Y \subseteq X$ such that for all n there is a subset J_n of I_n such that $Y_n = Y \cap X_n$ provides Y with a CW structure with characteristic maps $\{\beta_j : j \in J_n, n \geq 0\}$.

Example 15.2. $\mathrm{Sk}_n X \subseteq X$ is a subcomplex.

Proposition 15.3. *Let X be a CW complex with a chosen cell structure. Any compact subspace of X lies in some finite subcomplex.*

Proof. See [10], p. 196. □

Remark 15.4. For fixed cell structures, unions and intersections of subcomplexes are subcomplexes.

The n-sphere S^n (for $n > 0$) admits a very simple CW structure: Let $* = \mathrm{Sk}_0(S^n) = \mathrm{Sk}_1(S^n) = \cdots = \mathrm{Sk}_{n-1}(S^n)$, and attach an n-cell using the unique map $S^{n-1} \to *$. This is a minimal CW structure — you need at least two cells to build S^n.

This is great — much simpler than the simplest construction of S^n as a simplicial complex, for example — but it is not ideal for all applications.

Here's another CW structure on S^n. Regard S^n as the unit sphere in $\mathbb{R}^{n+1}$, filter the Euclidean space by leading subspaces $\mathbb{R}^k = \langle e_1, \dots, e_k \rangle$, and define

$$\mathrm{Sk}_k S^n = S^n \cap \mathbb{R}^{k+1} = S^k.$$

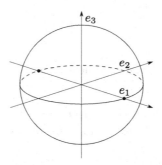

Now there are two k-cells for each k with $0 \le k \le n$, given by the two hemispheres of S^k. For each k there are two characteristic maps,

$$u_k, \ell_k : D^k \to S^k$$

defining the upper and lower hemispheres:

$$u_k(x) = (x, \sqrt{1 - |x|^2}), \quad \ell_k(x) = (x, -\sqrt{1 - |x|^2}).$$

Note that if $|x| = 1$ then $|u_k(x)| = |\ell_k(x)| = 1$, so each characteristic map restricts on the boundary to a map to S^{k-1} — to the same map, in fact! — and these restrictions serve as attaching maps. This cell structure has the advantage that S^{n-1} is a subcomplex of S^n.

The case $n = \infty$ is allowed here. Then $\mathbb{R}^\infty$ denotes the countably infinite dimensional inner product space that is the topological union of the leading subspaces $\mathbb{R}^n$. The CW complex S^∞ is of finite type but not finite dimensional. It has the following interesting property. We know that S^n is not contractible for any finite n (because $\widetilde{H}_*(S^n) \ne 0$), but:

Proposition 15.5. S^∞ *is contractible.*

Proof. This is an example of a "swindle" (another concept introduced by Eilenberg), making use of infinite dimensionality. Let $T : \mathbb{R}^\infty \to \mathbb{R}^\infty$ send $(x_1, x_2, \dots)$ to $(0, x_1, x_2, \dots)$. This sends S^∞ to itself. The location of the leading nonzero entry is different for x and Tx, so the line segment joining x to Tx doesn't pass through the origin. Therefore

$$x \mapsto \frac{tx + (1-t)Tx}{|tx + (1-t)Tx|}$$

provides a homotopy $1 \simeq T$. On the other hand, T is homotopic to the constant map with value $(1, 0, 0, \ldots)$, again by an affine homotopy. $\qquad\square$

This "inefficient" CW structure on S^n has a second advantage: it's *equivariant* with respect to the antipodal involution. This provides us with a CW structure on the orbit space for this action. This is the *real projective space* of dimension k: $\mathbb{RP}^k = S^k / \sim$ where $x \sim -x$. The quotient map $\pi : S^k \to \mathbb{RP}^k$ is a double cover, identifying upper and lower hemispheres. The inclusion of one sphere in the next is compatible with this equivalence relation, and gives us "linear" embeddings $\mathbb{RP}^{k-1} \subseteq \mathbb{RP}^k$. This suggests that

$$\varnothing \subseteq \mathbb{RP}^0 \subseteq \mathbb{RP}^1 \subseteq \cdots \subseteq \mathbb{RP}^n$$

might serve as a CW filtration. Indeed, for each k,

$$
\begin{array}{ccc}
S^{k-1} & \longrightarrow & D^k \\
\downarrow{\scriptstyle \pi} & & \downarrow{\scriptstyle u} \\
\mathbb{RP}^{k-1} & \longrightarrow & \mathbb{RP}^k
\end{array}
$$

is a pushout: A line in $\mathbb{R}^{k+1}$ either lies in $\mathbb{R}^k$ or is determined by a unique point in the upper hemisphere of S^k.

A CW structure on the Grassmannian

More generally, the Grassmannian $\mathrm{Gr}_k(\mathbb{R}^n)$ of k dimensional vector subspaces of $\mathbb{R}^n$ admits a beautiful and explicit CW structure, which we describe following [46].

The Grassmannian is very symmetric — it has a transitive action by the Lie group $SO(n)$ of rotations in $\mathbb{R}^n$ — but to define a CW structure on it we must break this symmetry. This symmetry breaking occurs by picking a complete flag in $\mathbb{R}^n$. Any one will do (and the space of complete flags is acted on freely and transitively by $SO(n)$), so let's just agree to use the flag determined by the standard ordered basis: so

$$0 = \mathbb{R}^0 \subset \mathbb{R}^1 \subset \mathbb{R}^2 \subset \cdots \subset \mathbb{R}^n$$

where $\mathbb{R}^i$ is the "leading subspace," spanned by $\{e_1, \ldots, e_i\}$. The ith coordinate function x_i maps $\mathbb{R}^i$ surjectively to $\mathbb{R}$ with kernel $\mathbb{R}^{i-1}$.

Let $V \in \mathrm{Gr}_k(\mathbb{R}^n)$. Intersecting V with this flag gives a filtration of V by vector subspaces,

$$0 = V_0 \subseteq V_1 \subseteq \cdots \subseteq V, \quad V_i = V \cap \mathbb{R}^i.$$

It's still the case that $V_{i-1} = \ker(x_i : V_i \to \mathbb{R})$, but the restriction of x_i to V_i is no longer necessarily surjective. When it's zero, $V_{i-1} = V_i$. When it's surjective, $\dim V_i = 1 + \dim V_{i-1}$.

The k-plane V determines a weakly increasing sequence of integers

$$0 \le a_1 \le a_2 \le \cdots \le a_k \le n - k$$

by requiring that $i + a_i$ is the smallest index j for which $\dim V_j = i$. This sequence $a = (a_1, \ldots, a_k)$ is the *type* of V. Write $\mathring{e}(a)$ for the subset of $\mathrm{Gr}_k(\mathbb{R}^n)$ consisting of all k-planes of type a. They will be the cell interiors in a CW decomposition of $\mathrm{Gr}_k(\mathbb{R}^n)$.

The flag determines a section of the projection $V_k(\mathbb{R}^n) \to \mathrm{Gr}_k(\mathbb{R}^n)$ (which, to be sure, is only continuous when restricted to a fixed $\mathring{e}(a)$), by assigning to a k-plane V the frame $v_1, \ldots, v_k$, described as follows. The first vector, v_1, is the unit basis vector for V_{1+a_1} (which is one-dimensional) with positive $(1+a_1)$st entry; the next vector, v_2, is the unit vector in V_{2+a_2} that is orthogonal to v_1 and has positive $(2 + a_2)$nd entry; etc. This is the "column reduction" process of linear algebra.

For example, each 2-plane in $\mathbb{R}^4$ is spanned by a unique (orthonormal) frame forming the columns in a matrix of exactly one of the shapes

$$\begin{bmatrix} p & * \\ & p \\ & \\ & \end{bmatrix}, \begin{bmatrix} p & * \\ & * \\ & p \\ & \end{bmatrix}, \begin{bmatrix} p & * \\ & * \\ & * \\ & p \end{bmatrix}, \begin{bmatrix} * & * \\ p & * \\ & p \\ & \end{bmatrix}, \begin{bmatrix} * & * \\ p & * \\ & * \\ & p \end{bmatrix}, \begin{bmatrix} * & * \\ * & * \\ p & * \\ & p \end{bmatrix},$$

where p denotes a positive real, $*$ any real, and empty spaces are 0. The types are

$$(0,0), (0,1), (0,2), (1,1), (1,2), (2,2).$$

As V runs over k-planes in $\mathring{e}(a)$, the vector v_1 has a_1 degrees of freedom: the first a_1 entries can make up any vector of norm less than 1, and the bottom entry (marked p above) is then determined. The vector v_2 has $2+a_2$ nonzero entries, but is subject to the two conditions that it be orthogonal to v_1 and of unit length; so it enjoys a_2 degrees of freedom. We thus expect $\dim \mathring{e}(a) = \|a\|$, where

$$\|a\| = a_1 + \cdots + a_k.$$

Therefore we define

$$\mathrm{Sk}_d \mathrm{Gr}_k(\mathbb{R}^n) = \bigcup_{\|a\| \le d} \mathring{e}(a).$$

Theorem 15.6. *This filtration defines a finite CW structure on* $\mathrm{Gr}_k(\mathbb{R}^n)$.

Proof. We will construct a pushout diagram

$$
\begin{array}{ccc}
\amalg_{\|a\|=d}\, \partial e(a) & \longrightarrow & \amalg_{\|a\|=d}\, e(a) \\
\downarrow{\scriptstyle \alpha} & & \downarrow{\scriptstyle \beta} \\
\mathrm{Sk}_{d-1} & \longrightarrow & \mathrm{Sk}_d
\end{array}
$$

and show that for each a, $(e(a), \partial e(a)) \cong (D^d, S^{d-1})$.

Define $e(a)$ to be the closure of $\mathring{e}(a)$ regarded as a subset of the Stiefel variety $V_k(\mathbb{R}^n)$:

$$
e(a) = \{(v_1, \ldots, v_k) : v_i \in \mathbb{R}^n, v_i \cdot v_j = \delta_{i,j}, v_i \cdot b_i \geq 0\},
$$

where $b_i = e_{i+a_i}$ is the ith "pivotal" basis vector. The subspace $\partial e(a)$ is the subset where some $v_i \cdot b_i = 0$. The map β is given on $e(a)$ by sending $(v_1, \ldots, v_k)$ to its span.

What remains to be proved is that $e(a) \cong D^d$. We will prove that $e(a)$ is homeomorphic to a product of the disks

$$
D_i(a) = \{v \in \mathbb{R}^{i+a_i} : \|v\| = 1, b_j \cdot v = 0 \text{ for } j < i, b_i \cdot v \geq 0\}.
$$

Do this by induction on k. When $k = 1$, $e(a_1) = D_1(a_1)$. So we now need to construct a homeomorphism

$$
e(a') \times D_k(a) \to e(a),
$$

where $a' = (a_1, \ldots, a_{k-1})$. View an element of the source space here as an $n \times k$ matrix (V', v). In it, the last column, v, is zero in the pivotal entries — that is, it's orthogonal to the pivotal basis vectors b_i for $i < k$ — but is not orthogonal to the columns of V'; while in the target the last column has become orthogonal to the columns in V'. This homeomorphism will have the form

$$
(V', v) \mapsto (V', T_{V'}v), \quad V' = (v_1, \ldots, v_{k-1}),
$$

where $T_{V'}$ is an $n \times n$ rotation matrix such that

$$
T_{V'}b_i = v_i \text{ for } i < k \quad \text{and} \quad T_{V'}x - x \in \mathbb{R}^{(k-1)+a_{k-1}} \text{ for } x \in \mathbb{R}^{k+a_k}.
$$

Then for $i < k$

$$
v_i \cdot T_{V'}v = T_{V'}b_i \cdot T_{V'}v = b_i \cdot v = 0,
$$

and, since b_k is orthogonal to $\mathbb{R}^{(k-1)+a_{k-1}}$,

$$
v_k \cdot T_{V'}v = b_k \cdot v \geq 0,
$$

so this is a homeomorphism to $e(a)$.

The operator $T_{V'}$ is constructed as the composite

$$
T_{V'} = T(b_{k-1}, v_{k-1}) \cdots T(b_2, v_2)T(b_1, v_1)
$$

where, for any two unit vectors $b, v \in \mathbb{R}^n$ with $b + v \neq 0$, $T(b, v)$ is the rotation matrix that sends b to v and is the identity on the orthogonal complement of the span of b and v. $\square$

Exercises

Exercise 15.7. Complex projective space $\mathbb{CP}^n$ admits a CW structure in which $\mathrm{Sk}_{2k} = \mathrm{Sk}_{2k+1} = \mathbb{CP}^k$ for $0 \leq k \leq n$. Verify this by describing characteristic maps $D^{2k} \to \mathbb{CP}^k$.

Exercise 15.8. Provide a cell structure on the unit sphere $S^{2n-1} \subset \mathbb{C}^n$ that is equivariant with respect to the action of the subgroup μ_p of pth roots of unity in $\mathbb{C}^\times$. By this we mean: Each skeleton is closed under the group action, and the characteristic maps can be grouped as equivariant maps $D \to \mathrm{Sk}_k S^{2n-1}$, where D is a coproduct of copies of $\mu_p \times D^k$ in which μ_p acts trivially on the disk. This is an example of a "G-CW complex," a topic we will develop further in Lecture 56.

16 Homology of CW complexes

The skeleton filtration of a CW complex leads to a long exact sequence in homology, showing that the relative homology $H_*(X_k, X_{k-1})$ controls how the homology changes when you pass from X_{k-1} to X_k. What is this relative homology? If we pick a set of characteristic maps, we get the following diagram.

$$
\begin{array}{ccccc}
\coprod_{i \in I_k} S_i^{k-1} & \hookrightarrow & \coprod_{i \in I_k} D_i^k & \longrightarrow & \bigvee_{i \in I_k} S_i^k \\
\downarrow{\scriptstyle \alpha} & & \downarrow & & \vdots \\
X_{k-1} & \hookrightarrow & X_k & \longrightarrow & X_k/X_{k-1}
\end{array}
$$

where $\bigvee$ is the wedge sum (disjoint union with all basepoints identified): $\bigvee_i S_i^k$ is a "bouquet of spheres." The dotted map exists and is a homeomorphism.

Luckily, the inclusion $X_{k-1} \subseteq X_k$ satisfies what's needed (Corollary 10.3) to conclude that

$$
H_q(X_k, X_{k-1}) \to H_q(X_k/X_{k-1}, *) = \overline{H}_q(X_k/X_{k-1})
$$

is an isomorphism. After all, X_{k-1} is a deformation retract of the space you get from X_k by deleting the center of each k-cell.

We know $\overline{H}_q(X_k/X_{k-1})$ very well:

$$\overline{H}_q(\bigvee_{i \in I_k} S_i^k) \cong \begin{cases} \mathbb{Z}\langle I_k \rangle & q = k \\ 0 & q \neq k. \end{cases}$$

Moral: The relative homology $H_k(X_k, X_{k-1})$ keeps track of the k-cells of X.

Definition 16.1. The group of *cellular n-chains* in a CW complex X is

$$C_k(X) = H_k(X_k, X_{k-1}).$$

A choice of k-cells determines a basis for this free abelian group:

$$C_k(X) \cong \mathbb{Z}\langle I_k \rangle.$$

If we put the fact that $H_q(X_k, X_{k-1}) = 0$ for $q \neq k$ into the homology long exact sequence of the pair, we find first that

$$H_q(X_{k-1}) \xrightarrow{\cong} H_q(X_k) \quad \text{for} \quad q \neq k, k-1,$$

and then that there is an exact sequence

$$0 \to H_k(X_k) \to C_k(X) \to H_{k-1}(X_{k-1}) \to H_{k-1}(X_k) \to 0.$$

So if we fix a dimension q, and watch how H_q varies as we move through the skelata of X, we find the following picture. First, $H_0(X_0)$ surjects onto $H_0(X_1)$, which is thereafter unchanged as you go to larger skelata. Now fix $q > 0$. Since X_0 is discrete, $H_q(X_0) = 0$. Then $H_q(X_k)$ continues to be 0 till you get up to X_q. $H_q(X_q)$ is a subgroup of the free abelian group $C_q(X)$ and hence is free abelian. Relations may get introduced into it when we pass to X_{q+1}; but thereafter all the maps

$$H_q(X_{q+1}) \to H_q(X_{q+2}) \to \cdots$$

are isomorphisms. All the q-dimensional homology of X is created on X_q, and all the relations in $H_q(X)$ occur by X_{q+1}.

This stable value of $H_q(X_k)$ maps isomorphically to $H_q(X)$, even if X is infinite dimensional. This is because the union of the images of any finite set of singular simplices in X is compact and so lies in a finite subcomplex (Proposition 15.3) and in particular lies in a finite skeleton. So any chain in X is the image of a chain in some skeleton. Since $H_q(X_k) \xrightarrow{\cong} H_q(X_{k+1})$ for $k > q$, we find that $H_q(X_q) \to H_q(X)$ is surjective. Similarly, if $c \in S_q(X_k)$ is a boundary in X, then it's a boundary in X_ℓ for some $\ell \geq k$. This shows that the map $H_q(X_{q+1}) \to H_q(X)$ is injective. We summarize:

Proposition 16.2. *Let $k, q \geq 0$. Then*

$$H_q(X_k) = 0 \quad \text{for } k < q$$

and

$$H_q(X_k) \xrightarrow{\cong} H_q(X) \quad \text{for } k > q.$$

In particular, $H_q(X) = 0$ if q exceeds the dimension of X.

We have defined the cellular n-chains of a CW complex X,

$$C_n(X) = H_n(X_n, X_{n-1}),$$

and found that it is the free abelian group on the set of n cells. We claim that these abelian groups are related to each other; they form the groups in a chain complex.

What should the boundary of an n-cell be? The n-cell is represented by a characteristic map $D^n \to X_n$ whose boundary is the attaching map $\alpha : S^{n-1} \to X_{n-1}$. This is a lot of information, and hard to interpret because X_{n-1} is itself potentially a complicated space. But things get much simpler if I pinch X_{n-2} to a point. This suggests defining

$$
\begin{array}{ccc}
C_n(X) & \xrightarrow{\quad\quad\quad d \quad\quad\quad} & C_{n-1}(X) \\
\| & & \| \\
H_n(X_n, X_{n-1}) & \xrightarrow{\partial_{n-1}} H_{n-1}(X_{n-1}) \xrightarrow{j_{n-1}} & H_{n-1}(X_{n-1}, X_{n-2}).
\end{array}
$$

The fact that $d^2 = 0$ is embedded in the following large diagram, in which the two columns and the central row are exact.

$$
\begin{array}{ccc}
C_{n+1}(X) = H_{n+1}(X_{n+1}, X_n) & & 0 = H_{n-1}(X_{n-2}) \\
\bigg\downarrow \partial_n \quad\quad \searrow d & & \bigg\downarrow \\
0 \longrightarrow H_n(X_n) \xrightarrow{j_n} C_n(X) = H_n(X_n, X_{n-1}) \xrightarrow{\partial_{n-1}} H_{n-1}(X_{n-1}) \\
\bigg\downarrow \quad\quad \searrow d & & \bigg\downarrow j_{n-1} \\
H_n(X_{n+1}) & & C_{n-1}(X) = H_{n-1}(X_{n-1}, X_{n-2}) \\
\bigg\downarrow & & \\
0 = H_n(X_{n+1}, X_n) & &
\end{array}
$$

Now, $\partial_{n-1} \circ j_n = 0$. So the composite of the diagonals is zero, i.e. $d^2 = 0$, and we have a chain complex! This is the *cellular chain complex* of X.

We should compute the homology of this chain complex, $H_n(C_*(X)) = \ker d / \operatorname{im} d$. Now

$$\ker d = \ker(j_{n-1} \circ \partial_{n-1}).$$

But j_{n-1} is injective, so

$$\ker d = \ker \partial_{n-1} = \operatorname{im} j_n = H_n(X_n).$$

On the other hand

$$\operatorname{im} d = j_n(\operatorname{im} \partial_n) = \operatorname{im} \partial_n \subseteq H_n(X_n).$$

So

$$H_n(C_*(X)) = H_n(X_n)/\operatorname{im} \partial_n = H_n(X_{n+1})$$

by exactness of the left column; but as we know this is exactly $H_n(X)$! We have proven the following result.

Theorem 16.3. *For a CW complex X, there is an isomorphism*

$$H_*(C_*(X)) \cong H_*(X)$$

natural with respect to filtration-preserving maps between CW complexes.

This has an immediate and surprisingly useful corollary.

Corollary 16.4. *Suppose that the CW complex X has only even cells — that is, $X_{2k} \hookrightarrow X_{2k+1}$ is an isomorphism for all k. Then*

$$H_*(X) \cong C_*(X).$$

That is, $H_n(X)$ is trivial for n odd, it is free abelian for all n, and the rank of $H_n(X)$ for n even is the number of n-cells.

Example 16.5. Complex projective space $\mathbb{CP}^n$ has a CW structure in which

$$\mathrm{Sk}_{2k}\mathbb{CP}^n = \mathrm{Sk}_{2k+1}\mathbb{CP}^n = \mathbb{CP}^k.$$

The attaching $S^{2k-1} \to \mathbb{CP}^k$ sends $v \in S^{2k-1} \subseteq \mathbb{C}^k$ to the complex line through v. So

$$H_k(\mathbb{CP}^n) = \begin{cases} \mathbb{Z} & \text{for } 0 \leq k \leq 2n, \ k \text{ even} \\ 0 & \text{otherwise}. \end{cases}$$

Remark 16.6. Notice that in our proof of Theorem 16.3 we used only properties contained in the Eilenberg-Steenrod axioms. As a result, any construction of an ordinary homology theory satisfying the Eilenberg-Steenrod axioms gives you the same values on CW complexes as singular homology.

Exercises

Exercise 16.7. Compute the homology of $\mathrm{Gr}_2(\mathbb{C}^4)$.

Exercise 16.8. Let $p, q \in \mathbb{Z}$, and let X be the 2-dimensional CW complex obtained by attaching two 2-cells to S^1 using maps of degree p and q. Compute $\pi_1(X)$ and $H_*(X)$.

17 Real projective space

Let's try to compute $H_*(\mathbb{RP}^n)$. This computation will invoke a second way to think of the cellular chain group $C_n(X)$. Each cell has a characteristic map $D^n \to X_n$, and we have the diagram

$$\coprod(D^n, S^{n-1}) \longrightarrow (X_n, X_{n-1})$$
$$\searrow \qquad \downarrow$$
$$(\bigvee S^n, *).$$

We've shown that the vertical map induces an isomorphism in homology, and the diagonal does as well. (For example, note that $\coprod D^n$ has a CW structure in which the $(n-1)$-skeleton is $\coprod S^{n-1}$.) So

$$H_n(\coprod(D^n, S^{n-1})) \xrightarrow{\cong} C_n(X).$$

We have a CW structure on $\mathbb{RP}^n$ with $\mathrm{Sk}_k(\mathbb{RP}^n) = \mathbb{RP}^k$; there is one k-cell — which we'll denote by e_k — for each k between 0 and n. The attaching map of the n-cell is the double cover $\pi : S^{n-1} \to \mathbb{RP}^{n-1}$. The cellular chain complex looks like this:

$$0 \leftarrow C_0(\mathbb{RP}^n) \leftarrow C_1(\mathbb{RP}^n) \leftarrow \cdots \leftarrow C_n(\mathbb{RP}^n) \leftarrow 0$$
$$\| \qquad\qquad \| \qquad\qquad\qquad \|$$
$$0 \longleftarrow \mathbb{Z}\langle e_0 \rangle \xleftarrow{d=0} \mathbb{Z}\langle e_1 \rangle \longleftarrow \cdots \longleftarrow \mathbb{Z}\langle e_n \rangle \longleftarrow 0.$$

The first differential is zero because $H_0(\mathbb{RP}^n) = \mathbb{Z}$. For $n > 1$, the differential in the cellular chain complex is given by the top row in the following commutative diagram, whose bottom row we will proceed to explain.

$$C_n = H_n(\mathbb{RP}^n, \mathbb{RP}^{n-1}) \xrightarrow{\ \partial\ } H_{n-1}(\mathbb{RP}^{n-1}) \longrightarrow H_{n-1}(\mathbb{RP}^{n-1}, \mathbb{RP}^{n-2}) = C_{n-1}$$
$$\uparrow \cong \qquad\qquad\qquad \uparrow \pi_* \qquad\qquad\qquad \uparrow \cong$$
$$H_n(D^n, S^{n-1}) \xrightarrow[\cong]{\ \partial\ } H_{n-1}(S^{n-1}) \longrightarrow H_{n-1}(D^{n-1}/S^{n-2}, *).$$

The right square arises from the commutative diagram

$$\begin{array}{ccc} \mathbb{RP}^{n-1} & \xrightarrow{\quad\text{pinch}\quad} & \mathbb{RP}^{n-1}/\mathbb{RP}^{n-2} \\ \pi \uparrow & & \uparrow \cong \\ S^{n-1} \longrightarrow & S^{n-1}/S^{n-2} = S^{n-1} \vee S^{n-1} \longrightarrow S^{n-1} & = D^{n-1}/S^{n-2}. \end{array}$$

One of the maps $S^{n-1} \to S^{n-1}$ from the wedge is the identity, and the other map is the antipodal map $\alpha : S^{n-1} \to S^{n-1}$. Write σ for a generator of $H_{n-1}(S^{n-1})$. Then H_{n-1} applied to the bottom path sends $\sigma \mapsto (\sigma, \sigma) \mapsto \sigma + \alpha_* \sigma$. So we need to know the degree of the antipodal map on S^{n-1}. The antipodal map reverses all n coordinates in $\mathbb{R}^n$. Each reversal is a reflection, and acts on S^{n-1} by a map of degree -1. So

$$\deg \alpha = (-1)^n.$$

Therefore the cellular complex of $\mathbb{RP}^n$ is as follows:

dim	-1	0	1	$\cdots$	n	$n+1$	$\cdots$

$$0 \longleftarrow \mathbb{Z} \xleftarrow{\ 0\ } \mathbb{Z} \xleftarrow{\ 2\ } \cdots \xleftarrow{\ 2\text{ or }0\ } \mathbb{Z} \longleftarrow 0 \longleftarrow \cdots$$

The homology is then easy to read off.

Proposition 17.1. *The homology of real projective space is as follows.*

$$H_k(\mathbb{RP}^n) = \begin{cases} \mathbb{Z} & k = 0 \\ \mathbb{Z} & k = n \ odd \\ \mathbb{Z}/2\mathbb{Z} & k \ odd, \ 0 < k < n \\ 0 & otherwise. \end{cases}$$

Here's a table. Missing entries are 0.

dim	0	1	2	3	4	5	$\cdots$
$\mathbb{RP}^0$	$\mathbb{Z}$						
$\mathbb{RP}^1$	$\mathbb{Z}$	$\mathbb{Z}$					
$\mathbb{RP}^2$	$\mathbb{Z}$	$\mathbb{Z}/2$					
$\mathbb{RP}^3$	$\mathbb{Z}$	$\mathbb{Z}/2$	0	$\mathbb{Z}$			
$\mathbb{RP}^4$	$\mathbb{Z}$	$\mathbb{Z}/2$	0	$\mathbb{Z}/2$			
$\mathbb{RP}^5$	$\mathbb{Z}$	$\mathbb{Z}/2$	0	$\mathbb{Z}/2$	0	$\mathbb{Z}$	
$\vdots$	$\vdots$	$\vdots$	$\vdots$	$\vdots$	$\vdots$	$\vdots$	

Summary: In real projective space, odd cells create new generators; even cells (except for the zero-cell) create torsion in the previous dimension.

This example illustrates the significance of cellular homology, and, therefore, of singular homology. A CW structure involves attaching maps

$$\coprod S^{n-1} \to \mathrm{Sk}_{n-1}X\,.$$

Knowing these, up to homotopy, determines the full homotopy type of the CW complex. Homology does not record all this information. Instead, it records only information about the composite obtained by pinching out $\mathrm{Sk}_{n-2}X$.

In H_{n-1}, the diagonal map induces a map $\partial : \mathbb{Z}I_n \to \mathbb{Z}I_{n-1}$ that is none other than the differential in the cellular chain complex.

The moral: homology picks off only the "first order" structure of a CW complex.

Exercises

Exercise 17.2 (Lens spaces). Let p and q be relatively prime positive integers. Define a space $L(p,q)$ as the quotient of S^3, the unit sphere in $\mathbb{C}^2$, by the action of the group of pth roots of unity given by

$$\zeta \cdot (z_1, z_2) = (\zeta z_1, \zeta^q z_2)\,.$$

Impose on $L(p,q)$ the structure of a finite cell complex with one cell in each dimension between 0 and 3. The cell complex structure is just the filtration, but you should specify the characteristic maps as well. Then compute the homology of $L(p,q)$.

18   Euler characteristic and homology approximation

Theorem 18.1. *Let X be a finite CW complex with a_k k-cells. Then*

$$\chi(X) := \sum (-1)^k a_k$$

depends only on the homotopy type of X; it is independent of the choice of CW structure.

This integer $\chi(X)$ is called the *Euler characteristic* of X. We will prove this theorem by showing that $\chi(X)$ equals a number computed from the homology groups of X, which are themselves homotopy invariants.

We'll need a little bit of information about the structure of finitely generated abelian groups.

Let A be an abelian group. The set of *torsion* elements of A,

$$\text{Tors}(A) = \{a \in A : na = 0 \text{ for some } n \neq 0\},$$

is a subgroup of A. A group is *torsion free* if $\text{Tors}(A) = 0$. For any A the quotient group $A/\text{Tors}(A)$ is torsion free.

For a general abelian group, that's about all you can say. But now assume A is finitely generated. Then $\text{Tors}(A)$ is a finite abelian group and $A/\text{Tors}(A)$ is a finitely generated free abelian group, isomorphic to $\mathbb{Z}^r$ for some integer r called the *rank* of A. Pick elements of A that map to a set of generators of $A/\text{Tors}(A)$, and use them to define a map $A/\text{Tors}A \to A$ splitting the projection map. This shows that if A is finitely generated then

$$A \cong \text{Tors}(A) \oplus \mathbb{Z}^r.$$

Lemma 18.2. *Let $0 \to A \to B \to C \to 0$ be a short exact sequence of finitely generated abelian groups. Then*

$$\text{rank}\, A - \text{rank}\, B + \text{rank}\, C = 0.$$

Lemma 18.3. *A finite abelian group A is isomorphic to*

$$\mathbb{Z}/n_1\mathbb{Z} \oplus \mathbb{Z}/n_2\mathbb{Z} \oplus \cdots \oplus \mathbb{Z}/n_t\mathbb{Z}$$

for a uniquely defined sequence of natural numbers $n_1, \ldots, n_t$ such that $2 \leq n_1 | n_2 | \cdots | n_t$.

The n_i are the "torsion coefficients" of A. The abelian group A cannot be generated by fewer than t elements.

Theorem 18.4. *Let X be a finite CW complex. Then*

$$\chi(X) = \sum_k (-1)^k \text{rank}\, H_k(X).$$

Proof. Pick a CW structure with, say, a_k k-cells for each k. We have the cellular chain complex C_*. Write H_*, Z_*, and B_* for the homology, the cycles, and the boundaries, in this chain complex. From the definitions, we have two families of short exact sequences:

$$0 \to Z_k \to C_k \to B_{k-1} \to 0$$

and

$$0 \to B_k \to Z_k \to H_k \to 0 \,.$$

Let's use them and facts about rank to rewrite the alternating sum:

$$\sum_k (-1)^k a_k = \sum_k (-1)^k \mathrm{rank}(C_k)$$

$$= \sum_k (-1)^k (\mathrm{rank}\,(Z_k) + \mathrm{rank}\,(B_{k-1}))$$

$$= \sum_k (-1)^k (\mathrm{rank}\,(B_k) + \mathrm{rank}\,(H_k) + \mathrm{rank}\,(B_{k-1})) \,.$$

The terms $\mathrm{rank}\,B_k + \mathrm{rank}\,B_{k-1}$ cancel because it's an alternating sum. This leaves $\sum_k (-1)^k \mathrm{rank}\,H_k$. But $H_k \cong H_k^{\mathrm{sing}}(X)$. $\qquad\square$

In the early part of the 20th century, "homology groups" were not discussed. It was Emmy Noether who first described things that way. Instead, people worked mainly with the sequence of ranks,

$$\beta_k = \mathrm{rank}\,H_k(X) \,,$$

which are known (following Poincaré) as the *Betti numbers* of X.

Given a CW complex X of finite type, can we give a lower bound on the number of k-cells in terms of the homology of X? Let's see. $H_k(X)$ is finitely generated because $C_k(X) \hookleftarrow Z_k(X) \twoheadrightarrow H_k(X)$. Thus

$$H_k(X) \cong \mathbb{Z}^{r(k)} \oplus \bigoplus_{i=1}^{t(k)} \mathbb{Z}/n_i(k)\mathbb{Z}$$

where the $n_1(k), \ldots, n_{t(k)}(k)$ are the torsion coefficients of $H_k(X)$ and $r(k)$ is the rank. Note that $r(0)$ is the number of components, and $t(k) = 0$ for $k \le 0$.

The minimal chain complex with $H_k = \mathbb{Z}^r$ and $H_q = 0$ for $q \neq k$ is just the chain complex with 0 everywhere except for $\mathbb{Z}^r$ in the kth degree. The minimal chain complex *of free abelian groups* with $H_k = \mathbb{Z}/n\mathbb{Z}$ and $H_q = 0$ for $q \neq k$ is the chain complex with 0 everywhere except in dimensions $k+1$ and k, where we have $\mathbb{Z} \xrightarrow{n} \mathbb{Z}$. These small complexes are called *elementary chain complexes*.

This implies that a lower bound on the number of k-cells is

$$r(k) + t(k) + t(k-1) \,.$$

The first two terms correspond to generators for H_k, and the last to relations for H_{k-1}.

These elementary chain complexes can be realized as the reduced cellular chains of CW complexes (at least if $k > 0$). A wedge of r copies of S^k has a CW structure with one 0-cell and r k-cells, so its cellular chain complex has $\mathbb{Z}^r$ in dimension k and 0 in other positive dimensions. To construct a CW complex with cellular chain complex given by $\mathbb{Z} \xrightarrow{n} \mathbb{Z}$ in dimensions $k+1$ and k and 0 in other positive dimensions, start with S^k as k-skeleton and attach a $(k+1)$-cell by a map of degree n. For example, when $k = 1$ and $n = 2$, you have $\mathbb{RP}^2$. These CW complexes are called "Moore spaces." (John Moore (1923–2016) worked at Princeton. He was an MIT alumnus and my PhD thesis advisor.)

This maximally efficient construction of a CW complex in a homotopy type can in fact be achieved, at least in the simply connected case:

Theorem 18.5 (Homology approximation: Wall, [75]; see Exercise 51.8). *Let X be a simply connected CW complex of finite type. Then there exists a CW complex Y with $r(k) + t(k) + t(k-1)$ k-cells, for all k, and a homotopy equivalence $Y \to X$.*

The construction of Moore spaces can be generalized:

Proposition 18.6. *For any graded abelian group A_* with $A_k = 0$ for $k \leq 1$, there exists a simply connected CW complex X with $\widetilde{H}_*(X) = A_*$.*

Proof. Let A be any abelian group. Pick generators for A. They determine a surjection from a free abelian group F_0. The kernel F_1 of that surjection is free, being a subgroup of a free abelian group. Write G_0 for a basis of F_0, and G_1 for a basis of F_1.

Let $k \geq 2$. Define X_k to be the wedge of $|G_0|$ copies of S^k, so $H_k(X_k) = \mathbb{Z}G_0$. Now define an attaching map

$$\alpha : \coprod_{b \in G_1} S_b^k \to X_k$$

by specifying it on each summand S_b^k. The generator $b \in G_1$ is given by a linear combination of the generators of F_0, say

$$b = \sum_{i=1}^{s} n_i a_i .$$

We want to mimic this in topology. To do this, first map $S^k \to \bigvee^s S^k$ by pinching $(s - 1)$ tangent $(k - 1)$-spheres to a point. In homology, this map takes a generator of $H_k(S^k)$ to the sum of the generators of the k-dimensional homology of the various spheres in the bouquet. Map the ith sphere in the wedge to $S^k_{a_i} \subseteq X_k$ by a map of degree n_i. The map on the summand S^k_b is then the composite of these two maps,

$$S^k_b \to \bigvee_{i=1}^{s} S^k \to \bigvee_{a \in G_0} S^k_a = X_k .$$

Altogether, we get a map α that realizes $F_1 \to F_0$ in H_k. So using it as an attaching map produces a CW complex X with $\widetilde{H}_q(X) = A$ for $q = k$ and 0 otherwise. Write $M(A, k)$ for a CW complex produced in this way.

Finally, given a graded abelian group A_*, form the wedge over k of the spaces $M(A_k, k)$. ☐

Such a space $M(A, k)$, simply connected with $\widetilde{H}_q(M(A, k)) = A$ for $q = k$ and 0 otherwise, is called a *Moore space of type* (A, k) [49]. The notation is a bit deceptive, since (as we'll see) there is no functor $M(-, k) :$ **Ab** $\to$ HoTop$_*$ splitting $\overline{H}_k$, for any $k \geq 2$ [66]. In fact at this point it's not clear that these conditions determine the homotopy type of a CW complex, though this will follow from the Whitehead theorem 46.8.

On the other hand, if π is any group, this construction (with groups rather than abelian groups) may be used to construct a connected 2-dimensional CW complex with fundamental group π, by the van Kampen Theorem.

Exercises

Exercise 18.7. Show that the Euler characteristic is a "cut-and-paste" invariant, in the following sense. Let X and Y be subcomplexes of the finite CW complex $X \cup Y$. Show that

$$\chi(X \cup Y) = \chi(X) + \chi(Y) - \chi(X \cap Y) .$$

19 Coefficients

Abelian groups can be quite complicated, even finitely generated ones. Vector spaces over a field are so much simpler! A vector space is determined up to isomorphism by a single cardinality, its dimension. Wouldn't it be great to have a version of homology that took values in the category of vector spaces over a field?

We can do this, and more. Let R be any commutative ring at all. Instead of forming the free abelian group on $\mathrm{Sin}_*(X)$, we could just as well form the free R-module:

$$S_*(X; R) = R\mathrm{Sin}_*(X).$$

This gives, first, a simplicial object in the category of R-modules. Forming the alternating sum of the face maps produces a chain complex of R-modules: $S_n(X; R)$ is an R-module for each n, and $d : S_n(X; R) \to S_{n-1}(X; R)$ is an R-module homomorphism. The homology groups are then again R-modules:

$$H_n(X; R) = \frac{\ker(d : S_n(X; R) \to S_{n-1}(X; R))}{\mathrm{im}(d : S_{n+1}(X; R) \to S_n(X; R))} .$$

This is the *singular homology of X with coefficients in the commutative ring R*. It satisfies all the Eilenberg-Steenrod axioms, with

$$H_n(*; R) = \begin{cases} R & \text{for} \quad n = 0 \\ 0 & \text{otherwise} . \end{cases}$$

(We could actually have replaced the ring R by any abelian group here, but this will become much clearer after we have the tensor product as a tool.) This means that all the work we have done for "integral homology" carries over to homology with any coefficients. In particular, if X is a CW complex we have the cellular homology with coefficients in R, $C_*(X; R)$, and its homology is isomorphic to $H_*(X; R)$.

The coefficient rings that are most important in algebraic topology are simple ones: the integers and the prime fields $\mathbb{F}_p$ and $\mathbb{Q}$; almost always a PID.

As an experiment, let's compute $H_*(\mathbb{RP}^n; R)$ for various rings R. Let's start with $R = \mathbb{F}_2$, the field with 2 elements. This is a favorite among algebraic topologists, because using it for coefficients eliminates all sign issues. The cellular chain complex has $C_k(\mathbb{RP}^n; \mathbb{F}_2) = \mathbb{F}_2$ for $0 \leq k \leq n$, and the differential alternates between multiplication by 2 and by 0. But in $\mathbb{F}_2$, $2 = 0$: so $d = 0$, and the cellular chains coincide with the homology:

$$H_k(\mathbb{RP}^n; \mathbb{F}_2) = \begin{cases} \mathbb{F}_2 & \text{for} \quad 0 \leq k \leq n \\ 0 & \text{otherwise} . \end{cases}$$

On the other hand, suppose that R is a ring in which 2 is invertible. The universal case is $\mathbb{Z}[1/2]$, but any subring of the rationals containing $1/2$ would do just as well, as would $\mathbb{F}_p$ for p odd. Now the cellular chain complex (in dimensions 0 through n) looks like

$$R \xleftarrow{0} R \xleftarrow{\cong} R \xleftarrow{0} R \xleftarrow{\cong} \cdots \xleftarrow{\cong} R$$

for n even, and

$$R \xleftarrow{0} R \xleftarrow{\cong} R \xleftarrow{0} R \xleftarrow{\cong} \cdots \xleftarrow{0} R$$

for n odd. Therefore for n even

$$H_k(\mathbb{RP}^n; R) = \begin{cases} R & \text{for} \quad k = 0 \\ 0 & \text{otherwise} \end{cases}$$

and for n odd

$$H_k(\mathbb{RP}^n; R) = \begin{cases} R & \text{for} \quad k = 0 \\ R & \text{for} \quad k = n \\ 0 & \text{otherwise} . \end{cases}$$

You get a much simpler result: Away from 2, even-dimensional projective spaces look like points and odd-dimensional projective spaces look like spheres!

One can generalize this process a little bit, and allow coefficients not just in a commutative ring, but more generally in a module M over a commutative ring; in particular, in any abelian group. This is most cleanly done using the mechanism of the tensor product. That mechanism will also let us address the following natural question:

Question 19.1. Given $H_*(X; R)$, can we deduce $H_*(X; M)$ for an R-module M?

The answer is called the "universal coefficient theorem." We'll spend a few days developing what we need to talk about this.

Exercises

Exercise 19.2. Let X be a finite CW complex. Show that for any field F,

$$\chi(X) = \sum (-1)^k \dim_F H_k(X; F) .$$

Exercise 19.3. Let p be a prime number. Give an example of two maps $f, g : X \to Y$ inducing the same map on integral homology but not homology with coefficients in $\mathbb{F}_p$ (and that are therefore not homotopic).

20 Tensor product

The category of R-modules is what might be called a "categorical ring," in which addition corresponds to the direct sum, the zero element is the zero module, 1 is R itself, and multiplication is ... well, the subject for today. We care about the tensor product for two reasons: First, it allows us to deal smoothly with bilinear maps such as the cross-product, and ultimately compute the homology of a product in terms of the homology of the factors. Second, and of more immediate interest, it will allow us relate homology with coefficients in an any R-module to homology with coefficients in the PID R; for example, relate $H_*(X; M)$ to $H_*(X)$, where M is any abelian group.

Let's begin by recalling the definition of a bilinear map over a commutative ring R.

Definition 20.1. Given three R-modules, M, N, P, a *bilinear map* (or, to be explicit, an *R-bilinear map*) is a function $\beta : M \times N \to P$ that is R-linear separately in each variable; that is:

$$\beta(x + x', y) = \beta(x, y) + \beta(x', y), \quad \beta(x, y + y') = \beta(x, y) + \beta(x, y'),$$

and

$$\beta(rx, y) = r\beta(x, y), \quad \beta(x, ry) = r\beta(x, y),$$

for $x, x' \in M$, $y, y' \in N$, and $r \in R$.

Of course, if R is the ring of integers then the second pair of identities follows from the first pair.

Example 20.2. $\mathbb{R}^n \times \mathbb{R}^n \to \mathbb{R}$ given by the dot product is an $\mathbb{R}$-bilinear map. The cross product $\mathbb{R}^3 \times \mathbb{R}^3 \to \mathbb{R}^3$ is $\mathbb{R}$-bilinear. If R is any ring, the multiplication $R \times R \to R$ is $\mathbb{Z}$-bilinear, but is R-bilinear if and only if the ring is commutative.

Wouldn't it be great to reduce stuff about bilinear maps to linear maps? We're going to do this by means of a universal property.

Definition 20.3. Let M, N be R-modules. A *tensor product* of M and N is an R-module P and a bilinear map $\beta_0 : M \times N \to P$ such that for every R-bilinear map $\beta : M \times N \to Q$ there is a unique factorization

$$
\begin{array}{ccc}
M \times N & \xrightarrow{\beta_0} & P \\
& \searrow{\scriptstyle \beta} & \downarrow{\scriptstyle f} \\
& & Q
\end{array}
$$

through an R-module homomorphism f.

We should have pointed out that a composition like $f \circ \beta_0$ is indeed again R-bilinear; but this is easy to check.

So β_0 is a "universal" bilinear map out of $M \times N$. Instead of β_0 we're going to write $\otimes : M \times N \to P$. This means that $\beta(x, y) = f(x \otimes y)$ in the above diagram. There are lots of things to say about this. When you have something that is defined by means of a universal property, you know that it's unique ... but you still have to check that it exists!

Construction 20.4. Let's think about how to construct a univeral R-bilinear map out of $M \times N$. Let $\beta : M \times N \to Q$ be any R-bilinear map. This β isn't linear. Maybe we should first extend it to a linear map. There is a unique R-linear extension over the free R-module $R\langle M \times N \rangle$ generated by the set $M \times N$:

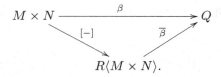

The map $[-]$, including a basis, isn't bilinear. We should quotient $R\langle M \times N \rangle$ by a submodule S of relations to make it bilinear. So S is the sub R-module generated by the four families of elements (corresponding to the four relations in the definition of R-bilinearity):

(1) $[(x + x', y)] - [(x, y)] - [(x', y)]$
(2) $[(x, y + y')] - [(x, y)] - [(x, y')]$
(3) $[(rx, y)] - r[(x, y)]$
(4) $[(x, ry)] - r[(x, y)]$

for $x, x' \in M$, $y, y' \in N$, and $r \in R$. Now the composite $M \times N \to R\langle M \times N \rangle / S$ *is* R-bilinear — we've quotiented out by all things that prevented it from being so! And the map $\overline{\beta} : R\langle M \times N \rangle \to Q$ factors as $R\langle M \times N \rangle \to R\langle M \times N \rangle / S \xrightarrow{f} Q$, where f is R-linear, and uniquely because the map to the quotient is surjective. This completes the construction.

If you find yourself using this construction, stop and think about what you're doing. You're never going to use this construction to compute anything. Here's an example: For any abelian group A,

$$A \times \mathbb{Z}/n\mathbb{Z} \to A/nA, \quad (a, b) \mapsto ba \mod nA$$

is clearly bilinear, and is universal as such. Just look: If $\beta : A \times \mathbb{Z}/n\mathbb{Z} \to Q$ is bilinear then $\beta(na, b) = n\beta(a, b) = \beta(a, nb) = \beta(a, 0) = 0$, so β factors through A/nA; and $A \times \mathbb{Z}/n\mathbb{Z} \to A/nA$ is surjective. So $A \otimes \mathbb{Z}/n\mathbb{Z} = A/nA$.

Remark 20.5. The image of $M \times N$ in $R\langle M \times N\rangle/S$ generates it as an R-module. These elements $x \otimes y$ are called "decomposable tensors."

What are the properties of such a universal bilinear map?

Property 20.6 (Uniqueness). Suppose $\beta_0 : M \times N \to P$ and $\beta_0' : M \times N \to P'$ are both universal. Then there's a linear map $f : P \to P'$ such that $\beta_0' = f\beta_0$ and a linear map $f' : P' \to P$ such that $\beta_0 = f'\beta_0'$. The composite $f'f : P \to P$ is a linear map such that $f'f\beta_0 = f'\beta_0' = \beta_0$. The identity map is another. But by universality, there's only one such linear map, so $f'f = 1_P$. An identical argument shows that $ff' = 1_{P'}$ as well, so they are inverse linear isomorphisms. Both f and f' are unique, so we may say:

> The universal R-bilinear map $\beta_0 : M \times N \to P$ is unique up to a unique R-linear isomorphism.

This entitles us to speak of "the" universal bilinear map out of $M \times N$, and give the target a symbol: $M \otimes_R N$. If R is the ring of integers, or otherwise understood, we will drop it from the notation.

Property 20.7 (Functoriality). Suppose $f : M \to M'$ and $g : N \to N'$. Study the diagram

$$
\begin{array}{ccc}
M \times N & \xrightarrow{\ \otimes\ } & M \otimes N \\
\downarrow{\scriptstyle f\times g} & \searrow & \vdots\ {\scriptstyle f\otimes g} \\
M' \times N' & \xrightarrow{\ \otimes\ } & M' \otimes N'.
\end{array}
$$

There is a unique R-linear map $f \otimes g$ making the diagram commute, because the diagonal map is R-bilinear and the map $M \times N \to M \otimes N$ is the universal R-bilinear map out of $M \times N$. You are invited to show that this makes $- \otimes_R - : \mathbf{Mod}_R \times \mathbf{Mod}_R \to \mathbf{Mod}_R$ into a functor.

Property 20.8 (Unitality, associativity, commutativity). I said that the tensor product was going to render Mod_R a "categorical ring," so we should check various properties of the tensor product. For example, $R \otimes_R M$ should be isomorphic to M. Let's think about this for a minute. We have an R-bilinear map $R \times M \to M$, given by multiplication. We just need to check the universal property. Suppose we have an R-bilinear map $\beta : R \times M \to P$. We have to construct a map $f : M \to P$ such that $\beta(r, x) = f(rx)$ and show it's unique. Our only choice is $f(x) = \beta(1, x)$, and that works.

Similarly, we should check that there's a unique isomorphism $L \otimes (M \otimes N) \xrightarrow{\cong} (L \otimes M) \otimes N$ that's compatible with $L \times (M \times N) \cong (L \times M) \times N$, and that there's a unique isomorphism $M \otimes N \to N \otimes M$ that's compatible with the switch map $M \times N \to N \times M$. There are a few other things to check, too: Have fun!

Property 20.9 (Sums). What happens with $M \otimes (\bigoplus_{i \in I} N_i)$? Here I can be any set, not just a finite set. How does this tensor product relate to $\bigoplus_{j \in I} (M \otimes N_j)$? Let's construct a map

$$f : \bigoplus_{i \in I} (M \otimes N_i) \to M \otimes \left(\bigoplus_{j \in I} N_j \right) .$$

We just need to define maps $M \otimes N_i \to M \otimes (\bigoplus_{j \in I} N_j)$ because the direct sum is the coproduct. We can use $1 \otimes \mathrm{in}_i$ where $\mathrm{in}_i : N_i \to \bigoplus_{j \in I} N_j$ is the defining inclusion. These give you a map f.

What about a map the other way? We'll define a map out of the tensor product using the universal property. So we need to define a bilinear map out of $M \times (\bigoplus_{j \in I} N_j)$. By linearity in the second factor, it will suffice to say where to send an element of the form $x \otimes y \in M \otimes N_j$. Just send it to $\mathrm{in}_j (x \otimes y)$! Now it's up to you to check that these are inverses.

Property 20.10 (Distributivity). Suppose $f : M' \to M$, $r \in R$, and $g, g_0, g_1 : N' \to N$. Then

$$f \otimes (g_0 + g_1) = f \otimes g_0 + f \otimes g_1 : M' \otimes N' \to M \otimes N$$

and

$$f \otimes rg = r(f \otimes g) = (rf) \otimes g : M' \otimes N' \to M \otimes N .$$

Again I'll leave this to you to check. Of course you can take sums in the first variable as well.

Our immediate use of this construction is to give a clean definition of "homology with coefficients in M," where M is any abelian group. First, endow singular chains with coefficients in M like this:

$$S_*(X; M) = S_*(X) \otimes M .$$

The boundary maps are just $d \otimes 1 : S_n(X) \otimes M \to S_{n-1}(X) \otimes M$. Then we define

$$H_n(X; M) = H_n(S_*(X; M)) .$$

Since $S_n(X) = \mathbb{Z}\mathrm{Sin}_n(X)$, $S_n(X; M)$ is a direct sum of copies of M indexed by the n-simplices in X.

If M happens to be a ring, this coincides with the notation used in the last lecture. In fact, if M is a module over a commutative ring R, we might as well have defined $S_*(X; M)$ as

$$S_*(X; M) = S_*(X; R) \otimes_R M.$$

One virtue of this way of thinking of it is that it makes it clear that in this case the differential in $S_*(X; M)$ is R-linear.

These definitions extend to relative homology. As we have noted, if $A \subseteq X$ then the map $S_n(A) \to S_n(X)$ is a split monomorphism. The same argument shows that if R is a commutative ring then $S_n(A; R) \to S_n(X; R)$ is a split monomorphism of free R-modules. So the quotient R-module $S_n(X, A; R)$ is again free,

$$0 \to S_n(A; R) \to S_n(X; R) \to S_n(X, A; R) \to 0$$

is a split exact sequence of R-modules, and applying the functor $- \otimes_R M$ to it gives us another short exact sequence of R-modules

$$0 \to S_n(A; M) \to S_n(X; M) \to S_n(X, A; M) \to 0.$$

In any case, $S_*(X, A; M)$ is a chain complex of R-modules and we define

$$H_n(X, A; M) = H_n(S_*(X, A; M)).$$

This gives a functor to R-modules, and the boundary map

$$\partial : H_n(X, A; M) \to H_{n-1}(A; M)$$

is an R-module homomorphism. Notice that

$$H_n(*; M) = \begin{cases} M & \text{for } n = 0 \\ 0 & \text{otherwise}. \end{cases}$$

The following result is immediate:

Proposition 20.11. *For any abelian group M, $(X, A) \mapsto H_*(X, A; M)$ provides a homology theory satisfying the Eilenberg-Steenrod axioms with $H_0(*; M) = M$. This construction is natural in the abelian group M. If M is an R-module, for a commutative ring R, then the homology theory takes values in the category of R-modules.*

We have said that the homology theory is natural in the coefficient module: an R-module homomorphism $M' \to M$ induces chain maps $S_*(X, A; M') \to S_*(X, A; M)$. Moreover, if

$$0 \to M' \to M \to M'' \to 0$$

is a short exact sequence of R-modules then the induced sequence

$$0 \to S_*(X; M') \to S_*(X; M) \to S_*(X; M'') \to 0$$

is again short exact, and we arrive at a long exact sequence in homology, the "coefficient long exact sequence":

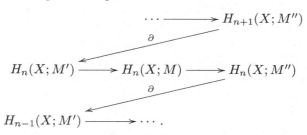

A particularly important case of all this is when R is a field; then $S_*(X; R)$ is a chain complex of vector spaces over R, and $H_*(X; R)$ is a graded vector space over R.

We motivated consideration of the tensor product by suggesting that it will be useful in answering questions like: If you know $H_*(X)$, can you compute $H_*(X; M)$ for other abelian coefficient groups? For example, is it possible that the latter is just $H_*(X) \otimes M$? We have already seen examples where this fails! The problem is that tensoring with a general abelian group fails to preserve exactness. We were lucky in forming $S_n(X, A; M)$, because we were tensoring M with a *split* short exact sequence. But on the level of homology, we have to work harder.

Exercises

Exercise 20.12. Let m, n be positive integers and consider the cyclic groups $\mathbb{Z}/m$ and $\mathbb{Z}/n$. Compute the tensor product $\mathbb{Z}/m \otimes \mathbb{Z}/n$.

Exercise 20.13. Let $A \subseteq X$ and $B \subseteq Y$ be subsets. Construct a natural chain map

$$S_*(X, A) \otimes S_*(Y, B) \to S_*(X \times Y, A \times Y \cup X \times B)$$

that is a homology isomorphism if A and B are open. (Hint: Exercise 9.10 might be useful.) So there is a natural "relative cross product" map

$$H_*(X, A; R) \otimes_R H_*(Y, B; R) \to H_*(X \times Y, A \times Y \cup X \times B; R)$$

that is an isomorphism if A and B are open, R is a PID, and either $H_*(X, A; R)$ or $H_*(Y, B; R)$ is free over R.

21 Tensor and Tor

We continue to study properties of the tensor product. Recall that

$$\mathbb{Z}/m\mathbb{Z} \otimes N = N/mN \, .$$

Consider the exact sequence

$$0 \to \mathbb{Z} \xrightarrow{2} \mathbb{Z} \to \mathbb{Z}/2\mathbb{Z} \to 0 \, .$$

Let's tensor it with $\mathbb{Z}/2\mathbb{Z}$. We get

$$0 \to \mathbb{Z}/2\mathbb{Z} \to \mathbb{Z}/2\mathbb{Z} \to \mathbb{Z}/2\mathbb{Z} \to 0 \, .$$

This cannot be a short exact sequence! This is a major tragedy: tensoring doesn't preserve exact sequences; one says that the functor $\mathbb{Z}/m\mathbb{Z} \otimes -$ is not "exact." This is why we can't form homology with coefficients in M by simply tensoring homology with M.

Tensoring does respect certain exact sequences:

Proposition 21.1. *The functor $N \mapsto M \otimes_R N$ preserves cokernels; it is right exact.*

Proof. Suppose that $N' \to N \to N'' \to 0$ is exact and let $f : M \otimes N \to Q$. We wish to show that there is a unique factorization as shown in the diagram

$$
\begin{array}{ccccccc}
M \otimes N' & \longrightarrow & M \otimes N & \longrightarrow & M \otimes N'' & \longrightarrow & 0 \\
 & {}_0 \searrow & \downarrow {}^f & \swarrow & & & \\
 & & Q \, . & & & &
\end{array}
$$

This is equivalent to asking whether there is a unique factorization of the corresponding diagram of bilinear maps,

$$
\begin{array}{ccccccc}
M \times N' & \longrightarrow & M \times N & \longrightarrow & M \times N'' & \longrightarrow & 0 \\
 & {}_0 \searrow & \downarrow {}^\beta & \swarrow & & & \\
 & & Q & & & &
\end{array}
$$

— uniqueness of the linear factorization is guaranteed by the fact that $M \times N''$ generates $M \otimes N''$. This unique factorization reflects the fact that $M \times -$ preserves cokernels. $\qquad\square$

Question 21.2. What about this argument fails if we try to use it to show that $M \otimes_R -$ preserves kernels?

Failure of exactness is bad, so let's try to repair it. A key observation is that if M is *free*, then $M \otimes_R -$ is exact. If $M = RI$, the free R-module on a set I, then $M \otimes_R N = \oplus_{i \in I} N$, since tensoring distributes over direct sums. Then we remember the following "obvious" fact:

Lemma 21.3. *If $M_i' \to M_i \to M_i''$ is exact for all $i \in I$, then so is*

$$\bigoplus M_i' \to \bigoplus M_i \to \bigoplus M_i'' \,.$$

Proof. Clearly the composite is zero. Let $(x_i \in M_i, i \in I) \in \bigoplus M_i$ and suppose it maps to zero. That means that each x_i maps to zero in M_i'' and hence is in the image of some $x_i' \in M_i'$. Just make sure to take $x_i' = 0$ if $x_i = 0$, so that it's still the case that only finitely many entries are nonzero. $\square$

To exploit this observation, we'll "resolve" M by free modules. For a start, this means: find a surjection from a free R-module,

$$\varepsilon : F_0 \to M \,.$$

This amounts to specifying a set of R-module generators. For a general ring R, the kernel of ε may not be free. For the moment, let's make sure that it is by assuming that R is a PID, and write F_1 for the kernel. The failure of $M \otimes_R -$ to be exact is measured, at least partially, by the leftmost term (defined as the kernel) in the exact sequence

$$0 \to \mathrm{Tor}_1^R(M, N) \to F_1 \otimes_R N \to F_0 \otimes_R N \to M \otimes_R N \to 0 \,.$$

The notation suggests that this Tor term is independent of the resolution. This is indeed the case, as we shall show presently. But before we do, let's compute some Tor groups.

Example 21.4. For any PID R, if $M = F$ is free over R we can take $F_0 = F$ and $F_1 = 0$, and discover that then $\mathrm{Tor}_1^R(F, N) = 0$ for any N.

Example 21.5. Let $R = \mathbb{Z}$, $M = \mathbb{Z}/n\mathbb{Z}$, and be N any abelian group. When $R = \mathbb{Z}$ it is often omitted from the notation for Tor. There is a nice free resolution staring at us: $F_0 = F_1 = \mathbb{Z}$, and $F_1 \to F_0$ given by multiplication by n. The sequence defining Tor_1 looks like

$$0 \to \mathrm{Tor}_1(\mathbb{Z}/n\mathbb{Z}, N) \to \mathbb{Z} \otimes N \xrightarrow{n \otimes 1} \mathbb{Z} \otimes N \to \mathbb{Z}/n\mathbb{Z} \otimes N \to 0 \,,$$

so

$$\mathbb{Z}/n\mathbb{Z} \otimes N = N/nN \,, \quad \mathrm{Tor}_1(\mathbb{Z}/n\mathbb{Z}, N) = \ker(n|N) \,.$$

The torsion in this case is the "n-torsion" in N. This accounts for the name.

Functors like Tor_1 can be usefully defined for any ring, and moving to that general case makes their significance clearer and illuminates the reason why Tor_1 is independent of choice of generators.

So let R be any ring and M a module over it. By picking R-module generators I can produce a surjection from a free R-module, $F_0 \to M$. Write K_0 for the kernel of this map. It is the module of relations among the generators. We can no longer guarantee that it's free, but we can at least find a set of module generators for it, and construct a surjection from a free R-module, $F_1 \to K_0$. Continuing in this way, we get a diagram like this —

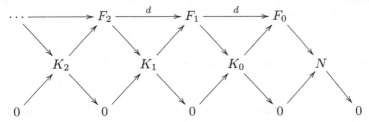

— in which the Λ-shaped subdiagrams are short exact sequences and F_s is free for all s. Splicing these short exact sequences gives you an exact sequence in the top row. This is a *free resolution of N*. The top row, F_*, is a chain complex. It maps to the very short chain complex with N in degree 0 and 0 elsewhere, and this chain map is a homology isomorphism (or "quasi-isomorphism"). We have in effect replaced N with this chain complex of free modules. The module N may be very complicated, with generators, relations, relations between relations .... All this is laid out in front of us by the free resolution. Generators of F_0 map to generators for N, and generators for F_1 map to relations among those generators, and so on.

Now we can try to define higher Tor functors by tensoring F_* with M and taking homology. If R is a PID and the resolution is just $F_1 \to F_0$, forming homology is precisely taking cokernel and kernel, as we did above. In general, we define

$$\text{Tor}_n^R(M, N) = H_n(M \otimes_R F_*).$$

In the next lecture we will check that this is well-defined — independent of free resolution, and functorial in the arguments. For the moment, notice that

$$\text{Tor}_n^R(M, F) = 0 \quad \text{for} \quad n > 0 \quad \text{if } F \text{ is free},$$

since I can take $F \xleftarrow{\cong} F \leftarrow 0 \leftarrow \cdots$ as a free resolution; and that

$$\mathrm{Tor}_0^R(M, N) = M \otimes_R N$$

since $M \otimes_R -$ is right-exact.

22 Fundamental theorem of homological algebra

We will now show that the R-modules $\mathrm{Tor}_n^R(M, N)$ are well-defined and functorial. This will be an application of a very general principle.

Theorem 22.1 (Fundamental Theorem of Homological Algebra). *Let N and M be R-modules; let*

$$\cdots \to F_1 \to F_0 \to N \to 0$$

be a sequence of R-modules in which each F_n is free; let

$$\cdots \to E_1 \to E_0 \to M \to 0$$

be an exact sequence of R-modules; and let $f : N \to M$ be a homomorphism. Then f lifts to a chain map $f_* : F_* \to E_*$, uniquely up to chain homotopy.

Proof. Let's try to construct f_0. We know that $F_0 = RS_0$ for some set S_0. Map the generators of F_0 into N via ε_N and then into M via f, and then lift the images to E_0 via ε_M (which is possible because ε_M is surjective). Then extend to an R-module homomorphism, to get f_0.

Now restrict f_0 to kernels to get g_0:

$$
\begin{array}{ccccccc}
0 & \longrightarrow & L_0 = \ker(\varepsilon_N) & \longrightarrow & F_0 & \xrightarrow{\ \varepsilon_N\ } & N \\
 & & \Big\downarrow{\scriptstyle g_0} & & \Big\downarrow{\scriptstyle f_0} & & \Big\downarrow{\scriptstyle f} \\
0 & \longrightarrow & K_0 = \ker(\varepsilon_M) & \longrightarrow & E_0 & \xrightarrow{\ \varepsilon_M\ } & M & \longrightarrow & 0 .
\end{array}
$$

Now the map $d : F_1 \to F_0$ satisfies $\varepsilon_N \circ d = 0$, and so factors through a map to $L_0 = \ker \varepsilon_N$. Similarly, $d : E_1 \to E_0$ factors through a map $E_1 \to K_0$, and this map must be surjective because the sequence $E_1 \to E_0 \to N$ is exact. We find ourselves in exactly the same situation as before:

$$
\begin{array}{ccccccc}
0 & \longrightarrow & L_1 & \longrightarrow & F_1 & \longrightarrow & L_0 \\
 & & \Big\downarrow{\scriptstyle g_1} & & \Big\downarrow{\scriptstyle f_1} & & \Big\downarrow{\scriptstyle g_0} \\
0 & \longrightarrow & K_1 & \longrightarrow & E_1 & \longrightarrow & K_0 & \longrightarrow & 0
\end{array}
$$

So we construct f_* by induction.

Now we need to prove the chain homotopy claim. So suppose I have $f_*, f'_* : F_* \to E_*$, both lifting $f : N \to M$. Then $f'_n - f_n$ (which we'll rename ℓ_n) is a chain map lifting $0 : N \to M$. We want to construct a chain null-homotopy of ℓ_*; that is, we want a sequence of maps $h : F_n \to E_{n+1}$ such that $dh + hd = \ell_*$. At the bottom, $F_{-1} = 0$, so we want $h : F_0 \to E_1$ such that $dh = \ell_0$. This factorization happens in two steps.

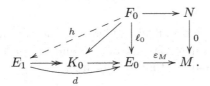

First, $\varepsilon_M \ell_0 = 0$ implies that ℓ_0 factors through $K_0 = \ker \varepsilon_M$. Next, $E_1 \to K_0$ is surjective, by exactness, and F_0 is free, so we can lift generators and extend R-linearly to get $h : F_0 \to E_1$.

The next step is organized by the diagram

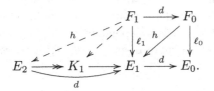

This diagram doesn't commute; $dh = \ell_0$, but the (d, h, ℓ_1) triangle doesn't commute. Rather, we want to construct $h : F_1 \to E_2$ such that $dh = \ell_1 - hd$. Since

$$d(\ell_1 - hd) = \ell_0 d - dhd = (\ell_0 - dh)d = 0$$

the map $\ell_1 - hd$ lifts to $K_1 = \ker d$. But then it lifts through E_2, since $E_2 \to K_1$ is surjective and F_1 is free.

Exactly the same process continues. $\qquad\square$

This proof uses a property of freeness that is shared by a broader class of modules.

Definition 22.2. An R-module P is *projective* if any map out of P factors through any surjection:

Every free module is projective, and this is the property of freeness that we have been using; the Fundamental Theorem of Homological Algebra holds under the weaker assumption that each F_n is projective.

Any direct summand in a projective is also projective. Any projective module is a direct summand of a free module. Over a PID, every projective is free, because any submodule of a free is free. But there are examples of nonfree projectives:

Example 22.3. Let k be a field and let R be the product ring $k \times k$. Then R acts on k in two ways, via $(a, b)c = ac$ and via $(a, b)c = bc$. These are both projective R-modules that are not free.

Now we will apply Theorem 22.1 to verify that our proposed construction of Tor is independent of free (or projective!) resolution, and is functorial.

Suppose we have an R-module homomorphism $f : N' \to N$. Pick arbitrary free resolutions $N' \leftarrow F'_*$ and $N \leftarrow F_*$, and pick any chain map $f_* : F'_* \to F_*$ lifting f. We claim that the map induced in homology by $1 \otimes f_* : M \otimes_R F'_* \to M \otimes_R F_*$ is independent of the choice of lift. Suppose f'_* is another lift, and pick a chain homotopy $h : f_* \simeq f'_*$. Since $M \otimes_R -$ is additive, the relation

$$1 \otimes h : 1 \otimes f_* \simeq 1 \otimes f'_*$$

still holds. So $1 \otimes f_*$ and $1 \otimes f'_*$ induce the same map in homology.

For example let $f : N' \to N$ and $g : N \to N''$; let $F'_* \to N'$, $F_* \to N$, and $F''_* \to N''$ be projective resolutions, and pick chain lifts f_* of f and g_* of g. The $g_* \circ f_*$ is a chain lift of $g \circ f$, and $(1 \otimes g_*) \circ (1 \otimes f_*)$ is a chain lift of $1 \otimes (g \circ f) = (1 \otimes g) \circ (1 \otimes f)$. So the map it induces in homology is the map induced by $1 \otimes (g \circ f)$. By functoriality of the formation of homology, this is also the composite of the maps induced by $1 \otimes g$ and $1 \otimes f$.

This argument establishes that $\operatorname{Tor}^R_*(M, N)$ is uniquely functorial in N. In particular $\operatorname{Tor}^R_*(M, N)$ is independent of choice of projective resolution. Functoriality in M is also easy to see.

My last general comment about Tor is that there's a symmetry here. Of course, $M \otimes_R N \cong N \otimes_R M$. This uses the fact that R is commutative. This leads right on to saying that $\operatorname{Tor}^R_n(M, N) \cong \operatorname{Tor}^R_n(N, M)$. We've been computing Tor by taking a resolution of the second variable. But I could equally have taken a resolution of the first variable. This follows from Theorem 22.1.

Example 22.4. I want to give an example in which you do have higher Tor modules. Let k be a field, and let $R = k[d]/(d^2)$. This is sometimes called the "dual numbers," or the exterior algebra over k. What is an R-module? It's just a k-vector space M with an operator d (given by multiplication by d) that satisfies $d^2 = 0$. Even though there's no grading around, I can still define the "homology" of M:

$$H(M; d) = \frac{\ker d}{\operatorname{im} d}.$$

The k-algebra R is "augmented" by an algebra map $\varepsilon : R \to k$ splitting the unit: $\varepsilon(d) = 0$. This renders k an R-module. Let's construct a free R-module resolution of this module. Here's a picture.

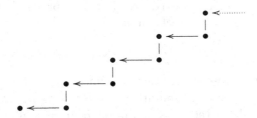

The vertical lines indicate multiplication by d. We could write this as

$$0 \leftarrow k \overset{\varepsilon}{\leftarrow} R \overset{d}{\leftarrow} R \overset{d}{\leftarrow} R \leftarrow \cdots.$$

Now tensor this over R with an R-module M; so M is a vector space equipped with an operator d with $d^2 = 0$. Each copy of R gets replaced by a copy of M, and the differential gives multiplication by d on M. So taking homology gives

$$\operatorname{Tor}_n^R(k, M) = \begin{cases} k \otimes_R M = M/dM & \text{for } n = 0 \\ H(M; d) & \text{for } n > 0. \end{cases}$$

So for example

$$\operatorname{Tor}_n^R(k, k) = k \quad \text{for } n \geq 0.$$

Exercises

Exercise 22.5. Let $0 \to M' \to M \to M'' \to 0$ be a short exact sequence of R-modules. Construct from it a natural long exact sequence of the form

$$\cdots \longrightarrow \mathrm{Tor}_2^R(M'', N)$$

$$\mathrm{Tor}_1^R(M', N) \longrightarrow \mathrm{Tor}_1^R(M, N) \longrightarrow \mathrm{Tor}_1^R(M'', N)$$

$$M' \otimes_R N \longrightarrow M \otimes_R N \longrightarrow M'' \otimes_R N \longrightarrow 0 \,.$$

23   Hom and Lim

We will now develop more properties of the tensor product: its relationship to homomorphisms and to direct limits.

Hom

The tensor product arose in our study of bilinear maps. Even more natural are *linear maps*. Given a commutative ring R and two R-modules M and N, we can think about the collection of all R-linear maps from M to N. Not only does this set form an abelian group (under pointwise addition of homomorphisms); it forms an R-module, with

$$(rf)(y) = f(ry) = rf(y), \quad r \in R, \, y \in M \,.$$

The check that this is again an R-module homomorphism uses commutativity of R. We will write $\mathrm{Hom}_R(M, N)$, or just $\mathrm{Hom}(M, N)$, for this R-module.

Since $\mathrm{Hom}(M, N)$ is an R-module, we are entitled to think about what an R-module homomorphism into it is. Given

$$f : L \to \mathrm{Hom}(M, N)$$

we can define a new function

$$\hat{f} : L \times M \to N, \quad \hat{f}(x, y) = (f(x))(y) \in N \,.$$

You should check that this new function $\hat{f}$ is R-bilinear! So we get a natural map

$$\mathrm{Hom}(L, \mathrm{Hom}(M, N)) \to \mathrm{Hom}(L \otimes M, N) \,.$$

Conversely, given a map $\hat{f} : L \otimes M \to N$ and $x \in L$, we can define $f(x) : M \to N$ by the same formula. These are inverse operations, so:

Lemma 23.1. *The natural map* $\mathrm{Hom}(L, \mathrm{Hom}(M, N)) \to \mathrm{Hom}(L \otimes M, N)$ *is an isomorphism.*

One says that $- \otimes M$ and $\mathrm{Hom}(M, -)$ are *adjoint*. The notion of adjoint functors will be taken up again in Lecture 39.

Lim

The second thing we will discuss is a generalization of one perspective on how the rational numbers are constructed from the integers — by a limit process: There are compatible maps in the diagram

$$
\begin{array}{ccccccccc}
\mathbb{Z} & \xrightarrow{\ 2\ } & \mathbb{Z} & \xrightarrow{\ 3\ } & \mathbb{Z} & \xrightarrow{\ 4\ } & \mathbb{Z} & \xrightarrow{\ 5\ } & \cdots \\
\downarrow{\scriptstyle 1} & & \downarrow{\scriptstyle 1/2} & & \downarrow{\scriptstyle 1/3!} & & \downarrow{\scriptstyle 1/4!} & & \\
\mathbb{Q} & \xrightarrow{=} & \mathbb{Q} & \xrightarrow{=} & \mathbb{Q} & \xrightarrow{=} & \mathbb{Q} & \xrightarrow{=} & \cdots
\end{array}
$$

and $\mathbb{Q}$ is the "universal," or "initial," abelian group that all these copies of the integers map to in a compatible way.

We will formalize this process, using partially ordered sets as indexing sets. Recall from Lecture 3 that a *partially ordered set*, or *poset*, is a small category $\mathcal{I}$ such that $\#\mathcal{I}(i, j) \leq 1$ and the only isomorphisms are the identity maps. Write $i \leq j$ if $\mathcal{I}(i, j) \neq \varnothing$. Then $i \leq i$, $i \leq j \leq k$ implies $i \leq k$, and if $i \leq j$ and $j \leq i$ then $i = j$. We will be interested in a particular class of posets.

Definition 23.2. A poset $(\mathcal{I}, \leq)$ is *directed* if for every $i, j \in \mathcal{I}$ there exists $k \in \mathcal{I}$ such that $i \leq k$ and $j \leq k$.

Example 23.3. This is a very common condition. A first example is the natural numbers $\mathbb{N}$ with $\leq$ as the order. Another example is the positive natural numbers, with $i \leq j$ if $i|j$. This is because $i, j|(ij)$. A topological example: if X is a space, A a subspace, we can consider the poset $\mathcal{U}_A$ whose elements are the open subsets of X containing A, ordered by saying that $U \leq V$ if $U \supseteq V$. This is directed because an intersection of two opens is again open. You should think of this directed set as capturing how A is approximated from above by open neighborhoods. It will play a big role in our treatment of Poincaré duality.

Definition 23.4. Let $\mathcal{I}$ be a directed set. An $\mathcal{I}$-*directed system* in a category $\mathcal{C}$ is a functor $\mathcal{I} \to \mathcal{C}$. This means that for every $i \in \mathcal{I}$ we are given an object $X_i \in \mathcal{C}$, and for every $i \leq j$ we are given a map $f_{i,j} : X_i \to X_j$, in such a way that $f_{i,i} = 1_{X_i}$ and if $i \leq j \leq k$ then $f_{i,k} = f_{j,k} \circ f_{i,j} : X_i \to X_k$.

Example 23.5. If $\mathcal{I} = (\mathbb{N}, \leq)$, then you get a "linear system" $X_0 \xrightarrow{f_{01}} X_1 \xrightarrow{f_{12}} X_2 \to \cdots$.

Example 23.6. Suppose $\mathcal{I} = (\mathbb{N}_{>0}, |)$, i.e., the second example above. You can consider $\mathcal{I} \to \mathbf{Ab}$, say assigning to each i the integers $\mathbb{Z}$, and $f_{ij} : \mathbb{Z} \xrightarrow{j/i} \mathbb{Z}$.

Example 23.7. There is an evident $\mathcal{U}_A$-directed system of spaces given by sending $U \subseteq A$ to the space $X - U$. We will encounter this example when we discuss Poincaré duality.

These directed systems can be a little complicated. But there's a simple one, namely the constant system.

Example 23.8. Let $\mathcal{I}$ be any directed set and $\mathcal{C}$ any category. An object $A \in \mathcal{C}$ determines an $\mathcal{I}$-directed set, namely the constant functor $c_A : \mathcal{I} \to \mathcal{C}$.

Not every directed system is constant, but we can try to find a best approximating constant system. To compare systems, we need morphisms. $\mathcal{I}$-directed systems in $\mathcal{C}$ are functors $\mathcal{I} \to \mathcal{C}$. They are related by natural transformations, and those are the morphisms in the category of $\mathcal{I}$-directed systems. That is to say, a morphism is a choice of map $g_i : X_i \to Y_i$, for each $i \in \mathcal{I}$, such that

$$
\begin{array}{ccc}
X_i & \longrightarrow & X_j \\
\downarrow{\scriptstyle g_i} & & \downarrow{\scriptstyle g_j} \\
Y_i & \longrightarrow & Y_j
\end{array}
$$

commutes for all $i \leq j$.

Definition 23.9. Let $X : \mathcal{I} \to \mathcal{C}$ be a directed system. A *direct limit* of X is an object L and a map $X \to c_L$ that is initial among maps to constant systems. This means that given any other map to a constant system, say $X \to c_A$, there is a unique map $f : L \to A$ such that

commutes.

This is a "universal property." Being solutions to a universal property, any two direct limits are canonically isomorphic; but a directed system may fail to have a direct limit. For example, the linear directed systems we used to create the rational numbers exists in the category of finitely generated abelian groups; but $\mathbb{Q}$ is not finitely generated, and there's no finitely generated group that will serve as a direct limit of this system in the category of finitely generated abelian groups.

Example 23.10. Suppose we have an increasing sequence of subspaces, $X_0 \subseteq X_1 \subseteq \cdots \subseteq X$. This gives us a directed system of spaces, directed by the poset $(\mathbb{N}, \leq)$. It's pretty clear that as a *set* the direct limit of this system is the union of the subspaces. Saying that X is the direct limit of this directed system of spaces is saying first that X is the union of the X_i's, and second that the topology on X is determined by the topology on the subspaces; it's the "weak topology," characterized by the property that a map $f : X \to Y$ is continuous if and only if the restriction of f to each X_n is continuous. This is saying that a subset of X is open if and only if its intersection with each X_n is open in X. For example, a CW complex is the direct limit of its skelata.

Direct limits may be constructed from the material of coproducts and quotients. So suppose $X : \mathcal{I} \to \mathcal{C}$ is a directed system. To attempt to construct the direct limit, begin by forming (if possible) the coproduct over the elements of $\mathcal{I}$,

$$\coprod_{i \in \mathcal{I}} X_i \, .$$

There are maps $\mathrm{in}_i : X_i \to \coprod X_i$, but they are not yet compatible with the order relation in $\mathcal{I}$. Form a quotient of the coproduct to enforce that compatibility:

$$\varinjlim_{i \in \mathcal{I}} X_i = \left(\coprod_{i \in \mathcal{I}} X_i \right) / \sim,$$

where $\sim$ is the equivalence relation generated by requiring that for any $i \in \mathcal{I}$ and any $x \in X_i$,

$$\mathrm{in}_i x \sim \mathrm{in}_j f_{ij}(x) \, .$$

The process of forming the coproduct and the quotient will depend upon the category you are working in, and may not be possible. In sets, coproduct is disjoint union and the quotient just forms equivalence classes. In abelian groups, the coproduct is the direct sum and to form the quotient you divide by the subgroup generated by differences. In general we really want to form a "coequalizer," as discussed later in Lecture 39.

The conclusion is that direct limits exist in **Set**, **Top**, **Ab**, **Mod**$_R$, and many other categories.

Here's a lemma that lets us identify when a map to a constant system is a direct limit.

Lemma 23.11. *Let* $X : \mathcal{I} \to \mathbf{Ab}$ *(or* $\mathbf{Mod}_R$*). A map* $f : X \to c_L$ *(given by* $f_i : X_i \to L$ *for* $i \in \mathcal{I}$*) is the direct limit if and only if:*

(1) For every $x \in L$*, there exists an* $i \in \mathcal{I}$ *and an* $x_i \in X_i$ *such that* $f_i(x_i) = x$.

(2) Let $x_i \in X_i$ *be such that* $f_i(x_i) = 0$ *in* L*. Then there exists some* $j \geq i$ *such that* $f_{ij}(x_i) = 0$ *in* X_j.

Proof. Homework.   □

Proposition 23.12. *The direct limit functor* $\varinjlim_{\mathcal{I}} : \mathrm{Fun}(\mathcal{I}, \mathbf{Ab}) \to \mathbf{Ab}$ *is exact. In other words, if* $X \xrightarrow{p} Y \xrightarrow{q} Z$ *is an exact sequence of* $\mathcal{I}$*-directed systems (meaning that at every degree we get an exact sequence of abelian groups), then* $\varinjlim_{\mathcal{I}} X \to \varinjlim_{\mathcal{I}} Y \to \varinjlim_{\mathcal{I}} Z$ *is exact.*

Proof. First of all, $qp : X \to Z$ is zero, which is to say that it factors through the constant zero object, so $\varinjlim_{\mathcal{I}} X \to \varinjlim_{\mathcal{I}} Z$ is certainly the zero map. Let $y \in \varinjlim_{\mathcal{I}} Y$, and suppose y maps to 0 in $\varinjlim_{\mathcal{I}} Z$. By condition (1) of Lemma 23.11, there exists i such that $y = f_i(y_i)$ for some $y_i \in Y_i$. Then $0 = q(y) = qf_i(y_i) = f_i q(y_i)$ because q is a map of direct systems. By condition (2), this means that there is $j \geq i$ such that $f_{ij}q(y_i) = 0$ in Z_j. So $qf_{ij}y_i = 0$, again because q is a map of direct systems. Then $f_{ij}y_i$ is an element in Y_j that maps to zero under q, so there is some $x_j \in X_j$ such that $p(x_j) = f_{ij}y_i$. Then $f_j(x_j) \in \varinjlim_{\mathcal{I}} X$ maps to y.   □

The exactness of the direct limit has many useful consequences. For example:

Corollary 23.13. *Let* $i \mapsto C(i)$ *be a directed system of chain complexes. Then there is a natural isomorphism*

$$\varinjlim_{i \in \mathcal{I}} H_*(C(i)) \to H_*(\varinjlim_{i \in \mathcal{I}} C(i)).$$

Lim and tensor

Direct limits and the tensor product are nicely related, and the way to see that is to use the adjunction with Hom that we started with today.

Proposition 23.14. *Let $\mathcal{I}$ be a direct set, and let $M : \mathcal{I} \to \mathbf{Mod}_R$ be a $\mathcal{I}$-directed system of R-modules. There is a natural isomorphism*

$$(\varinjlim_{i \in \mathcal{I}} M_i) \otimes_R N \cong \varinjlim_{i \in \mathcal{I}} (M_i \otimes_R N).$$

Proof. Let's verify that both sides satisfy the same universal property. A map from $(\varinjlim_{i \in \mathcal{I}} M_i) \otimes_R N$ to an R-module L is the same thing as a linear map $\varinjlim_{i \in \mathcal{I}} M_i \to \operatorname{Hom}_R(N, L)$. This is the same as a compatible family of maps $M_i \to \operatorname{Hom}_R(N, L)$, which in turn is the same as a compatible family of maps $M_i \otimes_R N \to L$, which is the same as a linear map $\varinjlim_{i \in \mathcal{I}} (M_i \otimes_R N) \to L$. $\square$

Putting together things we have just said:

Corollary 23.15. $H_*(X; \mathbb{Q}) = H_*(X) \otimes \mathbb{Q}$.

So for example we can redefine the Betti numbers of a space X as

$$\beta_n = \dim_{\mathbb{Q}} H_n(X; \mathbb{Q}).$$

Exercises

Exercise 23.16. Verify Lemma 23.11.

Exercise 23.17. (a) Embed $\mathbb{Z}/p^n\mathbb{Z}$ into $\mathbb{Z}/p^{n+1}\mathbb{Z}$ by sending 1 to p, and write $\mathbb{Z}_{p^\infty}$ for the union. It's called the *Prüfer group* (at p). Show that $\mathbb{Z}_{p^\infty} \cong \mathbb{Z}[1/p]/\mathbb{Z}$ and that

$$\mathbb{Q}/\mathbb{Z} \cong \bigoplus_p \mathbb{Z}_{p^\infty},$$

where the sum runs over the prime numbers.

(b) Compute $\mathbb{Z}_{p^\infty} \otimes_{\mathbb{Z}} A$ for A each of the following abelian groups: $\mathbb{Z}/n\mathbb{Z}$, $\mathbb{Z}[1/q]$ (for q a prime), and $\mathbb{Z}_{q^\infty}$ (for q a prime).

(c) Compute $\operatorname{Tor}_1^{\mathbb{Z}}(M, \mathbb{Z}[1/p])$ and $\operatorname{Tor}_1^{\mathbb{Z}}(M, \mathbb{Z}_{p^\infty})$, for any abelian group M in terms of the self-map $p : M \to M$.

(d) Show that if $f : X \to Y$ induces an isomorphism in homology with coefficients in the prime fields $\mathbb{F}_p$ (for all primes p) and $\mathbb{Q}$, then it induces an isomorphism in homology with coefficients in $\mathbb{Z}$.

24　Universal coefficient theorem

Finally, let's return to the question: Suppose that we are given $H_*(X;\mathbb{Z})$. Can we compute $H_*(X;\mathbb{Z}/2\mathbb{Z})$? This is non-obvious. For example, consider the map $\mathbb{RP}^2 \to S^2$ that pinches $\mathbb{RP}^1$ to a point. $H_2(\mathbb{RP}^2;\mathbb{Z}) = 0$, so in H_2 this map is certainly zero. But in $\mathbb{Z}/2\mathbb{Z}$-coefficients, in dimension 2, this map gives an isomorphism. This shows that there's no *functorial* determination of $H_*(X;\mathbb{Z}/2)$ in terms of $H_*(X;\mathbb{Z})$; the effect of a map in integral homology does not determine its effect in mod 2 homology. So how *do* we go between different coefficients?

Let R be a commutative ring and M an R-module, and suppose we have a chain complex C_* of R-modules. It could be the singular complex of a space, but it doesn't have to be. Let's compare $H_n(C_*) \otimes M$ with $H_n(C_* \otimes M)$. (Here and below we'll just write $\otimes$ for $\otimes_R$.) The latter thing gives homology with coefficients in M. Let's investigate how these two are related, and build up conditions on R and C_* as we go along.

First, there's a natural map

$$\alpha : H_n(C_*) \otimes M \to H_n(C_* \otimes M),$$

sending $[z] \otimes m$ to $[z \otimes m]$. We propose to find conditions under which it is injective. The map α fits into a commutative diagram with exact columns like this:

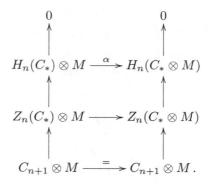

Now, $Z_n(C_* \otimes M)$ is a submodule of $C_n \otimes M$, but the map $Z_n(C) \otimes M \to C_n \otimes M$ need not be injective ... unless we impose more restrictions. If we can guarantee that it is, then the middle map is injective and a diagram chase shows that α is too.

So let's assume that R is a PID and that C_n is a free R-module for all n. Then the submodule $B_{n-1}(C_*) \subseteq C_{n-1}$ is again free, so the short exact

sequence in the top row of

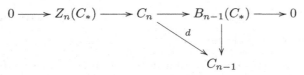

splits. So $Z_n(C_*) \to C_n$ is a split monomorphism, and hence $Z_n(C_*) \otimes M \to C_n \otimes M$ is too.

In fact, a little thought shows that this argument produces a splitting of the map α.

Now α is still not always an isomorphism. But it certainly is if $M = R$, and it's compatible with direct sums, so it certainly is if M is free. The idea is now to resolve M by frees, and see where that takes us.

So let

$$0 \to F_1 \to F_0 \to M \to 0$$

be a free resolution of M. Again, we're using the assumption that R is a PID, to guarantee that $\ker(F_0 \to M)$ is free. Again using the assumption that each C_n is free, we get a short exact sequence of chain complexes

$$0 \to C_* \otimes F_1 \to C_* \otimes F_0 \to C_* \otimes M \to 0.$$

In homology, this gives a long exact sequence. Unsplicing it gives the left-hand column in the following diagram.

$$
\begin{array}{ccc}
0 & & 0 \\
\downarrow & & \downarrow \\
\mathrm{coker}(H_n(C_* \otimes F_1) \to H_n(C_* \otimes F_0)) & \xrightarrow{\ \cong\ } & \mathrm{coker}(H_n(C_*) \otimes F_1 \to H_n(C_*) \otimes F_0)) \\
\downarrow & & \downarrow \\
H_n(C_* \otimes M) & \xrightarrow{\ =\ } & H_n(C_* \otimes M) \\
\downarrow{\scriptstyle\partial} & & \downarrow \\
\ker(H_{n-1}(C_* \otimes F_1) \to H_{n-1}(C_* \otimes F_0)) & \xrightarrow{\ \cong\ } & \ker(H_{n-1}(C_*) \otimes F_1 \to H_{n-1}(C_*) \otimes F_0) \\
\downarrow & & \downarrow \\
0 & & 0.
\end{array}
$$

The map α is an isomorphism when the module involved is free, and this lets us replace the left hand column with the indicated right hand column. But

$$\mathrm{coker}(H_n(C_*) \otimes F_1 \to H_n(C_*) \otimes F_0)) = H_n(C_*) \otimes M$$

and
$$\ker(H_{n-1}(C_*) \otimes F_1 \to H_{n-1}(C_*) \otimes F_0) = \mathrm{Tor}_1^R(H_{n-1}(C_*), M).$$

We have proved the following theorem.

Theorem 24.1 (Universal Coefficient Theorem). *Let R be a PID and let C_* a chain complex of R-modules such that C_n is free for all n and let M be an R-module. Then there is a natural short exact sequence of R-modules*

$$0 \to H_n(C_*) \otimes_R M \xrightarrow{\alpha} H_n(C_* \otimes_R M) \xrightarrow{\partial} \mathrm{Tor}_1^R(H_{n-1}(C_*), M) \to 0$$

that splits (but not naturally).

Example 24.2. The pinch map $\mathbb{RP}^2 \to S^2$ induces the following map of universal coefficient short exact sequences:

$$
\begin{array}{ccccccccc}
0 & \to & H_2(\mathbb{RP}^2) \otimes \mathbb{Z}/2\mathbb{Z} & \to & H_2(\mathbb{RP}^2; \mathbb{Z}/2\mathbb{Z}) & \xrightarrow{\cong} & \mathrm{Tor}_1(H_1(\mathbb{RP}^2), \mathbb{Z}/2\mathbb{Z}) & \to & 0 \\
& & \downarrow{\scriptstyle 0} & & \downarrow{\scriptstyle \cong} & & \downarrow{\scriptstyle 0} & & \\
0 & \to & H_2(S^2) \otimes \mathbb{Z}/2\mathbb{Z} & \xrightarrow{\cong} & H_2(S^2; \mathbb{Z}/2\mathbb{Z}) & \to & \mathrm{Tor}_1(H_1(S^2), \mathbb{Z}/2\mathbb{Z}) & \to & 0.
\end{array}
$$

This shows that the splitting of the universal coefficient short exact sequence cannot be made natural, and it explains the mystery that we began with.

Exercises

Exercise 24.3. Let X be a CW complex of finite type. Determine the dimension of $H_n(X; \mathbb{F}_p)$, for each n, in terms of the Betti numbers and torsion coefficients of $H_*(X; \mathbb{Z})$.

Exercise 24.4. The hypotheses of Theorem 24.1 are essential. Construct two counterexamples: one with $R = \mathbb{Z}$ but in which the groups in the chain complex are not free, and one in which $R = k[d]/d^2$ and the modules in C_* are free over R.

Exercise 24.5. Give an example of two Moore spaces that have the same homology with any field coefficients but that are not homotopy equivalent.

25   Künneth and Eilenberg-Zilber

We want to compute the homology of a product. Long ago, in Lecture 7, we constructed a bilinear map $S_p(X) \times S_q(Y) \to S_{p+q}(X \times Y)$, called the cross product. So we get a linear map $S_p(X) \otimes S_q(Y) \to S_{p+q}(X \times Y)$, and it satisfies the Leibniz formula, i.e. $d(x \times y) = dx \times y + (-1)^p x \times dy$. The method we used works with any coefficient ring, not just the integers.

Definition 25.1. Let C_*, D_* be two chain complexes. Their *tensor product* is the chain complex with

$$(C_* \otimes D_*)_n = \bigoplus_{p+q=n} C_p \otimes D_q.$$

The differential $(C_* \otimes D_*)_n \to (C_* \otimes D_*)_{n-1}$ sends $C_p \otimes D_q$ into the submodule $C_{p-1} \otimes D_q \bigoplus C_p \otimes D_{q-1}$ by

$$x \otimes y \mapsto dx \otimes y + (-1)^p x \otimes dy.$$

So the cross-product is a map of chain complexes $S_*(X) \otimes S_*(Y) \to S_*(X \times Y)$. There are two questions:
(1) Is this map an isomorphism in homology?
(2) How is the homology of a tensor product of chain complexes related to the tensor product of their homologies?

It's easy to see what happens in dimension zero, because $\pi_0(X) \times \pi_0(Y) = \pi_0(X \times Y)$ implies that $H_0(X) \otimes H_0(Y) \xrightarrow{\cong} H_0(X \times Y)$.

Let's dispose of the purely algebraic question (2) first.

Theorem 25.2. *Let R be a PID and C_*, D_* be chain complexes of R-modules. Assume that C_n is a free R-module for all n. There is a short exact sequence*

$$0 \to (H_*(C) \otimes_R H_*(D))_n \to H_n(C_* \otimes_R D_*) \to \mathrm{Tor}_1^R(H_*(C), H_*(D))_{n-1} \to 0$$

natural in these data, where

$$(H_*(C) \otimes_R H_*(D))_n = \bigoplus_{p+q=n} H_p(C) \otimes_R H_q(D)$$

and

$$\mathrm{Tor}_1^R(H_*(C), H_*(D))_{n-1} = \bigoplus_{p+q=n-1} \mathrm{Tor}_1^R(H_p(C), H_q(D)).$$

The sequence splits (but not naturally).

Proof. This is exactly the same as the proof for the UCT. It's a good idea to work through this on your own. □

Corollary 25.3. *Let R be a PID and assume C'_n and C_n are R free for all n. If $C'_* \to C_*$ and $D'_* \to D_*$ are homology isomorphisms then so is $C'_* \otimes D'_* \to C_* \otimes D_*$.*

Proof. Apply the five-lemma. □

Our attack on question (1) is via the method of "acyclic models." This is really a special case of the Fundamental Theorem of Homological Algebra, Theorem 22.1.

Definition 25.4. Let $\mathcal{C}$ be a category, and fix a set $\mathbb{M}$ of objects in $\mathcal{C}$, to be called the "models." A functor $F : \mathcal{C} \to \mathbf{Ab}$ is $\mathbb{M}$-*free* if it is the free abelian group functor applied to a coproduct of functors of the form $\mathcal{C}(M, -)$ for $M \in \mathbb{M}$.

Example 25.5. Since we are interested in the singular homology of a product of two spaces, it may be sensible to take as $\mathcal{C}$ the category of ordered pairs of spaces, $\mathcal{C} = \mathbf{Top}^2$, and for $\mathbb{M}$ the set of pairs of simplices, $\mathbb{M} = \{(\Delta^p, \Delta^q) : p, q \geq 0\}$. Then

$$S_n(X \times Y) = \mathbb{Z}[\mathbf{Top}(\Delta^n \times X) \times \mathbf{Top}(\Delta^n, Y)] = \mathbb{Z}\mathbf{Top}^2((\Delta^n, \Delta^n), (X, Y))$$

is $\mathbb{M}$-free; it's given by a single model, namely (Δ^n, Δ^n).

Example 25.6. With the same category and models,

$$(S_*(X) \otimes S_*(Y))_n = \bigoplus_{p+q=n} S_p(X) \otimes S_q(Y)$$

is $\mathbb{M}$-free, since the tensor product has as free basis the set

$$\coprod_{p+q=n} \mathrm{Sin}_p(X) \times \mathrm{Sin}_q(Y) = \coprod_{p+q=n} \mathbf{Top}^2((\Delta^p, \Delta^q), (X, Y)).$$

Definition 25.7. A natural transformation of functors $\theta : F \to G$ in $\mathrm{Fun}(\mathcal{C}, \mathbf{Ab})$ is an $\mathbb{M}$-*epimorphism* if $\theta_M : F(M) \to G(M)$ is a surjection of abelian groups for every $M \in \mathbb{M}$. A *sequence* of natural transformations is a composable pair $G' \to G \to G''$ with trivial composition. It is $\mathbb{M}$-*exact* if $G'(M) \to G(M) \to G''(M)$ is exact for all $M \in \mathbb{M}$.

Example 25.8. We claim that

$$\cdots \to S_n(X \times Y) \to S_{n-1}(X \times Y) \to \cdots \to S_0(X \times Y) \to H_0(X \times Y) \to 0$$

is $\mathbb{M}$-exact. Just plug in (Δ^p, Δ^q): you get an exact sequence, since $\Delta^p \times \Delta^q$ is contractible.

Example 25.9. The sequence

$$\cdots \to (S_*(X) \otimes S_*(Y))_n \to (S_*(X) \otimes S_*(Y))_{n-1} \to \cdots$$
$$\to S_0(X) \otimes S_0(Y) \to H_0(X) \otimes H_0(Y) \to 0$$

is also $\mathbb{M}$-exact, by Corollary 25.3.

The terms "M-free" and "M-exact" relate to each other in the expected way:

Lemma 25.10. *Let $\mathcal{C}$ be a category with a set of models $\mathbb{M}$ and let F, G, G' : $\mathcal{C} \to \mathbf{Ab}$ be functors. Suppose that F is $\mathbb{M}$-free, let $G' \to G$ be a $\mathbb{M}$-epimorphism, and let $f : F \to G$ be any natural transformation. Then there is a lifting:*

$$
\begin{array}{ccc}
 & & G' \\
 & \overline{f} \nearrow & \big\downarrow \\
F & \xrightarrow{\ f\ } & G.
\end{array}
$$

Proof. Clearly we may assume that $F(X) = \mathbb{Z}\mathcal{C}(M, X)$ for some $M \in \mathbb{M}$. Suppose first that $X = M \in \mathbb{M}$, so that

$$
\begin{array}{ccc}
 & & G'(M) \\
 & \overline{f}_M \nearrow & \big\downarrow \\
\mathbb{Z}\mathcal{C}(M, M) & \xrightarrow{\ f_M\ } & G(M).
\end{array}
$$

Consider $1_M \in \mathbb{Z}\mathcal{C}(M, M)$. Its image $f_M(1_M) \in G(M)$ is hit by some element in $c_M \in G'(M)$, since $G' \to G$ is an $\mathbb{M}$-epimorphism. Define $\overline{f}_M(1_M) = c_M$.

Now we exploit naturality! Any $\varphi : M \to X$ produces a commutative diagram

$$
\begin{array}{ccc}
\mathcal{C}(M, M) & \xrightarrow{\ \overline{f}_M\ } & G'(M) \\
\big\downarrow{\scriptstyle \varphi_*} & & \big\downarrow{\scriptstyle \varphi_*} \\
\mathcal{C}(M, X) & \xrightarrow{\ \overline{f}_X\ } & G'(X).
\end{array}
$$

Chase 1_M around the diagram, to see what the value of $\overline{f}_X(\varphi)$ must be:

$$\overline{f}_X(\varphi) = \overline{f}_X(\varphi_*(1_M)) = \varphi_*(\overline{f}_M(1_M)) = \varphi_*(c_M).$$

Now extend linearly. You should check that this does define a natural transformation. $\qquad\square$

This is precisely the condition required to prove the Fundamental Theorem of Homological Algebra. So we have:

Theorem 25.11 (Acyclic Models). *Let* $\mathbb{M}$ *be a set of models in a category* $\mathcal{C}$. *Let* $\theta : F \to G$ *be a natural transformation of functors from* $\mathcal{C}$ *to* **Ab**. *Let* F_* *and* G_* *be functors from* $\mathcal{C}$ *to chain complexes, with augmentations* $F_0 \to F$ *and* $G_0 \to G$. *Assume that* F_n *is* $\mathbb{M}$-*free for all* n, *and that* $G_* \to G \to 0$ *is an* $\mathbb{M}$-*exact sequence. Then there is a unique chain homotopy class of chain maps* $F_* \to G_*$ *covering* θ.

Corollary 25.12. *Suppose furthermore that* θ *is a natural isomorphism. If each* G_n *is* $\mathbb{M}$-*free and* $F_* \to F \to 0$ *is an* $\mathbb{M}$-*exact sequence, then any natural chain map* $F_* \to G_*$ *covering* θ *is a natural chain homotopy equivalence.*

Applying this to our category $\mathbf{Top}^2$ with models as before, we get the following theorem that completes work we did in Lecture 7.

Theorem 25.13 (Eilenberg-Zilber theorem). *Take coefficients in any commutative ring* R. *There are unique chain homotopy classes of natural chain maps:*

$$S_*(X) \otimes_R S_*(Y) \leftrightarrows S_*(X \times Y)$$

covering the usual isomorphism

$$H_0(X) \otimes_R H_0(Y) \cong H_0(X \times Y) \,,$$

and any such pair are natural chain homotopy inverses.

Corollary 25.14. *There is a canonical natural isomorphism* $H_*(S_*(X) \otimes_R S_*(Y)) \cong H_*(X \times Y)$.

Corollary 25.15. *Any choice of natural chain-level cross-product is naturally homotopy commutative.*

Proof. Let $\gamma_{X,Y} : S_*(X) \otimes S_*(Y) \to S_*(X \times Y)$ be any chain map covering $\times : H_0(X) \otimes H_0(Y) \to H_0(X \times Y)$. Then

$$
\begin{array}{ccc}
S_*(X) \times S_*(Y) & \xrightarrow{\gamma_{X,Y}} & S_*(X \times Y) \\
\downarrow{\scriptstyle \tau} & & \cong \downarrow{\scriptstyle T_*} \\
S_*(Y) \times S_*(X) & \xrightarrow{\gamma_{Y,X}} & S_*(Y \times X)
\end{array}
$$

is another; so it must be naturally homotopic to $\gamma_{X,Y}$. Similarly for a cross product in the other direction. $\qquad \square$

Combining this theorem with the algebraic Künneth theorem, we get:

Theorem 25.16 (Künneth theorem). *Take coefficients in a PID R. There is a short exact sequence*

$$0 \to (H_*(X) \otimes_R H_*(Y))_n \to H_n(X \times Y) \to \mathrm{Tor}_1^R(H_*(X), H_*(Y))_{n-1} \to 0$$

natural in X, Y, where

$$(H_*(X) \otimes_R H_*(Y))_n = \bigoplus_{p+q=n} H_p(X) \otimes_R H_q(Y)$$

and

$$\mathrm{Tor}_1^R(H_*(X), H_*(Y))_{n-1} = \bigoplus_{p+q=n-1} \mathrm{Tor}_1^R(H_p(X), H_q(Y)).$$

The sequence splits as R-modules, but not naturally.

Example 25.17. If $R = k$ is a field, every module is free, so the Tor term vanishes, and you get a Künneth *isomorphism*:

$$\times : H_*(X; k) \otimes_k H_*(Y; k) \xrightarrow{\cong} H_*(X \times Y; k).$$

This is rather spectacular. For example, what is $H_*(\mathbb{RP}^3 \times \mathbb{RP}^3; k)$, where k is a field? Well, if k has characteristic different from 2, $\mathbb{RP}^3$ has the same homology as S^3, so the product has the same homology as $S^3 \times S^3$: the dimensions are $1, 0, 0, 2, 0, 0, 1$. If char $k = 2$, on the other hand, the cohomology modules are either 0 or k, and we need to form the graded tensor product:

$$
\begin{array}{cccc}
k & k & k & k \\
k & k & k & k \\
k & k & k & k \\
k & k & k & k
\end{array}
$$

so the dimensions of the homology of the product are $1, 2, 3, 4, 3, 2, 1$.

The palindromic character of this sequence will be explained by Poincaré duality. Let's look also at what happens over the integers. Then we have the table of tensor products.

	$\mathbb{Z}$	$\mathbb{Z}/2\mathbb{Z}$	0	$\mathbb{Z}$
$\mathbb{Z}$	$\mathbb{Z}$	$\mathbb{Z}/2\mathbb{Z}$	0	$\mathbb{Z}$
$\mathbb{Z}/2\mathbb{Z}$	$\mathbb{Z}/2\mathbb{Z}$	$\mathbb{Z}/2\mathbb{Z}$	0	$\mathbb{Z}/2\mathbb{Z}$
0	0	0	0	0
$\mathbb{Z}$	$\mathbb{Z}$	$\mathbb{Z}/2\mathbb{Z}$	0	$\mathbb{Z}$

There is only one nonzero Tor group, namely

$$\mathrm{Tor}_1^{\mathbb{Z}}(H_1(\mathbb{RP}^3), H_1(\mathbb{RP}^3)) = \mathbb{Z}/2\mathbb{Z}.$$

Putting this together, we get the following groups.

$$
\begin{array}{c|c}
H_0 & \mathbb{Z} \\
H_1 & \mathbb{Z}/2\mathbb{Z} \oplus \mathbb{Z}/2\mathbb{Z} \\
H_2 & \mathbb{Z}/2\mathbb{Z} \\
H_3 & \mathbb{Z} \oplus \mathbb{Z} \oplus \mathbb{Z}/2\mathbb{Z} \\
H_4 & \mathbb{Z}/2\mathbb{Z} \oplus \mathbb{Z}/2\mathbb{Z} \\
H_5 & 0 \\
H_6 & \mathbb{Z}
\end{array}
$$

The failure of perfect symmetry here is interesting, and will also be explained by Poincaré duality.

Exercises

Exercise 25.18. An object A of a category $\mathcal{C}$ determines a functor $\mathcal{C}(A, -) : \mathcal{C} \to \mathbf{Set}$ sending X to $\mathcal{C}(A, X)$; a morphism $f : X \to Y$ sends $g \in \mathcal{C}(A, X)$ to $f \circ g \in \mathcal{C}(A, Y)$. This is the functor "co-represented" by A.

(a) Let $F : \mathcal{C} \to \mathbf{Set}$ be any functor, and consider the set (or possibly class) of natural transformations $\theta : \mathcal{C}(A, -) \to F$. Write down a (natural!) pair of inverse maps

$$
F(A) \leftrightarrows \mathrm{nt}(\mathcal{C}(A, -), F)
$$

(so that in this case the class of natural transformations is actually a set).

(b) Conclude that the obvious (natural!) map

$$
\mathcal{C}(B, A) \to \mathrm{nt}(\mathcal{C}(A, -), \mathcal{C}(B, -))
$$

is a bijection, and then that the set of natural isomorphisms from $\mathcal{C}(A, -)$ to $\mathcal{C}(B, -)$ is in bijection with the set of isomorphisms from B to A.

Exercise 25.19. Show that any chain map natural $\gamma_{X,Y}$ as in the Eilenberg-Zilber theorem is not only commutative up to natural chain homotopy, as in Corollary 25.15, but also "unital" and "associative" up to natural chain homotopy.

Chapter 3

Cohomology and duality

26 Coproducts, cohomology

The next topic is cohomology. This is like homology, but it's a contravariant rather than covariant functor of spaces. There are three reasons why you might like a contravariant functor.

(1) Many geometric contructions *pull back*; that is, they behave contravariantly. For example, if I have some covering space $\widetilde{X} \to X$ and a map $f : Y \to X$, I get a pullback covering space $f^*\widetilde{X}$. A better example is vector bundles (that we'll talk about in Chapter 6) — they don't push forward, they pull back. So if we want to study them by means of "natural" invariants, these invariants will have to lie in a (hopefully computable) group that also behaves contravariantly. This will lead to the theory of "characteristic classes."

(2) The structure induced by the diagonal map from a space to its square induces structure in contravariant functors that is more general and easier to study than what you get in homology.

(3) Cohomology turns out to be the target of the Poincaré duality map.

Coalgebras

Let's elaborate on point (2). Every space has a diagonal map
$$\Delta : X \to X \times X .$$
This induces a map $H_*(X; R) \to H_*(X \times X; R)$, for any coefficient group R. Now, if R is a ring, we get a cross product map
$$\times : H_*(X; R) \otimes_R H_*(X; R) \to H_*(X \times X; R) .$$

If R is a PID, the Künneth Theorem tells us that this map is a monomorphism. If the remaining term in the Künneth Theorem is zero, the cross product is an isomorphism. So if $H_*(X; R)$ is free over R (or even just flat over R), we get a "diagonal" or "coproduct"

$$\Delta : H_*(X; R) \to H_*(X; R) \otimes_R H_*(X; R).$$

If R is a field, this map is universally defined, and natural in X.

This kind of structure is unfamiliar, and at first seems a bit strange. After all, the tensor product is defined by a universal property for maps *out* of it; maps *into* it just are what they are.

Still, it's often useful, and we pause to fill in some of its properties.

Definition 26.1. Let R be a commutative ring. A *(graded) coalgebra* over R is a (graded) R-module M equipped with a "comultiplication" $\Delta : M \to M \otimes_R M$ and a "counit" map $\varepsilon : M \to R$ such that the following diagrams commute.

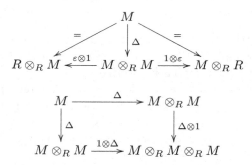

It is *commutative* if in addition

commutes, where $\tau(x \otimes y) = (-1)^{|x| \cdot |y|} y \otimes x$ is the twist map.

Using acyclic models, you saw in Exercise 25.19 and Corollary 25.15 that the cross product map is associative and commutative: The diagrams

$$
\begin{array}{ccc}
S_*(X) \otimes S_*(Y) \otimes S_*(Z) & \xrightarrow{\times \otimes 1} & S_*(X \times Y) \otimes S_*(Z) \\
\downarrow{\scriptstyle 1 \otimes \times} & & \downarrow{\scriptstyle \times} \\
S_*(X) \otimes S_*(Y \times Z) & \xrightarrow{\times} & S_*(X \times Y \times Z)
\end{array}
$$

and

$$S_*(X) \otimes S_*(Y) \xrightarrow{\ \tau\ } S_*(Y) \otimes S_*(X)$$

$$\downarrow \times \qquad\qquad \downarrow \times$$

$$S_*(X \times Y) \xrightarrow{\ T_*\ } S_*(Y \times X)$$

commute up to natural chain homotopy, where τ is as defined above on the tensor product and $T : X \times Y \to Y \times X$ is the swap map. Similar diagrams apply to the standard comparison map for the homology of tensor products of chain complexes,

$$\mu : H_*(C) \otimes H_*(D) \to H_*(C \otimes D),$$

and the result is this:

Corollary 26.2. *Suppose that R is a PID and that $H_*(X; R)$ is free over R. Then $H_*(X; R)$ has the natural structure of a commutative graded coalgebra over R.*

We could now just go on and talk about coalgebras. But they are less familiar, and available only if $H_*(X; R)$ is free over R. So instead we're going to dualize, talk about cohomology, and get an algebra structure. Some say that cohomology is better because you have algebras, but that's more of a sociological statement than a mathematical one.

Let's get on with it.

Cohomology

Definition 26.3. Let N be an abelian group. A *singular n-cochain* on X with values in N is a function $\mathrm{Sin}_n(X) \to N$.

If N is an R-module, then I can extend linearly to get an R-module homomorphism $S_n(X; R) \to N$.

Notation 26.4. Write

$$S^n(X; N) = \mathrm{Map}(\mathrm{Sin}_n(X), N) = \mathrm{Hom}_R(S_n(X; R), N).$$

It is naturally an R-module.

This is going to give us something contravariant, that's for sure. But we haven't quite finished dualizing. The differential $d : S_{n+1}(X; R) \to S_n(X; R)$ induces a "coboundary map"

$$d : S^n(X; N) \to S^{n+1}(X; N)$$

defined by

$$(df)(\sigma) = (-1)^{n+1} f(d\sigma).$$

The sign is a little strange, and we'll see an explanation in a minute. Anyway, we get a "cochain complex," with a differential that *increases* degree by 1. We still have $d^2 = 0$, since

$$(d^2 f)(\sigma) = \pm d(f(d\sigma)) = \pm f(d^2\sigma) = \pm f(0) = 0,$$

so we have *cocycles* and *coboundaries* and we can still take homology of this cochain complex.

Definition 26.5. The nth *singular cohomology group* of X with coefficients in an abelian group N is

$$H^n(X; N) = \frac{\ker(S^n(X; N) \to S^{n+1}(X; N))}{\operatorname{im}(S^{n-1}(X; N) \to S^n(X; N))}.$$

If N is an R-module, then $H^n(X; N)$ is again an R-module.

Let's first compute $H^0(X; N)$. A 0-cochain is a function $\mathrm{Sin}_0(X) \to N$; that is, a function (not required to be continuous!) $f : X \to N$. To compute df, take a 1-simplex $\sigma : \Delta^1 \to X$ and evaluate f on its boundary:

$$(df)(\sigma) = -f(d\sigma) = -f(\sigma(e_0) - \sigma(e_1)) = f(\sigma(e_1)) - f(\sigma(e_0)).$$

So f is a co*cycle* if it's constant on path components. That is to say:

Lemma 26.6. $H^0(X; N) = \mathrm{Map}(\pi_0(X), N)$.

Warning 26.7. $S^n(X; \mathbb{Z}) = \mathrm{Map}(\mathrm{Sin}_n(X); \mathbb{Z}) = \prod_{\mathrm{Sin}_n(X)} \mathbb{Z}$, which is probably an uncountable product. An awkward fact is that no infinite product of free abelian groups is free abelian.

The first thing a cohomology class does is to give a linear functional on homology, by "evaluation." Let's spin this out a bit.

We want to tensor together cochains and chains. But to do that we should make the differential in $S^*(X)$ go down, not up. Just as a notational matter, let's write

$$S^\vee_{-n}(X; N) = S^n(X; N)$$

and define a differential $d : S^\vee_{-n}(X) \to S^\vee_{-n-1}(X)$ to be the differential $d : S^n(X) \to S^{n+1}(X)$. Now $S^\vee_*(X)$ is a chain complex, albeit a nonpositively graded one. Form the graded tensor product, with

$$(S^\vee_*(X; N) \otimes S_*(X))_n = \bigoplus_{p+q=n} S^\vee_p(X; N) \otimes S_q(X).$$

Now evaluation is a map of graded abelian groups

$$\langle -, - \rangle : S_*^\vee(X; N) \otimes S_*(X) \to N \,,$$

where N is regarded as a chain complex concentrated in degree 0. We would like this map to be a chain map. So let $f \in S^n(X; N)$ and $\sigma \in S_n(X)$, and compute

$$0 = d\langle f, \sigma \rangle = \langle df, \sigma \rangle + (-1)^n \langle f, d\sigma \rangle \,.$$

This forces

$$(df)(\sigma) = \langle df, \sigma \rangle = -(-1)^n f(d\sigma) \,,$$

explaining the odd sign in our definition above.

Here's the payoff: There's a natural map

$$H_{-n}(S_*^\vee(X; N)) \otimes H_n(S_*(X)) \xrightarrow{\mu} H_0\left(S_*^\vee(X; N) \otimes S_*(X)\right) \to N \,.$$

This gives us the *Kronecker pairing*

$$\langle -, - \rangle : H^n(X; N) \otimes H_n(X) \to N \,.$$

We can develop the properties of cohomology in analogy with properties of homology. For example: If $A \subseteq X$, there is a restriction map $S^n(X; N) \to S^n(A; N)$, induced by the injection $\mathrm{Sin}_n(A) \hookrightarrow \mathrm{Sin}_n(X)$. And as long as A is nonempty, we can split this injection, so any function $\mathrm{Sin}_n(A) \to N$ extends to $\mathrm{Sin}_n(X) \to N$. This means that $S^n(X; N) \to S^n(A; N)$ is surjective. (This is the case if $A = \varnothing$, as well!)

Definition 26.8. The *relative n-cochain group* with coefficients in N is

$$S^n(X, A; N) = \ker\left(S^n(X; N) \to S^n(A; N)\right) \,.$$

This defines a sub cochain complex of $S^*(X; N)$, and we define

$$H^n(X, A; N) = H^n(S^*(X, A; N)) \,.$$

The short exact sequence of cochain complexes

$$0 \to S^*(X, A; N) \to S^*(X; N) \to S^*(A; N) \to 0$$

induces the *long exact cohomology sequence*

$$H^2(X, A; N) \longrightarrow \cdots$$

$$H^1(X, A; N) \longrightarrow H^1(X; N) \longrightarrow H^1(A; N) \quad \xrightarrow{\delta}$$

$$0 \longrightarrow H^0(X, A; N) \longrightarrow H^0(X; N) \longrightarrow H^0(A; N) \quad \xrightarrow{\delta}$$

Exercises

Exercise 26.9. The linear dual of a graded k-algebra of finite type (finitely generated in each degree) has a natural k-coalgebra structure. Determine the diagonal map in the dual of the polynomial algebra $k[x]$, $|x| = 2$. (It will transpire that the $H^*(\mathbb{CP}^\infty; k) = k[x]$, $|x| = 2$, so this is determining the coalgebra structure of $H_*(\mathbb{CP}^\infty; k)$.)

27   Ext and UCT

Let R be a commutative ring (probably a PID) and N an R-module. The singular cochains on X with values in N,

$$S^*(X; N) = \mathrm{Map}(\mathrm{Sin}_*(X), N),$$

then forms a cochain complex of R-modules. It is contravariantly functorial in X and covariantly functorial in N. The Kronecker pairing defines a map

$$H^n(X; N) \otimes_R H_n(X) \to N$$

whose adjoint

$$\beta : H^n(X; N) \to \mathrm{Hom}_R(H_n(X), N)$$

gives us an estimate of the cohomology in terms of the homology of X. Here's how well it does:

Theorem 27.1 (Mixed variance Universal Coefficient Theorem). *Let R be a PID and N an R-module, and let C_* be a chain-complex of free R-modules. Then there is a short exact sequence of R-modules,*

$$0 \to \mathrm{Ext}^1_R(H_{n-1}(C_*), N) \to H^n(\mathrm{Hom}_R(C_*, N)) \to \mathrm{Hom}_R(H_n(C_*), N) \to 0,$$

natural in C_* and N, that splits (but not naturally).

Taking $C_* = S_*(X; R)$, we have the short exact sequence

$$0 \to \mathrm{Ext}^1_R(H_{n-1}(X), N) \to H^n(X; N) \xrightarrow{\beta} \mathrm{Hom}_R(H_n(X), N) \to 0$$

that splits, but not naturally. This also holds for relative cohomology.

What is this Ext?

The problem that arises is that $\mathrm{Hom}_R(-, N) : \mathbf{Mod}_R \to \mathbf{Mod}_R$ is not exact. Suppose I have an injection $M' \to M$. Is $\mathrm{Hom}(M, N) \to \mathrm{Hom}(M', N)$ surjective? Does a map $M' \to N$ necessarily extend to a map $M \to N$? No! For example, $\mathbb{Z}/2\mathbb{Z} \hookrightarrow \mathbb{Z}/4\mathbb{Z}$ is an injection, but the identity map $\mathbb{Z}/2\mathbb{Z} \to \mathbb{Z}/2\mathbb{Z}$ does not extend over $\mathbb{Z}/4\mathbb{Z}$.

On the other hand, if $M' \xrightarrow{i} M \xrightarrow{p} M'' \to 0$ is an exact sequence of R-modules then

$$0 \to \operatorname{Hom}_R(M'', N) \to \operatorname{Hom}_R(M, N) \to \operatorname{Hom}_R(M', N)$$

is again exact. Check this statement, in which R is any commutative ring.

Now homological algebra comes to the rescue again to repair the failure of exactness. Fix a commutative ring R and an R-module M. Pick a projective resolution of M,

$$0 \leftarrow M \leftarrow F_0 \leftarrow F_1 \leftarrow F_2 \leftarrow \cdots .$$

Apply $\operatorname{Hom}_R(-, N)$ to get a cochain complex

$$0 \to \operatorname{Hom}_R(F_0, N) \to \operatorname{Hom}_R(F_1, N) \to \operatorname{Hom}_R(F_2, N) \to \cdots .$$

Definition 27.2. $\operatorname{Ext}_R^n(M, N) = H^n(\operatorname{Hom}_R(F_*, N))$.

Remark 27.3. Ext is well-defined and functorial, by the Fundamental Theorem of Homological Algebra, Theorem 22.1. If M is free (or projective) then $\operatorname{Ext}_R^n(M, -) = 0$ for $n > 0$, since we can take M as its own projective resolution. If R is a PID, then we can assume $F_1 = \ker(F_0 \to M)$ and $F_n = 0$ for $n > 1$, so $\operatorname{Ext}_R^n = 0$ if $n > 1$. If R is a field, then $\operatorname{Ext}_R^n = 0$ for $n > 0$.

Example 27.4. Let $R = \mathbb{Z}$ and take $M = \mathbb{Z}/k\mathbb{Z}$. This admits a simple free resolution: $0 \to \mathbb{Z} \xrightarrow{k} \mathbb{Z} \to \mathbb{Z}/k\mathbb{Z} \to 0$. Apply $\operatorname{Hom}(-, N)$ to it, and remember that $\operatorname{Hom}(\mathbb{Z}, N) = N$, to get the very short cochain complex, with entries in dimensions 0 and 1:

$$0 \to N \xrightarrow{k} N \to 0 .$$

Taking homology gives us

$$\operatorname{Hom}(\mathbb{Z}/k\mathbb{Z}, N) = \ker(k|N) , \quad \operatorname{Ext}^1(\mathbb{Z}/k\mathbb{Z}, N) = N/kN .$$

Proof of Theorem 27.1. First of all, we can't just copy the proof (in Lecture 24) of the homology universal coefficient theorem, since $\operatorname{Ext}_R^1(-, R)$ is not generally trivial.

Instead, we start by thinking about what an n-cocycle in $\operatorname{Hom}_R(C_*, N)$ is: It's a homomorphism $C_n \to N$ such that the composite $C_{n+1} \to C_n \to N$ is trivial. Write $B_n \subseteq C_n$ for the submodule of boundaries. We have a homomorphism that kills B_n; that is,

$$Z^n(\operatorname{Hom}_R(C_*, N)) \xrightarrow{\cong} \operatorname{Hom}_R(C_n/B_n, N) .$$

Now $H_n(C_*)$ (which we'll abbreviate as H_n) is the submodule Z_n/B_n of C_n/B_n; we have an exact sequence

$$0 \to H_n \to C_n/B_n \to B_{n-1} \to 0.$$

Apply $\mathrm{Hom}_R(-, N)$ to this short exact sequence. The result is again short exact, because the original sequence is split short exact since B_{n-1} is a submodule of the free R-module C_{n-1} and hence free. This gives us the bottom line in the map of short exact sequences

$$0 \twoheadrightarrow B^n \mathrm{Hom}_R(C_*, N) \to Z^n \mathrm{Hom}_R(C_*, N) \to H^n(\mathrm{Hom}_R(C_*, N)) \twoheadrightarrow 0$$

$$0 \to \mathrm{Hom}_R(B_{n-1}, N) \to \mathrm{Hom}_R(C_n/B_n, N) \longrightarrow \mathrm{Hom}_R(H_n, N) \longrightarrow 0.$$

The map β is the one we started with. The Snake Lemma 9.2 now shows that it is surjective and that

$$\ker \beta \cong \mathrm{coker}(B^n \mathrm{Hom}_R(C_*, N) \to \mathrm{Hom}_R(B_{n-1}, N)).$$

An element of $B^n \mathrm{Hom}_R(C_*, N)$ is a map $C_n \to N$ that factors as $C_n \xrightarrow{d} C_{n-1} \to N$. The observation is now that this is the same as requiring a factorization $C_n \xrightarrow{d} Z_{n-1} \to N$; once this factorization has been achieved, the map $Z_{n-1} \to N$ automatically extends to all of C_{n-1}. This is because $Z_{n-1} \subseteq C_{n-1}$ as a direct summand: the short exact sequence

$$0 \to Z_{n-1} \to C_{n-1} \to B_{n-2} \to 0$$

splits since B_{n-2} is free. Consequently we can rewrite our formula for $\ker \beta$ as

$$\ker \beta \cong \mathrm{coker}(\mathrm{Hom}_R(Z_{n-1}, N) \to \mathrm{Hom}_R(B_{n-1}, N)).$$

But after all

$$0 \leftarrow H_{n-1} \leftarrow Z_{n-1} \leftarrow B_{n-1} \leftarrow 0$$

is a free resolution, so this cokernel is precisely $\mathrm{Ext}_R^1(H_{n-1}(C_*), N)$. $\quad \square$

Question 27.5. Why is Ext called Ext?

Answer: It classifies extensions. Let R be a commutative ring, and let M, N be two R-modules. I can think about "extensions of M by N," that is, short exact sequences of the form

$$0 \to N \to L \to M \to 0.$$

We'll say that two extensions are *equivalent* if there's a map of short exact sequences between them that is the identity on N and on M. The map in the middle is necessarily then an isomorphism as well, by the 5-lemma. An alternate and equivalent definition of $\mathrm{Ext}_R^1(M, N)$ is as the set of extensions like this modulo this notion of equivalence. The zero in this group is the split extension. I invite you to think about why this is isomorphic to the definition using projective resolutions, and to construct the addition ("Baer sum") in this version of $\mathrm{Ext}_R^1(M, N)$ [33].

For example, with $R = \mathbb{Z}$, I have two extensions of $\mathbb{Z}/2\mathbb{Z}$ by $\mathbb{Z}/2\mathbb{Z}$:

$$0 \to \mathbb{Z}/2\mathbb{Z} \to \mathbb{Z}/2\mathbb{Z} \oplus \mathbb{Z}/2\mathbb{Z} \to \mathbb{Z}/2\mathbb{Z} \to 0$$

and

$$0 \to \mathbb{Z}/2\mathbb{Z} \to \mathbb{Z}/4\mathbb{Z} \to \mathbb{Z}/2\mathbb{Z} \to 0 \,.$$

The first represents $0 \in \mathrm{Ext}^1(\mathbb{Z}/2\mathbb{Z}, \mathbb{Z}/2\mathbb{Z})$, and the second represents the nonzero element in that group.

Eilenberg-Steenrod axioms for cohomology

The universal coefficient theorem is useful in transferring properties of homology to cohomology. For example, if $f : X \to Y$ is a map that induces an isomorphism in $H_*(-; R)$, then it induces an isomorphism in $H^*(-; N)$ for any R-module N, at least provided that R is a PID. (This is in fact true in general.)

Cohomology satisfies the appropriate analogues of the Eilenberg-Steenrod axioms.

Homotopy invariance: If $f_0 \simeq f_1 : (X, A) \to (Y, B)$, then

$$f_0^* = f_1^* : H^*(Y, B; N) \to H^*(X, A; N) \,.$$

I can't use the UCT to address this. But we established a chain homotopy $f_{0,*} \simeq f_{1,*} : S_*(X, A) \to S_*(Y, B)$, and applying Hom converts chain homotopies to cochain homotopies.

Excision: If $U \subseteq A \subseteq X$ such that $\overline{U} \subseteq \mathrm{Int}(A)$, then $H^*(X, A; N) \to H^*(X - U, A - U; N)$ is an isomorphism. This follows from excision in homology and the mixed variance UCT.

Milnor axiom: The inclusions induce an isomorphism

$$H^*(\textstyle\coprod_\alpha X_\alpha; N) \to \textstyle\prod_\alpha H^*(X_\alpha; N) \,.$$

As a result, it enjoys the fruit of these axioms, such as:

The Mayer-Vietoris sequence: If $A, B \subseteq X$ are such that their interiors cover X, then there is a long exact sequence

$$H^{n+1}(X; N) \longrightarrow \cdots$$

$$H^n(X; N) \longrightarrow H^n(A; N) \oplus H^n(B; N) \longrightarrow H^n(A \cap B; N)$$

$$\cdots \longrightarrow H^{n-1}(A \cap B; N).$$

Exercises

Exercise 27.6. Compute $H^*(\mathbb{RP}^n; R)$ where $R = \mathbb{Z}$, $R = \mathbb{Z}[1/2]$, and $R = \mathbb{F}_2$.

Exercise 27.7. Verify the Milnor axiom for singular cohomology.

Exercise 27.8. Check that our proof of the UCT (Theorem 24.1) only really required that $\operatorname{Tor}_1^R(C_n, M) = 0$ for all n. Use this observation to prove that if M is a finitely generated R-module then there is a natural short exact sequence

$$0 \to H^n(X) \otimes_R M \to H^n(X; M) \to \operatorname{Tor}_1^R(H^{n+1}(X), M) \to 0.$$

28   Products in cohomology

In Lecture 25 we used acyclic models to construct a natural chain homotopy equivalence

$$S_*(X) \otimes S_*(Y) \leftrightarrows S_*(X \times Y)$$

lifting the isomorphism $H_0(X) \otimes H_0(Y) \cong H_0(X \times Y)$. We used a choice of the bottom arrow (and gave the Eilenberg-Zilber map as an example) to produce a homology cross product $\times : H_*(X) \otimes H_*(Y) \to H_*(X \times Y)$. Now we want to use a map going the other way, in order to construct a cohomology cross product map $\times : H^*(X) \otimes H^*(Y) \to H^*(X \times Y)$.

There is an attractive choice in this case too: the *Alexander-Whitney map*. For each pair of natural numbers p, q, we will define a natural homomorphism

$$\alpha : S_{p+q}(X \times Y) \to S_p(X) \otimes S_q(Y).$$

It suffices to define this on simplices, so let $\sigma : \Delta^{p+q} \to X \times Y$ be a singular $(p+q)$-simplex in the product. Let

$$\sigma_1 = \mathrm{pr}_1 \circ \sigma : \Delta^{p+q} \to X \quad \text{and} \quad \sigma_2 = \mathrm{pr}_2 \circ \sigma : \Delta^{p+q} \to Y$$

be the two coordinates of σ. I have to produce a p-simplex in X and a q-simplex in Y.

First define two maps in the simplex category:

— the "front face" $\alpha_p : [p] \to [p+q]$, sending i to i for $0 \le i \le p$, and

— the "back face" $\omega_q : [q] \to [p+q]$, sending j to $p+j$ for $0 \le j \le q$.

Use the same symbols for the affine extensions to maps $\Delta^p \to \Delta^{p+q}$ and $\Delta^q \to \Delta^{p+q}$. Now let

$$\alpha(\sigma) = (\sigma_1 \circ \alpha_p) \otimes (\sigma_2 \circ \omega_q).$$

This seems like a very random construction; but it works! It's named after two great early algebraic topologists, James W. Alexander and Hassler Whitney. Both worked at the Institute for Advanced Study in Princeton. Alexander (1888–1971) is known for his polynomial and his horned sphere, as well as his duality theorem we will come to presently. To Whitney (1907–1989) we owe the characteristic classes and sum formula we will see later, as well as fundamental work embeddings and stratified spaces. For homework, you will show that these maps assemble into a chain map

$$\alpha : S_*(X \times Y) \to S_*(X) \otimes S_*(Y)$$

lifting the isomorphism $H_0(X \times Y) \to H_0(X) \otimes H_0(Y)$.

This works over any ring R. To get a map in cohomology, we should form a composite

$$S^p(X) \otimes S^q(Y) \xrightarrow{\hspace{3cm}} S^{p+q}(X \times Y)$$

$$\downarrow \qquad\qquad\qquad\qquad \|$$

$$\mathrm{Hom}_R(S_p(X) \otimes S_q(Y), R) \xrightarrow{\alpha^*} \mathrm{Hom}_R(S_{p+q}(X \times Y), R).$$

The first map goes like this: Given chain complexes C_* and D_*, we can consider the dual cochain complexes $\mathrm{Hom}_R(C_*, R)$ and $\mathrm{Hom}_R(D_*, R)$, and construct a chain map

$$\mathrm{Hom}_R(C_*, R) \otimes_R \mathrm{Hom}_R(D_*, R) \to \mathrm{Hom}_R(C_* \otimes_R D_*, R)$$

by

$$f \otimes g \mapsto \begin{cases} (x \otimes y \mapsto (-1)^{pq} f(x)g(y)) & |x| = |f| = p, \ |y| = |g| = q \\ 0 & \text{otherwise.} \end{cases}$$

Again, I leave it to you to check that this is a cochain map.

Altogether, we have constructed a natural cochain map

$$\times : S^p(X) \otimes S^q(Y) \to S^{p+q}(X \times Y).$$

From this, we get a homomorphism

$$H^*(S^*(X) \otimes S^*(Y)) \to H^*(X \times Y).$$

I'm not quite done! As in the Künneth theorem, there is an evident natural map

$$\mu : H^*(X) \otimes H^*(Y) \to H^*(S^*(X) \otimes S^*(Y)).$$

The composite

$$\times : H^*(X) \otimes H^*(Y) \to H^*(S^*(X) \otimes S^*(Y)) \to H^*(X \times Y)$$

is the *cohomology cross product*.

It's not very easy to do computations with this, directly. We'll find indirect means. Let me make some points about this construction, though.

Definition 28.1. The *cup product* is the map obtained by taking $X = Y$ and composing with the map induced by the diagonal $\Delta : X \to X \times X$:

$$\cup : H^p(X) \otimes H^q(X) \xrightarrow{\times} H^{p+q}(X \times X) \xrightarrow{\Delta^*} H^{p+q}(X).$$

These definitions make good sense with any ring for coefficients.

Let's explore this definition in dimension zero. I claim that $H^0(X; R) \cong \mathrm{Map}(\pi_0(X), R)$ as rings. When $p = q = 0$, both α_0 and ω_0 are the identity maps, so we are just forming the pointwise product of functions.

There's a distinguished element in $H^0(X)$, namely the the function $\pi_0(X) \to R$ that takes on the value 1 on every path component. This is the identity for the cup product. This comes about because when $p = 0$ in our above story, then α_0 is just including the 0-simplex, and ω_q is the identity.

The cross product is also associative. This follows again from acyclic models, but if we use the Alexander-Whitney map it holds even on the chain level.

Proposition 28.2. Let $f \in S^p(X)$, $g \in S^q(Y)$, and $h \in S^r(Z)$, and let $\sigma : \Delta^{p+q+r} \to X \times Y \times Z$ be any simplex. Then the Alexander-Whitney cross product satisfies:

$$((f \times g) \times h)(\sigma) = (f \times (g \times h))(\sigma).$$

Proof. Write σ_{12} for the composite of σ with the projection map $X \times Y \times Z \to X \times Y$, and so on. Then

$$((f \times g) \times h)(\sigma) = (-1)^{(p+q)r}(f \times g)(\sigma_{12} \circ \alpha_{p+q})h(\sigma_3 \circ \omega_r).$$

But

$$(f \times g)(\sigma_{12} \circ \alpha_{p+q}) = (-1)^{pq}f(\sigma_1 \circ \alpha_p)g(\sigma_2 \circ \mu_q),$$

where μ_q is the "middle face," sending ℓ to $\ell + p$ for $0 \le \ell \le q$. In other words,

$$((f \times g) \times h)(\sigma) = (-1)^{pq+qr+rp}f(\sigma_1 \circ \alpha_p)g(\sigma_2 \circ \mu_q)h(\sigma_3 \circ \omega_r).$$

I've used associativity of the coefficient ring. You get exactly the same thing when you expand $(f \times (g \times h))(\sigma)$, so the cross product is associative. $\square$

Of course the diagonal map is "associative," too, and we find that the cup product is associative:

$$(\alpha \cup \beta) \cup \gamma = \alpha \cup (\beta \cup \gamma).$$

Exercises

Exercise 28.3. Verify that the Alexander-Whitney map is a chain map lifting the isomorphism $H_0(X \times Y) \to H_0(X) \otimes H_0(y)$.

29 Cup product, continued

We have constructed an explicit map $\times : S^p(X) \otimes S^q(Y) \to S^{p+q}(X \times Y)$ via:

$$(f \times g)(\sigma) = (-1)^{pq}f(\sigma_1 \circ \alpha_p)g(\sigma_2 \circ \omega_q)$$

where $\alpha_p : \Delta^p \to \Delta^{p+q}$ sends i to i for $0 \le i \le p$ and $\omega_q : \Delta^q \to \Delta^{p+q}$ sends j to $p + j$ for $0 \le j \le q$. This is a cochain map; it induces a "cross product" $\times : H^p(X) \otimes H^q(Y) \to H_{p+q}(X \times Y)$, and, by composing with the map induced by the diagonal embedding, the "cup product"

$$\cup : H^p(X) \otimes H^q(X) \to H^{p+q}(X).$$

We formalize the structure that this product imposes on cohomology.

Definition 29.1. Let R be a commutative ring. A *graded R-algebra* is a graded R-module $\ldots, A_{-1}, A_0, A_1, A_2, \ldots$ equipped with maps $A_p \otimes_R A_q \to A_{p+q}$ and a map $\eta : R \to A_0$ that make the following diagram commute.

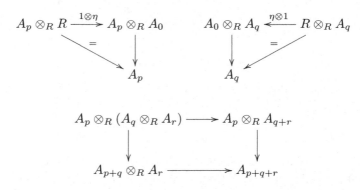

A graded R-algebra A is *commutative* if the following diagram commutes:

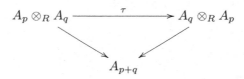

where $\tau(x \otimes y) = (-1)^{pq} y \otimes x$.

We claim that $H^*(X; R)$ forms a commutative graded R-algebra under the cup product. From what we did in the last lecture, it is clearly a graded R-algebra. But commutativity is nontrivial. On the cochain level, this is clearly not graded commutative. We're going to have to work hard — in fact, so hard that you're going to do it for homework. What needs to be checked is that the following diagram commutes up to natural chain homotopy.

$$
\begin{array}{ccc}
S_*(X \times Y) & \xrightarrow{\;T_*\;} & S_*(Y \times X) \\
{\scriptstyle \alpha_{X,Y}}\downarrow & & \downarrow{\scriptstyle \alpha_{Y,X}} \\
S_*(X) \otimes_R S_*(Y) & \xrightarrow{\;\tau\;} & S_*(Y) \otimes_R S_*(X)
\end{array}
$$

Acyclic models helps us prove things like this. Let us summarize:

Theorem 29.2. *With any commutative ring of coefficients, $H^*(X)$ is naturally a commutative graded R-algebra.*

So, for example, if $|x|$ is odd then $x^2 = -x^2$, or $2x^2 = 0$. If 2 is invertible in the coefficient ring, this implies that $x^2 = 0$.

You might hope that there is some way to produce a commutative product on a chain complex modeling $H^*(X)$. With coefficients in $\mathbb{Q}$, this is possible, by a construction due to Dennis Sullivan. With coefficients in a field of nonzero characteristic, it is not possible. Steenrod operations provide the obstruction.

My goal now is to compute the cohomology algebras of some spaces. Some spaces are easy! There is only one possible product structure on $H^*(S^n)$, for example. (When $n = 0$, we get a free module of rank 2 in dimension 0. This admits a variety of commutative algebra structures; but we have already seen that $H^0(S^0) = \mathbb{Z} \times \mathbb{Z}$ as an algebra.) Maybe the next thing to try is a product of spheres. More generally, we should ask whether there is an algebra structure on $H^*(X) \otimes H^*(Y)$ making the cross product an algebra map. If A and B are two graded algebras, there *is* a natural algebra structure on $A \otimes B$, given by $1 = 1 \otimes 1$ and

$$(a' \otimes b')(a \otimes b) = (-1)^{|b'| \cdot |a|} a'a \otimes b'b.$$

If A and B are commutative, then so is $A \otimes B$ with this algebra structure.

Proposition 29.3. *The cohomology cross product*

$$\times : H^*(X) \otimes H^*(Y) \to H^*(X \times Y)$$

is an R-algebra homomorphism.

Proof. We have diagonal maps $\Delta_X : X \to X \times X$ and $\Delta_Y : Y \to Y \times Y$. The diagonal on $X \times Y$ factors as

$$
\begin{array}{ccc}
X \times Y & \xrightarrow{\ \ \ \Delta_{X \times Y}\ \ \ } & X \times Y \times X \times Y \\
& \searrow{\scriptstyle \Delta_X \times \Delta_Y} \qquad \nearrow{\scriptstyle 1 \times T \times 1} & \\
& X \times X \times Y \times Y. &
\end{array}
$$

Let $\alpha_1, \alpha_2 \in H^*(X)$ and $\beta_1, \beta_2 \in H^*(Y)$. Then $\alpha_1 \times \beta_1, \alpha_2 \times \beta_2 \in H^*(X \times Y)$, and I want to calculate $(\alpha_1 \times \beta_1) \cup (\alpha_2 \times \beta_2)$. Let's see:

$$
\begin{aligned}
(\alpha_1 \times \beta_1) \cup (\alpha_2 \times \beta_2) &= \Delta^*_{X \times Y}(\alpha_1 \times \beta_1 \times \alpha_2 \times \beta_2) \\
&= (\Delta_X \times \Delta_Y)^*(1 \times T \times 1)^*(\alpha_1 \times \beta_1 \times \alpha_2 \times \beta_2) \\
&= (\Delta_X \times \Delta_Y)^*(\alpha_1 \times T^*(\beta_1 \times \alpha_2) \times \beta_2) \\
&= (-1)^{|\alpha_2| \cdot |\beta_1|}(\Delta_X \times \Delta_Y)^*(\alpha_1 \times \alpha_2 \times \beta_1 \times \beta_2).
\end{aligned}
$$

Naturality of the cross product asserts that the diagram

$$
\begin{array}{ccc}
H^*(X \times Y) & \xleftarrow{\quad \times_{X \times Y} \quad} & H^*(X) \otimes_R H^*(Y) \\
\uparrow{\scriptstyle (\Delta_X \times \Delta_Y)^*} & & \uparrow{\scriptstyle \Delta_X^* \otimes \Delta_Y^*} \\
H^*(X \times X \times Y \times Y) & \xleftarrow{\quad \times_{X \times X, Y \times Y} \quad} & H^*(X \times X) \otimes H^*(Y \times Y)
\end{array}
$$

commutes. We learn:

$$
\begin{aligned}
(\alpha_1 \times \beta_1) \cup (\alpha_2 \times \beta_2) &= (-1)^{|\alpha_2| \cdot |\beta_1|} (\Delta_X \times \Delta_Y)^*(\alpha_1 \times \alpha_2 \times \beta_1 \times \beta_2) \\
&= (-1)^{|\alpha_2| \cdot |\beta_1|} (\alpha_1 \cup \alpha_2) \times (\beta_1 \cup \beta_2).
\end{aligned}
$$

That's exactly what we wanted.   □

We will see later, in Theorem 33.3, that the cross product map is often an isomorphism.

Example 29.4. How about $H^*(S^p \times S^q)$? I'll assume that p and q are both positive, and leave the other cases to you. The Künneth theorem guarantees that $\times : H^*(S^p) \otimes H^*(S^q) \to H^*(S^p \times S^q)$ is an isomorphism. Write α for a generator of S^p and β for a generator of S^q; and use the same notations for the pullbacks of these elements to $S^p \times S^q$ under the projections. Then

$$
H^*(S^p \times S^q) = \mathbb{Z}\langle 1, \alpha, \beta, \alpha \cup \beta \rangle,
$$

and

$$
\alpha^2 = 0, \quad \beta^2 = 0, \quad \alpha\beta = (-1)^{pq} \beta\alpha.
$$

This calculation is useful! For example:

Corollary 29.5. *Let $p, q > 0$. Any map $S^{p+q} \to S^p \times S^q$ induces the zero map in $H^{p+q}(-)$.*

Proof. Let $f : S^{p+q} \to S^p \times S^q$ be such a map. It induces an algebra map $f^* : H^*(S^p \times S^q) \to H^*(S^{p+q})$. This map must kill α and β, for degree reasons. But then it also kills their product, since f^* is multiplicative.   □

The space $S^p \vee S^q \vee S^{p+q}$ has the same homology and cohomology groups as $S^p \times S^q$. Both are built as CW complexes with cells in dimensions $0, p, q$, and $p + q$. But they are not homotopy equivalent. We can see this now because there *is* a map $S^{p+q} \to S^p \vee S^q \vee S^{p+q}$ inducing an *isomorphism* in $H^{p+q}(-)$, namely, the inclusion of that summand.

Exercises

Exercise 29.6. Verify that the cup product renders $H^*(X)$ a commutative graded ring.

Exercise 29.7. (a) Let $n, k > 0$. Compute $H^*((S^k)^n; R)$ as an R-algebra. When k is odd, this is an "exterior algebra."

(b) $\mathbb{R}^n$ is the universal cover of $(S^1)^n = \mathbb{R}^n / \mathbb{Z}^n$. Let M be an $n \times n$ matrix with entries in $\mathbb{Z}$. It defines a linear map $\mathbb{R}^n \to \mathbb{R}^n$ in the usual way. Show that this map descends to a self-map of $(S^1)^n$. Compute the effect of this map on $H_1((S^1)^n)$ and on $H_n((S^1)^n)$.

30 Surfaces and nondegenerate symmetric bilinear forms

We are aiming towards a proof of a fundamental cohomological property of manifolds.

Definition 30.1. A (topological) manifold is a Hausdorff space such that every point has an open neighborhood that is homeomorphic to some finite dimensional Euclidean space.

If all these Euclidean spaces can be chosen to be $\mathbb{R}^n$, we have an n-manifold. This definition makes space for some very "large" manifolds: the "long line," for example, or an uncountable discrete set. Whenever necessary we will feel free to add hypotheses to eliminate these "non-geometric" examples. Compactness certainly avoids them! In fact any compact manifold is of the homotopy type of a finite CW complex [78].

In this lecture we will state a case of the Poincaré duality theorem and study some consequences of it, especially for compact 2-manifolds. Henri Poincaré (1854–1912), engineer, theoretical physicist, mathematician, Professor at the Sorbonne, could be said to be the founding father of the subject of topology.

This whole lecture will be happening with coefficients in $\mathbb{F}_2$.

Theorem 30.2. *Let M be a compact manifold of dimension n. There exists a unique class $[M] \in H_n(M)$, called the* fundamental class, *such that for every p, q with $p + q = n$ the pairing*

$$H^p(M) \otimes H^q(M) \xrightarrow{\cup} H^n(M) \xrightarrow{\langle -, [M] \rangle} \mathbb{F}_2$$

is perfect.

A bilinear map $V \otimes W \to R$ of modules over a PID R is a *perfect pairing* if the adjoint maps

$$\alpha : V \to \operatorname{Hom}_R(W, R) = W^\vee, \quad \beta : W \to \operatorname{Hom}_R(V, R) = V^\vee$$

are both isomorphisms. Since $\operatorname{Hom}_R(W, R)$ is torsion-free, both V and W must be torsion free. The maps α and β determine each other: $\alpha = \beta^\vee \circ \iota_V$ and $\beta = \alpha^\vee \circ \iota_W$, where $\iota_V : V \to V^{\vee\vee}$ is the canonical map. Since α and β are isomorphisms by assumption, the maps ι_V and ι_W are isomorphisms as well. This forces V and W to be of finite rank: they are both finitely generated free R-modules.

So we learn that $H^p(M)$ is finite-dimensional and vanishes for $p > n$.

Combining this pairing with the universal coefficient theorem, we get isomorphisms

$$H^p(M) \xrightarrow{\cong} \operatorname{Hom}(H^q(M), \mathbb{F}_2) \xleftarrow{\cong} H_q(M).$$

The homology and cohomology classes corresponding to each other under this isomorphism are said to be "Poincaré dual."

Using these isomorphisms, the cup product pairing can be rewritten as a homology pairing:

$$
\begin{array}{ccc}
H_p(M) \otimes H_q(M) & \xrightarrow{\;\pitchfork\;} & H_{n-p-q}(M) \\
\Big\downarrow{\scriptstyle \cong} & & \Big\downarrow{\scriptstyle \cong} \\
H^{n-p}(M) \otimes H^{n-q}(M) & \xrightarrow{\;\cup\;} & H^{2n-p-q}(M).
\end{array}
$$

This is the *intersection pairing*. Here's how to think of it. Take homology classes $\alpha \in H_p(M)$ and $\beta \in H_q(M)$ and represent them (if possible!) as the image of the fundamental classes of submanifolds of M, of dimensions p and q. Move them if necessary to make them intersect "transversely." Then their intersection will be a submanifold of dimension $n - p - q$, and it will represent the homology class $\alpha \pitchfork \beta$.

This relationship between the cup product and the intersection pairing is the source of the symbol for the cup product.

Example 30.3. Let $M = T^2 = S^1 \times S^1$. We know that

$$H^1(M) = \mathbb{F}_2\langle a, b \rangle$$

and $a^2 = b^2 = 0$, while $ab = ba$ generates $H^2(M)$. The Poincaré duals of these classes are represented by cycles α and β wrapping around one or the other of the two factor circles. They can be made to intersect in a single point. This reflects the fact that

$$\langle a \cup b, [M] \rangle = 1.$$

Similarly, the fact that $a^2 = 0$ reflects the fact that its Poincaré dual cycle α can be moved so as not to intersect itself. The picture below shows two possible α's.

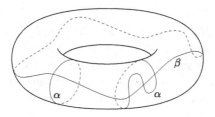

This example exhibits a particularly interesting fragment of the statement of Poincaré duality: In an even dimensional manifold — say $n = 2k$ — the cup product pairing gives us a nondegenerate symmetric bilinear form on $H^k(M)$. As indicated above, this can equally well be considered a bilinear form on $H_k(M)$, and it is then to be thought of as describing the number of points (mod 2) two k-cycles intersect in, when put in general position relative to one another. It's called the *intersection form*. We'll denote it by

$$\alpha \cdot \beta = \langle a \cup b, [M] \rangle,$$

where again a and α are Poincaré dual, and b and β are dual.

Example 30.4. In terms of the basis α, β, the intersection form for T^2 has matrix

$$\begin{bmatrix} 0 & 1 \\ 1 & 0 \end{bmatrix}.$$

This is a "hyperbolic form."

Let's discuss finite dimensional nondegenerate symmetric bilinear forms over $\mathbb{F}_2$ in general. A form on V restricts to a form on any subspace $W \subseteq V$, but the restricted form may be degenerate. Any subspace has an *orthogonal complement*

$$W^\perp = \{v \in V : v \cdot w = 0 \text{ for all } w \in W\}.$$

Lemma 30.5. *The restriction of a nondegenerate bilinear form on V to a subspace W is nondegenerate exactly when $W \cap W^\perp = 0$. In that case $W^\perp$ is also nondegenerate, and the splitting*

$$V \cong W \oplus W^\perp$$

respects the forms.

Using this easy lemma, we may inductively decompose a general finite dimensional symmetric bilinear form. First, suppose that there is a vector $v \in V$ such that $v \cdot v = 1$. It generates a nondegenerate subspace and

$$V = \langle v \rangle \oplus \langle v \rangle^{\perp}.$$

Continuing to split off one-dimensional subspaces brings us to the situation of a nondegenerate symmetric bilinear form such that $v \cdot v = 0$ for every vector. Unless $V = 0$ we can pick a nonzero vector v. Since the form is nondegenerate, we may find another vector w such that $v \cdot w = 1$. The two together generate a 2-dimensional hyperbolic subspace. Split it off and continue. We conclude:

Proposition 30.6. *Any finite dimensional nondegenerate symmetric bilinear form over $\mathbb{F}_2$ splits as an orthogonal direct sum of forms with matrices* $[1]$ *and* $\begin{bmatrix} 0 & 1 \\ 1 & 0 \end{bmatrix}$.

Let **Bil** be the set of isomorphism classes of finite dimensional nondegenerate symmetric bilinear forms over $\mathbb{F}_2$. We've just given a classification of these things. This is a commutative monoid under orthogonal direct sum. It can be regarded as the set of nonsingular symmetric matrices modulo the equivalence relation of "similarity": Two matrices M and N are *similar*, $M \sim N$, if $N = AMA^T$ for some nonsingular A.

Claim 30.7. $\begin{bmatrix} 1 & & \\ & 1 & \\ & & 1 \end{bmatrix} \sim \begin{bmatrix} & & 1 \\ & 1 & \\ 1 & & \end{bmatrix}$.

Proof. This is the same thing as saying that $\begin{bmatrix} & & 1 \\ & 1 & \\ 1 & & \end{bmatrix} = AA^T$ for some nonsingular A. Let $A = \begin{bmatrix} 1 & 1 & 1 \\ 1 & 0 & 1 \\ 0 & 1 & 1 \end{bmatrix}$. □

It's easy to see that there are no further relations; **Bil** is the commutative monoid with two generators I and H, subject to the relation $H + I = 3I$.

Let's go back to topology, and let M be a compact 2-manifold. Then you get an intersection pairing on $H_1(M)$. Consider $\mathbb{RP}^2$. We know that $H_1(\mathbb{RP}^2) = \mathbb{F}_2$. This must be the form we labeled I. This says that any time you have a nontrivial cycle on a projective plane, there's nothing you can do to remove its self intersections. You can see this. The projective plane is a Möbius band with a disk sewn on along the boundary. The waist of the Möbius band serves as a generating cycle. The observation is that if this cycle is moved to intersect itself transversely, it must intersect itself an odd number of times.

We can produce new surfaces from old by a process of "addition." Given two connected surfaces Σ_1 and Σ_2, cut a disk out of each one and sew them together along the resulting circles. This is the *connected sum* $\Sigma_1 \# \Sigma_2$. You showed in 12.2 that there is an isomorphism

$$H_1(\Sigma_1 \# \Sigma_2) \cong H_1(\Sigma_1) \oplus H_1(\Sigma_2).$$

I leave it to you to verify that the direct sum is orthogonal.

Write **Surf** for the set of homeomorphism classes of compact connected surfaces. Connected sum provides it with the structure of a commutative monoid. The classification (e.g. [35]) of surfaces may now be summarized as follows:

Theorem 30.8. *Formation of the intersection bilinear form gives an isomorphism of commutative monoids* **Surf** $\to$ **Bil**.

This is a kind of model result of algebraic topology! — a complete algebraic classification of a class of geometric objects. The oriented surfaces correspond to the bilinear forms of type gH; g is the *genus*. We must have a relation corresponding to $H \oplus I = 3I$, namely

$$T^2 \# \mathbb{RP}^2 \cong \mathbb{RP}^2 \# \mathbb{RP}^2 \# \mathbb{RP}^2.$$

You should verify this for yourself!

There's more to be said about this. Away from characteristic 2, symmetric bilinear forms and quadratic forms are interchangeable. But over $\mathbb{F}_2$ you can ask for a quadratic form q such that

$$q(x + y) = q(x) + q(y) + x \cdot y.$$

This is a "quadratic refinement" of the symmetric bilinear form. Of course it implies that $x \cdot x = 0$ for all x, so this will correspond to some further structure on an oriented surface. This structure is a "framing," a trivialization of the normal bundle of an embedding into a high dimensional Euclidean space. There are then further invariants of this framing; this is the story of the Kervaire invariant.

Exercises

Exercise 30.9. Construct a homeomorphism of surfaces corresponding to $H \oplus I = 3I$.

Exercise 30.10. Where does the Klein K^2 bottle figure in this classification of surfaces? Draw singular 1-simplices representing each of the four elements of $H_1(K^2)$, and verify that their intersections (including self-intersections) are determined by the intersection pairing.

Exercise 30.11. Show that any odd-dimensional compact manifold has vanishing Euler characteristic.

31   Local coefficients and orientations

The fact that a manifold is locally Euclidean puts surprising constraints on its cohomology, captured in the statement of Poincaré duality. To understand how this comes about, we have to find ways to promote "local information" — like the existence of Euclidean neighborhoods — to "global information" — like restrictions on the structure of the cohomology. Today we'll study the notion of an orientation, which is the first link between local and global.

The local-to-global device relevant to this is the notion of a "local coefficient system," which is based on the more primitive notion of a covering space. We merely summarize that theory, since it is a prerequisite for this course; see for example [35] for more details.

Covering spaces

Definition 31.1. A continuous map $p : E \to B$ is a *covering space* if
(1) every point pre-image is a discrete subspace of E, and
(2) every $b \in B$ has a neighborhood V admitting a map $p^{-1}(V) \to p^{-1}(b)$ such that the induced map

$$p^{-1}(V) \to V \times p^{-1}(b)$$

is a homeomorphism.

The space B is the "base," E the "total space."

Example 31.2. A first example is given by the projection map $\mathrm{pr}_1 : B \times F \to B$ where F is discrete. A covering space of this form is said to be *trivial*, so the covering space condition can be rephrased as "local triviality."

The first interesting example is the projection map $S^n \to \mathbb{RP}^n$ obtained by identifying antipodal maps on the sphere. This example generalizes in the following way.

Definition 31.3. An action of a group π on a space X is *principal* or *totally discontinuous* (terrible language, since we are certainly assuming that every group element acts by homeomorphisms) provided that every element $x \in X$ has a neighborhood U such that the only time U and gU intersect is when $g = 1$.

This is a strong form of "freeness" of the action. It is precisely what is needed to guarantee:

Lemma 31.4. *If π acts principally on X then the orbit projection map $X \to \pi \backslash X$ is a covering space.*

It is not hard to use local triviality to prove the following:

Theorem 31.5 (Unique path lifting). *Let $p : E \to B$ be a covering space, and $\omega : I \to B$ a path in the base. For any $e \in E$ such that $p(e) = \omega(0)$, there is a unique path $\widetilde{\omega} : I \to E$ in E such that $p\widetilde{\omega} = \omega$ and $\widetilde{\omega}(0) = e$.*

This theorem connects the theory of covering spaces with the "fundamental group" of a space B with distinguished basepoint b. It's denoted by $\pi_1(B, b)$, in homage to Henri Poincaré. It is the set of equivalence classes of loops at B, in which two loops are equivalent if they differ by a homotopy that fixes the endpoints. Loops can be added by juxtaposition:

$$(\alpha \cdot \omega)(t) = \begin{cases} \alpha(2t) & 0 \leq t \leq 1/2 \\ \omega(2t - 1) & 1/2 \leq t \leq 1. \end{cases}$$

A space B is *simply connected* if it is path connected and $\pi_1(B, b)$ is trivial for some (and hence any) $b \in B$.

The fundamental group is closely related to singular homology in dimension one. If we pick a generator σ for $H_1(S^1, *) = \mathbb{Z}$, we get a map $\pi_1(B, *) \to H_1(B, *)$ by sending the pointed homotopy class of $f : S^n \to X$ to $f_*(\sigma)$. This is the *Hurewicz map* (in dimension 1), and we have Poincaré's theorem:

Theorem 31.6 (e.g. [72, Theorem 9.2.1]). *Let B be a path-connected pointed space. The Hurewicz map factors through an isomorphism*

$$\pi_1(B, *)^{ab} \to H_1(B, *)$$

from the abelianization of the fundamental group.

This theorem will be generalized later; see Lectures 51 and 65. Combined with the Universal Coefficient Theorem 27.1, Poincaré's theorem implies:

Corollary 31.7. *Let B be a path-connected space and $b \in B$, and let N be any abelian group. There is a natural isomorphism*

$$\mathrm{Hom}(\pi_1(B, b), N) \to H^1(B; N).$$

Returning to the covering space story, Theorem 31.5 leads to a right action of $\pi_1(B, b)$ on $F = p^{-1}(b)$: Represent an element of $\pi_1(B, b)$ by a loop ω; for an element $e \in p^{-1}(b)$ let $\widetilde{\omega}$ be the lift of ω with $\widetilde{\omega}(0) = e$; and define

$$e \cdot [\omega] = \widetilde{\omega}(1) \in E.$$

This element lies in F because ω was a *loop*, ending at b. One must check that this action by $[\omega] \in \pi_1(B, b)$ does not depend upon the choice of representative ω, and that we do indeed get a *right action*:

$$e \cdot (ab) = (e \cdot a) \cdot b, \quad e \cdot 1 = e.$$

Given a principal π-action on X, with orbit space B, we can do more than just form the orbit space! If we also have a right action of π on a set F, we can form a new covering space over B with F as "generic" fiber. Write $F \times_\pi X$ for the quotient of the product space $F \times X$ by the equivalence relation

$$(s, gx) \sim (sg, x), \quad g \in \pi.$$

This is the "balanced product" or "Borel construction." The composite projection $F \times X \to X \to B$ factors through a map $p : F \times_\pi X \to B$, which is easily seen to be a covering space. Any element $x \in X$ determines a homeomorphism

$$F \to p^{-1}p(x) \quad \text{by} \quad s \mapsto [s, x].$$

Of course $* \times_\pi X = B$, and if we let π act on itself by right translation, $\pi \times_\pi X = X$.

Covering spaces of a fixed space B form a category $\mathbf{Cov}_B$, in which a morphism $E' \to E$ is "covering transformation," that is, a map $f : E' \to E$ making

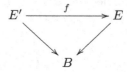

commute. Sending $p : E \to B$ to $p^{-1}(b)$ with its action by $\pi_1(B, b)$ gives a functor

$$\mathbf{Cov}_B \to \mathbf{Set}_{\pi_1(B,b)}$$

to the category of right actions of $\pi_1(B, b)$ on sets. For connected spaces, this is usually an equivalence of categories. The technical assumption required is this: A space B is *semilocally simply connected* if it is path connected and for every point b and every neighborhood U of b, there exists a smaller neighborhood V such that $\pi_1(V, b) \to \pi_1(X, b)$ is trivial. This is a very weak condition.

Theorem 31.8. *Assume that B is semi-locally simply connected. Then the functor $\mathbf{Cov}_B \to \mathbf{Set}_{\pi_1(B,b)}$ is an equivalence of categories.*

This "Galois correspondence" is another one of those perfect theorems in algebraic topology!

The covering space corresponding under this equivalence to the translation action of $\pi_1(B, b)$ on itself is the *universal cover* of B, denoted by $\widetilde{B} \to B$. $\widetilde{B}$ is simply connected. Since the automorphism group of π as a right π-set is π (acting by left translation), the automorphism group of $\widetilde{B} \to B$ as a covering space of B is $\pi_1(B, b)$. This action is principal, and the covering space corresponding to a $\pi_1(B, b)$-set S is given by the balanced product $S \times_{\pi_1(B,b)} \widetilde{B}$.

If $p : E \to B$ is a covering space, one of the things you may want to do is consider a *section* of p; that is, a continuous function $s : B \to E$ such that $p \circ s = 1_B$. Write $\Gamma(B; E)$ for the set of sections of $p : E \to B$. Under the correspondence of Theorem 31.8,

$$\Gamma(B; E) = (p^{-1}(b))^{\pi_1(B,b)},$$

the fixed point set for the action of $\pi_1(B, b)$ on $p^{-1}(b)$.

The local trivality implies:

Lemma 31.9. *Let $E \to B$ be a covering space, and suppose that s_1 and s_2 are sections. The subset of B on which these two sections coincide is open and closed.*

The following technical lemma will be important to us.

Proposition 31.10. *Let B be a normal space, $E \to B$ be a covering space, and K a compact subset of B. Any element of $\Gamma(K; E)$ extends to a section on an open neighborhood of K in B.*

Proof. (cp. [10, p. 334], [22, p. 150], [11, p. 66]) Let $\mathcal{U}$ be a set of open subsets of X that covers K and such that E is trivial over each element of $\mathcal{U}$. Since K is compact we may assume that $\mathcal{U}$ is finite: $\mathcal{U} = \{U_1, \ldots, U_n\}$. Let $U_0 = X - K$, so that the set $\{U_0, \ldots, U_n\}$ is an open cover of X. Any finite open cover of a normal space admits a *shrinking*: so there is a new open cover $\{V_0, \ldots, V_n\}$ such that $\overline{V_i} \subseteq U_i$ for all i. Since V_0 doesn't meet K, $\{V_1, \ldots, V_n\}$ covers K. Let $V = \bigcup_{i=1}^{n} V_i$, an open subset of X containing K.

Let s be a section of E over K. It certainly extends to a section over each U_i, say s_i. Now define

$$W = \{x \in V : \forall i, j \in I, \, x \in V_i \cap V_j \Rightarrow s_i(x) = s_j(x)\},$$

where $I = \{1, \ldots, n\}$. The subspace W inherits the finite open cover $\{V_i \cap W : i \in I\}$, and the gluing lemma for maps on open subsets shows that there is a continuous section of E on W restricting on $V_i \cap W$ to the same section that s_i does for each $i \in I$. Since $K \subseteq W$, it suffices to show that W contains an open neighborhood of K.

Let $x \in K$. We construct an open subset of X containing x and lying in W. Let $I(x)$ be the set of indices i such that $x \in \overline{V_i}$, and define

$$N_0(x) = \bigcap_{j \notin I(x)} (V - \overline{V_j}),$$

and for $i, j \in I(x)$ define

$$N_{i,j}(x) = \{y \in U_i \cap U_j \cap V : s_i(y) = s_j(y)\}.$$

Then $x \in N_{i,j}(x)$ since $x \in \overline{V_i} \subseteq U_i$ for all $i \in I(x)$ and the sections agree at x since $x \in K$. By Lemma 31.9, this is an open subset of V. Then put

$$N(x) = N_0(x) \cap \bigcap_{i,j \in I(x)} N_{i,j}(x).$$

This is certainly an open subset of X containing x. To see that it lies in W, let $y \in N(x) \subseteq V$. If $y \in V_i$ then $i \in I(x)$ since $N(x) \subseteq N_0(x)$, so if $y \in V_i \cap V_j$ then both i and j lie in $I(x)$. Then use the inclusion $N(x) \subseteq N_{i,j}(x)$ to see that $s_i(y) = s_j(y)$. $\qquad\square$

Local coefficient systems

Covering spaces come up naturally in our study of topological manifolds. For any space X, we can probe the structure of X in the neighborhood of $x \in X$ by studying the graded R-module $H_*(X, X - x; R)$, the *local* *homology of X at x*; where, to lighten notation, we have written $X - x$ in place of $X - \{x\}$. By excision, this group depends only on the structure of X "locally at x": For any neighborhood U of x, excising the complement of U gives an isomorphism

$$H_*(U, U - x) \xrightarrow{\cong} H_*(X, X - x).$$

When the space is an n-manifold — let's write M for it — the local homology is very simple. It's nonzero only in dimension n. This has a nice immediate consequence, by the way: there is a well-defined locally constant function dim : $M \to \mathbb{N}$, sending x to the dimension in which $H_*(M, M - x)$ is nontrivial. For an n-manifold, it's the constant function with value n.

In fact the whole family of homology groups $H_n(M, M - x)$ is "locally constant." This is captured in the statement that taken together, as x varies over M, they constitute a covering space over M. To see this, begin by defining

$$o_M = \coprod_{x \in M} H_n(M, M - x)$$

as sets. There is an evident projection map $p : o_M \to M$. We aim to put a topology on o_M with the property that this map is a covering space. This will use an important map $j_{A,x}$, defined for any closed set $A \subseteq M$ and $x \in A$ as the map induced by an inclusion of pairs:

$$j_{A,x} : H_n(M, M - A) \to H_n(M, M - x).$$

Define a basis of opens $V_{U,x,\alpha}$ in o_M indexed by triples (U, x, α) where U is open in M, $x \in U$, and $\alpha \in H_n(M, M - \overline{U})$:

$$V_{U,x,\alpha} = \{j_{\overline{U},x}(\alpha) : x \in U\}.$$

Each $\alpha \in H_n(M, M - \overline{U})$ thus defines a "sheet" of o_M over U. We leave it to you to check that this is indeed a covering space.

This covering space has more structure: each fiber is an abelian group, an infinite cyclic abelian group (or it's an R-module, free on one generator, if we have coefficients in a commutative ring R). These structures vary continuously as you move from one fiber to another. To illuminate this structure, observe that the category $\mathbf{Cov}_B$ has finite products; the product of $E' \downarrow B$ and $E \downarrow B$ is given by the fiber product or pullback, $E' \times_B E \downarrow B$. The terminal object is the identity map $B \downarrow B$. This lets us define an "abelian group object" in $\mathbf{Cov}_B$; it's an object $E \downarrow B$ together with maps $E \times_B E \to E$ and $B \to E$ over B, satisfying some evident conditions that are equivalent to requiring that they render each fiber an abelian group. If you have a ring around you can also ask for a map $(B \times R) \times_B E \to E$ making each fiber an R-module.

The structure we have defined is a *local coefficient system* (of R-modules). We already have an example; if M is an n-manifold, we have the *orientation local system* o_M over M.

It's useful to allow coefficients in a commutative ring R; so denote by

$$o_M \otimes R$$

the local system of R-modules obtained by tensoring each fiber with R.

The classification theorem for covering spaces has as a corollary:

Theorem 31.11. *Let B be semi-locally simply connected. Then forming the fiber over a point b gives an equivalence of categories from the category of local coefficient systems of R-modules over B and the category of modules over the group algebra $R\pi_1(B,b)$.*

Assume M is a connected n-manifold. The fibers of our local coefficient system o_M are quite simple: they are free of rank 1. Since any automorphism of such an R-module is given by multiplication by a unit in R, we find that the local coefficient system is defined by giving a homomorphism

$$\pi_1(M,b) \to R^\times$$

or, what is the same (by Corollary 31.7) an element of $H^1(M; R^\times)$.

The set of abelian group generators of the fibers of o_M form a sub covering space, a double cover of M, denoted by $o_M^\times$. It is the "orientation double cover." If M is orientable it is trivial; it consists of two copies of M. An orientation consists of a choice of one or the other of the components. If M is nonorientable the orientation double cover is again connected. An interesting and simple fact is that its total space is a manifold in its own right, and is orientable; in fact it carries a canonical orientation.

Similarly we can form the sub covering space of R-module generators of the fibers of $o_M \otimes R$; write $(o_M \otimes R)^\times$ for it.

Orientations

A "local R-orientation at x" is a choice of R-module generator of $H_n(M, M - x; R)$, and we make the following definition.

Definition 31.12. An R-*orientation* of an n-manifold M is a section of $(o_M \otimes R)^\times$. If an orientation exists, the manifold is R-*orientable*.

For example, when $R = \mathbb{F}_2$, every manifold is orientable, and uniquely so, since $\mathbb{F}_2^\times = \{1\}$. A $\mathbb{Z}$-orientation (or simply "orientation") is a section of the orientation double cover. A manifold is "R-orientable" if it admits an R-orientation.

This relates to the "globalization" project we started out talking about. A section over B is in fact called a "global section." In the case of the orientation local system, we have a canonical map

$$ j : H_n(M; R) \to \Gamma(M; o_M \otimes R) , $$

described as follows. The value of $j(a)$ at $x \in M$ is the restriction of a to $H_n(M, M - x)$. The first "local-to-global" theorem, a special case of Poincaré duality, is this:

Theorem 31.13 (Orientation Theorem). *If M is compact, the map j : $H_n(M; R) \to \Gamma(M; o_M \otimes R)$ is an isomorphism.*

We will prove this theorem in the next lecture.

When $R = \mathbb{Z}$ we will usually drop the "R-." A connected n-manifold is either non-orientable, or admits exactly two orientations. Euclidean space is orientable, and an orientation is determined by a choice of ordered basis.

Still with $R = \mathbb{Z}$, the homomorphism

$$ w_1 : \pi_1(M, b) \to \mathbb{Z}^\times = \{\pm 1\} , $$

regarded as an element of $H^1(M; \mathbb{F}_2)$, is the "first Stiefel-Whitney class." If it is trivial, you can pick consistent generators for $H_n(M, M - x; \mathbb{Z})$ as x runs over M: the manifold is "orientable," and is *oriented* by one of the two possible choices, which is then called an *orientation*. If the orientation local system is nontrivial, the manifold is *nonorientable*. I hope it's clear that the Möbius band is nonorientable, and hence any surface containing the Möbius band is as well.

For general R, the representation of $\pi_1(B)$ on the fiber of $o_M \otimes R$ over b is given by the composite $\pi_1(B) \to \{\pm 1\} \to R^\times$. If this is the trivial homomorphism, the fixed points of this representation on R form all of R. If not, the fixed points are the subgroup of R of elements of order 2, written $R[2]$.

Corollary 31.14. *If M is a compact connected n-manifold, then*

$$ H_n(M; R) \cong \begin{cases} R & \text{if } M \text{ is orientable} \\ R[2] & \text{if not.} \end{cases} $$

In the first case, a generator of $H_n(M; R)$ is a *fundamental class* for the manifold. You should think of the manifold itself as a cycle representing this homology class. It is characterized as the class that restricts to the chosen generator of $H_n(M, M - x)$ for all x; this is saying that the cycle "covers" each point x exactly once, with the correct orientation.

The first isomorphism in Corollary 31.14 depends upon this choice of fundamental class. But in the second case, the isomorphism is canonical. Over $\mathbb{F}_2$, any compact connected manifold has a unique fundamental class, the generator of $H_n(M; \mathbb{F}_2) = \mathbb{F}_2$.

Exercises

Exercise 31.15. Check that o_M is a covering space of the manifold M.

Exercise 31.16. Show that if $p : N \downarrow M$ is a covering space of manifolds then $p^* o_M \cong o_N$, where p^* denotes the pull-back functor $\mathbf{Cov}_M \to \mathbf{Cov}_N$. Define a canonical orientation on the total space of $o_M^\times$.

Exercise 31.17. Show that any Lie group is orientable.

32   Proof of the orientation theorem

We are studying the way in which local homological information gives rise to global information, especially on an n-manifold M. The tool is the map

$$j : H_n(M; R) \to \Gamma(M; o_M \otimes R)$$

sending a class c to the section of the orientation local coefficient system given at $x \in M$ by the restriction $j_x(c) \in H_n(M, M - x)$. We asserted that if M is compact then j is an isomorphism and that $H_q(M) = 0$ for $q > n$. The proof will be by an induction that requires us to formulate a more general statement.

Let $A \subseteq M$ be a compact subset. A class in $H_n(M, M - A)$ is represented by an n-chain whose boundary lies outside of A. Does it cover A evenly? We can give meaning to this question as follows. Let $x \in A$. Then $M - A \subseteq M - x$, so we have a restriction map

$$j_{A,x} : H_n(M, M - A) \to H_n(M, M - x)$$

that tests whether the chain covers x. As x ranges over A, these maps together give us a map to the group of sections of o_M over A,

$$j_A : H_n(M, M - A) \to \Gamma(A; o_M).$$

(The topology of o_M is set up so that $j_A(c)$ is continuous for any $c \in H_n(M, M - A)$.) Because $H_n(M, M - A)$ deals with chains that "stretch over A" — with boundary in $M - A$ — we will employ the following notation.

Notation 32.1. $H_n(M, M - A) = H_n(M|A)$.

Here is our more general assertion.

Theorem 32.2. *Let M be an n-manifold and let A be a compact subset of M. Then $H_q(M|A; R) = 0$ for $q > n$, and the map $j_A : H_n(M|A; R) \to \Gamma(A; o_M \otimes R)$ is an isomorphism.*

If M is compact, we may take $A = M$ and learn that $H_q(M; R) = 0$ for $q > n$ and

$$j_M : H_n(M; R) \xrightarrow{\cong} \Gamma(M; o_M \otimes R).$$

But the theorem covers much more exotic situations as well; perhaps A is a Cantor set in some Euclidean space, for example.

We follow [10] in proving this, and refer you to that reference for the modifications appropriate for the more general statement when A is assumed merely closed rather than compact.

First we establish two general facts about the statement of Theorem 32.2.

Proposition 32.3. *Let A and B be compact subspaces of the n-manifold M, and suppose the result holds for A, B, and $A \cap B$. Then it holds for $A \cup B$.*

Proof. The relative Mayer-Vietoris theorem (Exercise 11.9) and the hypothesis that $H_{n+1}(M|A \cap B) = 0$ gives us exactness of the top row in the (clearly commutative) ladder

$$
\begin{array}{ccccccc}
0 & \longrightarrow & H_n(M|A \cup B) & \longrightarrow & H_n(M|A) \oplus H_n(M|B) & \longrightarrow & H_n(M|A \cap B) \\
& & \downarrow{\scriptstyle j_{A \cup B}} & & \downarrow{\scriptstyle j_A \oplus j_B} & & \downarrow{\scriptstyle j_{A \cap B}} \\
0 & \longrightarrow & \Gamma(A \cup B; o_M) & \longrightarrow & \Gamma(A; o_M) \oplus \Gamma(B; o_M) & \longrightarrow & \Gamma(A \cap B; o_M).
\end{array}
$$

Exactness of the bottom row is clear: A section over $A \cup B$ is precisely a section over A and a section over B that agree on the intersection. So the five-lemma shows that $j_{A \cup B}$ is an isomorphism. Looking further back in the Mayer-Vietoris sequence gives the vanishing of $H_q(M|A \cup B)$ for $q > n$. $\square$

Proposition 32.4. *Let $A_1 \supseteq A_2 \supseteq \cdots$ be a decreasing sequence of compact subsets of M, and assume that the theorem holds for each A_i. Then it holds for the intersection $A = \bigcap A_i$.*

The proof of this proposition entails two lemmas, which we'll dispose of first.

Lemma 32.5. *Let $A_1 \supseteq A_2 \supseteq \cdots$ be a decreasing sequence of closed subsets of a space X, with intersection A. Then*

$$\varinjlim_i H_q(X, X - A_i) \xrightarrow{\cong} H_q(X, X - A).$$

Proof. Let $\sigma : \Delta^q \to X$ be any q-simplex in $X - A$. The subsets $X - A_i$ form an open cover of $\mathrm{im}(\sigma)$, so since it is compact it lies in some single $X - A_i$. The criterion of Lemma 23.11 shows that

$$\varinjlim_i S_q(X - A_i) \xrightarrow{\cong} S_q(X - A).$$

Thus

$$\varinjlim_i S_q(X, X - A_i) \xrightarrow{\cong} S_q(X, X - A_i)$$

by exactness of direct limit and the five-lemma. The claim then follows again using exactness of direct limit. $\square$

Lemma 32.6. *Let $A_1 \supseteq A_2 \supseteq \cdots$ be a decreasing sequence of compact subsets in a Hausdorff space X and let A be their intersection. For any open neighborhood U of A there exists i such that $A_i \subseteq U$.*

Proof. A is compact, being a closed subset of the compact Hausdorff space A_1. Since A is the intersection of the A_i, and $A \subseteq U$, the intersection of the decreasing sequence of compact sets $A_i - U$ is empty. Thus by the finite intersection property one of them, say $U - A_i$, must be empty; but that says that $A_i \subseteq U$. $\square$

Proof of Proposition 32.4. By Lemma 32.5, $H_q(M|A) = 0$ for $q > n$. In dimension n, we contemplate the commutative diagram

$$
\begin{array}{ccc}
\varinjlim_i H_n(M|A_i) & \xrightarrow{\cong} & H_n(M|A) \\
\Big\downarrow{\scriptstyle \cong} & & \Big\downarrow \\
\varinjlim_i \Gamma(A_i; o_M) & \xrightarrow{\cong} & \Gamma(A; o_M).
\end{array}
$$

The top map an isomorphism by Lemma 32.5, and the left one by the hypothesis of the Proposition.

To see that the bottom map is an isomorphism, we'll verify the two conditions for a map to be a direct limit from Lecture 23. Lemmas 31.10 and 32.5 imply that any section over A is the restriction of a section over some A_i. On the other hand, suppose that a section $\sigma \in \Gamma(A_i; o_M)$ vanishes on A. Then it vanishes on some open set containing A, again Lemma 31.10. Some A_j lies in that open set, again by Lemma 32.5. We may assume that $j \geq i$, and conclude that σ already vanishes on A_j.   □

Proof of Theorem 32.2. There are five steps. In describing them, we will call a subset of M "Euclidean" if it lies inside some open set homeomorphic to $\mathbb{R}^n$.

(1) $M = \mathbb{R}^n$, A a compact convex subset.
(2) $M = \mathbb{R}^n$, A a finite union of compact convex subsets.
(3) $M = \mathbb{R}^n$, A any compact subset.
(4) M arbitrary, A a finite union of compact Euclidean subsets.
(5) M arbitrary, A an arbitrary compact subset.

Notes on the proofs: (1) To be clear, "convex" implies nonempty. By translating A, we may assume that $0 \in A$. The compact subset A lies in some disk, and by a homothety we may assume that the disk is the unit disk D^n. Then we claim that the inclusion $i : S^{n-1} \to \mathbb{R}^n - A$ is a deformation retract. A retraction is given by $r(x) = x/||x||$, and a homotopy from ir to the identity is given by

$$h(x,t) = \left(t + \frac{1-t}{||x||} \right) x\,.$$

It follows that $H_q(\mathbb{R}^n, \mathbb{R}^n - A) \cong H_q(\mathbb{R}^n, \mathbb{R}^n - D^n)$ for all q. This group is zero for $q > n$. In dimension n, note that restricting to the origin gives an isomorphism $H_n(\mathbb{R}^n, \mathbb{R}^n - D^n) \to H_n(\mathbb{R}^n, \mathbb{R}^n - 0)$ since $\mathbb{R}^n - D^n$ is a deformation retract of $\mathbb{R}^n - 0$. The local system $o_{\mathbb{R}^n}$ is trivial, since $\mathbb{R}^n$ is simply connected, so restricting to the origin gives an isomorphism $\Gamma(D^n, o_{\mathbb{R}^n}) \to H_n(\mathbb{R}^n, \mathbb{R}^n - 0)$. This implies that $j_{D^n} : H_n(\mathbb{R}^n, \mathbb{R}^n - D^n) \to \Gamma(D^n, o_{\mathbb{R}^n})$ is an isomorphism. The restriction $\Gamma(D^n, o_{\mathbb{R}^n}) \to \Gamma(A, o_{\mathbb{R}^n})$ is also an isomorphism, since $A \to D^n$ is a deformation retract. So by the commutative diagram

$$H_n(\mathbb{R}^n, \mathbb{R}^n - D^n) \xrightarrow{\;\cong\;} H_n(\mathbb{R}^n, \mathbb{R}^n - A)$$

$$\cong \Big\downarrow j_{D^n} \qquad\qquad\qquad \Big\downarrow j_A$$

$$\Gamma(D^n, o_{\mathbb{R}^n}) \xrightarrow{\;\;\cong\;\;} \Gamma(A, o_{\mathbb{R}^n})$$

we find that $j_A : H_n(\mathbb{R}^n, \mathbb{R}^n - A) \to \Gamma(A; o_{\mathbb{R}^n})$ is an isomorphism.

(2) by Proposition 32.3.

(3) For each $j \geq 1$, let C_j be a finite subset of A such that

$$A \subseteq \bigcup_{x \in C_j} B_{1/j}(x).$$

Since any intersection of convex sets is either empty or convex,

$$A_k = \bigcap_{j=1}^{k} \bigcup_{x \in C_j} B_{1/j}(x)$$

is a union of finitely many convex sets, and since A is closed it is the intersection of this decreasing family. So the result follows from (1), (2), and Proposition 32.4.

(4) by (3) and (2).

(5) The compact subset A of M is the intersection of a decreasing family of finite unions of compact Euclidean subsets, so the result follows from Proposition 32.4. To see this it's convenient to use the fact that if a manifold admits a finite cover by Euclidean spaces then it is metrizable. (See for example [26, Theorem 2.68].) The union of elements in a finite Euclidean cover of A is thus metrizable. It also admits a countable dense subset, so A does as well; pick one, say S. For each $x \in S$ there is an $\epsilon(x) > 0$ such that the ball of radius $\epsilon(x)$ with center x is Euclidean. For any j, the collection of balls with center in S and radius the minimum of $1/j$ and $\epsilon(x)/2$ is a cover of A by opens whose closures are compact and Euclidean. For each j pick a finite subcover, and let B_j be the union of the closures. This sequence of unions of Euclidean compacts may not be decreasing, but the sequence $A_k = \bigcap_{j \leq k} B_k$ is. $\qquad\square$

33 A plethora of products

We are now heading towards a statement of Poincaré duality. It will use a certain action of cohomology on homology, which we set up now.

Kronecker pairing

Recall that we have the Kronecker pairing

$$\langle -, - \rangle : H^p(X; R) \otimes H_p(X; R) \to R.$$

It can't be "natural," because H^p is contravariant while homology is covariant. But given $f : X \to Y$, $b \in H^p(Y)$, and $x \in H_p(X)$, we can ask: How does $\langle f^*b, x \rangle$ relate to $\langle b, f_*x \rangle$?

Claim 33.1. $\langle f^*b, x \rangle = \langle b, f_*x \rangle$.

Proof. This is easy, and true on the level of chains! I find it useful to write out diagrams to show where things are. We want the diagram

$$
\begin{array}{ccc}
\mathrm{Hom}(S_p(Y), R) \otimes S_p(X) & \xrightarrow{\;1 \otimes f_*\;} & \mathrm{Hom}(S_p(Y), R) \otimes S_p(Y) \\
\downarrow{\scriptstyle f^* \otimes 1} & & \downarrow{\scriptstyle \langle -, - \rangle} \\
\mathrm{Hom}(S_p(X), R) \otimes S_p(X) & \xrightarrow{\quad \langle -, - \rangle \quad} & R
\end{array}
$$

to commute. Going to the right and then down gives

$$\beta \otimes \xi \mapsto \beta \otimes f_*(\xi) \mapsto \beta(f_*\xi).$$

The other way gives

$$\beta \otimes \xi \mapsto f^*(\beta) \otimes \xi = (\beta \circ f_*) \otimes \xi \mapsto (\beta \circ f_*)(\xi).$$

This is exactly $\beta(f_*\xi)$. $\qquad\square$

There's actually another product in play here:

$$\mu : H(C_*) \otimes H(D_*) \to H(C_* \otimes D_*)$$

given by $[c] \otimes [d] \mapsto [c \otimes d]$. I used it to pass from the chain level computation we did to the homology statement.

Cross products

We also have the two cross products:

$$\times : H_p(X) \otimes H_q(Y) \to H_{p+q}(X \times Y)$$

and

$$\times : H^p(X) \otimes H^q(Y) \to H^{p+q}(X \times Y).$$

You might think this is fishy because both maps are in the same direction, while homology and cohomology have opposite variances. But it's OK, because we used different things to make these constructions: for example, the Eilenberg-Zilber map ζ for homology and the Alexander-Whitney map α for cohomology. Still, they're related:

$$\zeta : S_*(X) \otimes S_*(Y) \leftrightarrows S_*(X \times Y) : \alpha$$

are homotopy inverse chain homotopy equivalence covering the obvious isomorphism $H_0(X) \otimes H_0(Y) \cong H_0(X \times Y)$ (Theorem 25.13). Moreover:

Lemma 33.2. *Let $a \in H^p(X), b \in H^q(Y), x \in H_p(X), y \in H_q(Y)$. Then:*

$$\langle a \times b, x \times y \rangle = (-1)^{pq} \langle a, x \rangle \langle b, y \rangle \, .$$

Proof. Say $a = [f], b = [g], x = [\xi], y = [\eta]$. Then $\langle a \times b, x \times y \rangle$ is represented by

$$(f \otimes g)\alpha\zeta(\xi \otimes \eta) \simeq (f \otimes g)(\xi \otimes \eta) = (-1)^{pq} f(\xi)g(\eta) \, . \qquad \square$$

We can use this to prove a restricted form of the Künneth theorem in cohomology.

Theorem 33.3. *Let R be a PID. Assume that $H_p(X)$ is a finitely generated free R-module for all p. Then*

$$\times : H^*(X; R) \otimes_R H^*(Y; R) \to H^*(X \times Y; R)$$

is an isomorphism.

Proof. Write $M^\vee$ for the R-linear dual of an R-module M. By our assumption about $H_p(X)$, the map

$$H_p(X)^\vee \otimes H_q(Y)^\vee \to (H_p(X) \otimes H_q(Y))^\vee \, ,$$

sending $f \otimes g$ to $(x \otimes y \mapsto (-1)^{pq} f(x)g(y))$, is an isomorphism. The homology Künneth theorem guarantees that the bottom map in the following diagram is an isomorphism.

$$
\begin{array}{ccc}
\displaystyle\bigoplus_{p+q=n} H^p(X) \otimes H^q(Y) & \xrightarrow{\quad\quad\times\quad\quad} & H^n(X \times Y) \\[2em]
\Big\| & & \Big\| \\[1em]
\displaystyle\bigoplus_{p+q=n} H_p(X)^\vee \otimes H_q(Y)^\vee & \xrightarrow{\cong} \left(\displaystyle\bigoplus_{p+q=n} H_p(X) \otimes H_q(Y)\right)^\vee \xleftarrow{\cong} & H_n(X \times Y)^\vee
\end{array}
$$

Commutativity of this diagram is exactly the content of Lemma 33.2. $\qquad \square$

We saw in Proposition 29.3 that the cohomology cross-product is an algebra map, so under the conditions of the theorem it is an isomorphism of algebras. You do need some finiteness assumption, even if you are working over a field. For example let T be an infinite set, regarded as a space with the discrete topology. Then $H^0(T; R) = \mathrm{Map}(T, R)$. But

$$\mathrm{Map}(T, R) \otimes \mathrm{Map}(T, R) \to \mathrm{Map}(T \times T, R)$$

sending $f \otimes g$ to $(s, t) \to f(s)g(t)$ is not surjective; the characteristic function of the diagonal is not in the image, for example (unless $R = 0$).

Cup and cap products

There are more products around. For example, there is a map

$$H^p(Y) \otimes H^q(X, A) \to H^{p+q}(Y \times X, Y \times A).$$

Constructing this is on your homework. Suppose $Y = X$. Then I get

$$\cup : H^*(X) \otimes H^*(X, A) \to H^*(X \times X, X \times A) \xrightarrow{\Delta^*} H^*(X, A),$$

where $\Delta : (X, A) \to (X \times X, X \times A)$ is the "relative diagonal." This *relative cup product* makes $H^*(X, A)$ into a module over the graded algebra $H^*(X)$. The relative cohomology is *not* a ring — it doesn't have a unit, for example — but it is a module. And the long exact sequence of the pair is a sequence of $H^*(X)$-modules.

I want to introduce you to one more product, one that will enter into our expression of Poincaré duality. This is the *cap product*. What can we do with $S^p(X) \otimes S_n(X)$? Well, we can form the composite:

$$\cap : S^p(X) \otimes S_n(X) \xrightarrow{1 \times (\alpha \circ \Delta_*)} S^p(X) \otimes S_p(X) \otimes S_{n-p}(X) \xrightarrow{\langle -, - \rangle \otimes 1} S_{n-p}(X).$$

If we choose to use for α the Alexander-Whitney map, we can write:

$$\cap : \beta \otimes \sigma \mapsto \beta \otimes (\sigma \circ \alpha_p) \otimes (\sigma \circ \omega_q) \mapsto (\beta(\sigma \circ \alpha_p))\,(\sigma \circ \omega_q).$$

We are evaluating the cochain on part of the chain, leaving a lower dimensional chain left over.

This composite is a chain map, and so induces a map in homology:

$$\cap : H^p(X) \otimes H_n(X) \to H_{n-p}(X).$$

Notice how the dimensions work. Long ago a bad choice was made: If cohomology were graded with negative integers, the way the gradations work here would look better.

There are also two slant products, but maybe I won't talk about them.

Exercises

Exercise 33.4. Construct a fully relative cup product

$$H^*(X, A) \otimes H^*(Y, B) \to H^*(X \times Y, X \times B \cup A \times Y)$$

under the hypothesis that A is open in X and B is open in Y. Conclude that if X and Y are pointed spaces whose basepoints have contractible open neighborhoods there is a "smash product" map

$$\wedge : \overline{H}^*(X) \otimes \overline{H}^*(Y) \to \overline{H}^*(X \wedge Y).$$

34 Cap product and Čech cohomology

We have a few more things to say about the cap product, and we will then use it to give a statement of Poincaré duality.

Proposition 34.1. *The cap product enjoys the following properties.*
(1) $(a \cup b) \cap x = a \cap (b \cap x)$ *and* $1 \cap x = x$: $H_*(X)$ *is a module for* $H^*(X)$.
(2) Given a map $f : X \to Y$, $b \in H^p(Y)$, *and* $x \in H_n(X)$,

$$f_*(f^*(b) \cap x) = b \cap f_*(x).$$

(3) Let $\varepsilon : H_*(X) \to R$ *be the augmentation. Then*

$$\varepsilon(b \cap x) = \langle b, x \rangle.$$

(4) Cap and cup are adjoint:

$$\langle a \cup b, x \rangle = \langle a, b \cap x \rangle.$$

Proof. (1) Easy.
(2) Let β be a cocycle representing b, and σ an n-simplex in X. Then

$$\begin{aligned}
f_*(f^*(\beta) \cap \sigma) &= f_*((f^*(\beta)(\sigma \circ \alpha_p)) \cdot (\sigma \circ \omega_q)) \\
&= f_*(\beta(f \circ \sigma \circ \alpha_p) \cdot (\sigma \circ \omega)) \\
&= \beta(f \circ \sigma \circ \alpha_p) \cdot f_*(\sigma \circ \omega_q) \\
&= \beta(f \circ \sigma \circ \alpha_p) \cdot (f \circ \sigma \circ \omega_q) \\
&= \beta \cap f_*(\sigma).
\end{aligned}$$

This formula goes by many names: the "projection formula," or "Frobenius reciprocity." Our proof shows that it holds on the chain level, provided you use the Alexander-Whitney map to construct the chain level cap product.
(3) We get zero unless $p = n$. Again let $\sigma \in \mathrm{Sin}_n(X)$, and compute:

$$\varepsilon(\beta \cap \sigma) = \varepsilon(\beta(\sigma) \cdot c^0_{\sigma(e_n)}) = \beta(\sigma)\varepsilon(c^0_{\sigma(e_n)}) = \beta(\sigma) = \langle \beta, \sigma \rangle.$$

(4) Homework. $\qquad\square$

Here now is a statement of Poincaré duality. It deals with the homological structure of compact topological manifolds. We recall the notion of an orientation, and Theorem 31.13 asserting the existence of a fundamental class $[M] \in H_n(M; R)$ in a compact R-oriented n-manifold.

Theorem 34.2 (Poincaré duality). *Let M be a topological n-manifold that is compact and oriented with respect to a PID R. Then there is a unique class $[M] \in H_n(M; R)$ that restricts to the orientation class in $H_n(M, M - a; R)$ for every $a \in M$. It has the property that*

$$- \cap [M] : H^p(M; R) \to H_q(M; R), \quad p + q = n,$$

is an isomorphism for all p.

You might want to go back to Lecture 25 and verify that $\mathbb{RP}^3 \times \mathbb{RP}^3$ satisfies this theorem.

Relative cap product

Our proof of Poincaré duality will be by induction. In order to make the induction go we will prove a substantially more general theorem, one that involves relative homology and cohomology. So we begin by understanding how the cap product behaves in relative homology.

Suppose $A \subseteq X$ is a subspace. We have:

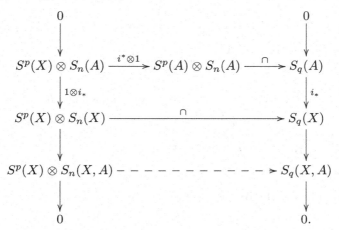

The left sequence is exact because $0 \to S_n(A) \to S_n(X) \to S_n(X, A) \to 0$ splits and tensoring with $S^p(X)$ (which is not free!) therefore leaves it exact. The solid arrow diagram commutes precisely by the chain-level projection formula. There is therefore a uniquely defined map on cokernels.

This chain map yields the *relative cap product*

$$\cap : H^p(X) \otimes H_n(X, A) \to H_q(X, A).$$

It renders $H_*(X, A)$ a module for the graded algebra $H^*(X)$.

I want to come back to an old question, about the significance of relative homology. Suppose that $K \subseteq X$ is a subspace, and consider the relative homology $H_*(X, X - K)$. Since the complement of $X - K$ in X is K, these groups should be regarded as giving information about K. If I enlarge K, I make $X - K$ smaller: $K \subseteq L$ induces $H_*(X, X - L) \to H_*(X - K)$; the relative homology is *contravariant* in the variable K (regarded as an object of the poset of subspaces of X).

Excision gives insight into how $H_*(X, X - K)$ depends on K. Suppose $K \subseteq U \subseteq X$ with $\overline{K} \subseteq \text{Int}(U)$. To simplify things, let's just suppose that K is closed and U is open. Then $X - U$ is closed, $X - K$ is open, and $X - U \subseteq X - K$, so excision asserts that the inclusion map

$$H_*(U, U - K) \to H_*(X, X - K)$$

is an isomorphism.

The cap product puts some structure on $H_*(X, X - K)$: it's a module over $H^*(X)$. But we can do better! We just decided that $H_*(X, X - K) = H_*(U, U - K)$, so the $H^*(X)$ action factors through an action by $H^*(U)$, for any open set U containing K. How does this refined action change when I decrease U?

Lemma 34.3. *Let* $K \subseteq V \subseteq U \subseteq X$, *with* K *closed and* U, V *open. Then:*

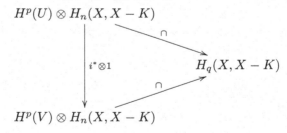

$$H^p(U) \otimes H_n(X, X - K)$$

$$i^* \otimes 1 \qquad\qquad \cap \qquad\qquad H_q(X, X - K)$$

$$\cap$$

$$H^p(V) \otimes H_n(X, X - K)$$

commutes.

Proof. This is just the projection formula again!   □

Čech cohomology

Let $\mathcal{U}_K$ be the set of open neighborhoods of K in X. It is partially ordered by reverse inclusion. This poset is directed, since the intersection of two opens is open.

Definition 34.4. The *Čech cohomology* of K is

$$\check{H}^p(K) = \varinjlim_{U \in \mathcal{U}_K} H^p(U).$$

I apologize for this bad notation; its possible dependence on the way K is sitting in X is not recorded. The maps in this directed system are all maps of graded algebras, so the direct limit is naturally a commutative graded algebra. (Eduard Čech (1893–1960) was a Czech topologist, working at Masaryk University in Brno and the Charles University in Prague.)

Since tensor product commutes with direct limits, we now get a cap product pairing

$$\cap : \check{H}^p(K) \otimes H_n(X, X - K) \to H_q(X, X - K)$$

satisfying the expected properties. This is the best you can do. It's the natural structure that this relative homology has: $H_*(X, X - K)$ is a module over $\check{H}^*(K)$.

There are compatible restriction maps $H^p(U) \to H^p(K)$, so there is a natural map

$$\check{H}^*(K) \to H^*(K).$$

This map is often an isomorphism. Suppose $K \subseteq X$ satisfies the following "regular neighborhood" condition: For every open $U \supseteq K$, there exists an open V with $U \supseteq V \supseteq K$ such that $K \hookrightarrow V$ is a homotopy equivalence (or actually just a homology isomorphism).

Lemma 34.5. *Under these conditions,* $\check{H}^*(K) \to H^*(K)$ *is an isomorphism.*

Proof. We will check that the map to $H^p(K)$ satisfies the conditions we established in Lemma 23.11 to be a direct limit.

So let $x \in H^p(K)$. Let U be a neighborood of K in X such that $H^p(U) \to H^p(K)$ is an isomorphism. Then indeed x is in the image of $H^p(U)$.

Then let U be a neighborhood of K and let $x \in H^p(U)$ restrict to 0 in $H^p(K)$. Let V be a sub-neighborhood such that $H^p(V) \to H^p(K)$ is an isomorphism. Then x restricts to 0 in $H^p(V)$. □

On the other hand, here's an example that distinguishes $\check{H}^*$ from H^*. This is a famous example. The "topologist's sine curve" or "Warsaw circle" is the subspace of $\mathbb{R}^2$ defined as follows. It is union of three subsets, A, B, and C. A is the graph of $\sin(2\pi/x)$ where $0 < x < 1$. B is the interval

$0 \times [-1, 1]$. C is a continuous curve from $(0, -1)$ to $(1, 0)$ and meeting $A \cup B$ only at its endpoints. This is a counterexample for a lot of things; you've probably seen it in your point-set topology course.

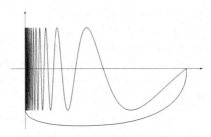

What is the singular homology of the topologist's sine curve? Use Mayer-Vietoris! I can choose V to be some connected portion of the continuous curve from $(0, -1)$ to $(1, 0)$, and U to contain the rest of the space in a way that intersects V in two open intervals. Then V is contractible, and U is made up of two contractible path components. (This space is not locally path connected, and one of these path components is not closed.)

The Mayer-Vietoris sequence looks like

$$0 \to H_1(X) \xrightarrow{\partial} H_0(U \cap V) \to H_0(U) \oplus H_0(V) \to H_0(X) \to 0.$$

The two path components of $U \cap V$ do not become connected in U, so $\partial = 0$ and we find that $\varepsilon : H_*(X) \xrightarrow{\cong} H_*(*)$ and hence $H^*(X) \cong H^*(*)$.

How about $\check{H}^*$? Let $X \subset U$ be an open neighborhood. The interval $0 \times [-1, 1]$ has an ϵ-neighborhood, for some small ϵ, that's contained in U. This implies that there exists a neighborhood $X \subseteq V \subseteq U$ such that $V \simeq S^1$. Consequently

$$\check{H}^*(X) = \varinjlim_{U \in \mathcal{U}_X} H^*(U) \cong H^*(S^1)$$

by a cofinality argument that we will detail in Lemma 35.6. So $\check{H}^*(X) \neq H^*(X)$.

Nevertheless, under quite general conditions the Čech cohomology of a compact Hausdorff space is a topological invariant. The "Čech construction" provides an intrinsic definition of Čech cohomology, one that is a topological invariant by construction. See Dold's beautiful book [15] for this and other topics discussed in this lecture.

Exercises

Exercise 34.6. Verify the remaining parts of Proposition 34.1.

35 Čech cohomology as a cohomology theory

Let X be any space, and let $K \subseteq X$ be a closed subspace. We've defined the Čech cohomology of K as the direct limit of $H^*(U)$ as U ranges over the poset $\mathcal{U}_K$ of open neighborhoods of K. This often coincides with $H^*(K)$ but will not be the same in general. Nevertheless it behaves like a cohomology theory. To expand on this claim, we should begin by defining a relative version.

Suppose $L \subseteq K$ is a pair of closed subsets of a space X. Let (U, V) be a "neighborhood pair" for (K, L):

$$
\begin{array}{ccc}
L & \subseteq & K \\
\cap & & \cap \\
V & \subseteq & U
\end{array}
$$

with U and V open. These again form a directed set $\mathcal{U}_{K,L}$, with partial order given by reverse inclusion of pairs. Then define

$$\check{H}^p(K, L) = \varinjlim_{(U,V) \in \mathcal{U}_{K,L}} H^p(U, V).$$

We will want to verify versions of the Eilenberg-Steenrod axioms for these functors. For a start, I have to explain how maps induce maps.

Let $\mathcal{I}$ be a directed set and $A : \mathcal{I} \to \mathbf{Ab}$ a functor. If we have an order-preserving map — a functor — $\varphi : \mathcal{J} \to \mathcal{I}$ from another directed set, we get $A\varphi : \mathcal{J} \to \mathbf{Ab}$; so $(A\varphi)_j = A_{\varphi(j)}$. I can form two direct limits: $\varinjlim_{\mathcal{J}} A\varphi$ and $\varinjlim_{\mathcal{I}} A$. I claim that they are related by a map

$$\varinjlim_{\mathcal{J}} A\varphi \to \varinjlim_{\mathcal{I}} A.$$

Using the universal property of direct limits, we need to come up with compatible maps $f_j : A_{\varphi(j)} \to \varinjlim_{\mathcal{I}} A$. We have compatible maps $\mathrm{in}_i : A_i \to \varinjlim_{\mathcal{I}} A$ for $i \in \mathcal{I}$, so we can take $f_j = \mathrm{in}_{\varphi(j)}$.

These maps are compatible under composition of order-preserving maps.

Example 35.1. A closed inclusion $i : L \subseteq K$ induces an order-preserving map $\varphi : \mathcal{U}_K \to \mathcal{U}_L$. The functor $H^p : \mathcal{U}_L \to \mathbf{Ab}$ restricts to $H^p : \mathcal{U}_K \to \mathbf{Ab}$, so we get maps

$$\varinjlim_{\mathcal{U}_K} H^p = \varinjlim_{\mathcal{U}_K} H^p \varphi \to \varinjlim_{\mathcal{U}_L} H^p$$

i.e.

$$i^* : \check{H}^p(K) \to \check{H}^p(L).$$

This makes $\check{H}^p$ into a contravariant functor on the partially ordered set of closed subsets of X.

I can do the same thing for relative cohomology, and get the maps involved in the following two theorems, whose proofs will come in due course.

Theorem 35.2 (Long exact sequence). *Let* (K, L) *be a closed pair in* X. *There is a long exact sequence*

$$\cdots \to \check{H}^p(K, L) \to \check{H}^p(K) \to \check{H}^p(L) \xrightarrow{\delta} \check{H}^{p+1}(K, L) \to \cdots$$

that is natural in the pair.

Theorem 35.3 (Excision). *Suppose A and B are closed subsets of a normal space, or compact subsets of a Hausdorff space. Then the map*

$$\check{H}^p(A \cup B, A) \xrightarrow{\cong} \check{H}^p(B, A \cap B)$$

induced by the inclusion is an isomorphism.

Each of these theorems relates direct limits defined over different directed sets. To prove them, I will want to rewrite the various direct limits as direct limits over the same directed set. This raises the following ...

Question 35.4. When does $\varphi : \mathcal{J} \to \mathcal{I}$ induce an isomorphism $\varinjlim_{\mathcal{J}} A\varphi \to \varinjlim_{\mathcal{I}} A$?

This is a lot like taking a sequence and a subsequence and asking when they have the same limit. There's a cofinality condition in analysis, that has a similar expression here.

Definition 35.5. $\varphi : \mathcal{J} \to \mathcal{I}$ is *cofinal* if for all $i \in \mathcal{I}$, there exists $j \in \mathcal{J}$ such that $i \leq \varphi(j)$.

For example, any surjective order-preserving map is cofinal. For another example, let $(\mathbb{N}_{>0}, <)$ be the positive integers with their usual order, and $(\mathbb{N}_{>0}, |)$ the same set but with the divisibility order. There is an order-preserving map $\varphi : (\mathbb{N}_{>0}, <) \to (\mathbb{N}_{>0}, |)$ given by $n \mapsto n!$. This map is far from surjective, but any integer n divides some factorial (n divides $n!$, for example), so φ is cofinal. We claimed that both these systems produce $\mathbb{Q}$ as direct limit. Here's why.

Lemma 35.6. *If $\varphi : \mathcal{J} \to \mathcal{I}$ is cofinal then $\varinjlim_{\mathcal{J}} A\varphi \to \varinjlim_{\mathcal{I}} A$ is an isomorphism.*

Proof. Check that $\{A_{\varphi(j)} \to \varinjlim_{\mathcal{I}} A\}$ satisfies the necessary and sufficient conditions to be $\varinjlim_{\mathcal{J}} A\varphi$.

(1) Let $a \in \varinjlim_{\mathcal{I}} A$. We know that there exists some $i \in \mathcal{I}$ and $a_i \in A$ such that $a_i \mapsto a$. Pick j such that $i \leq \varphi(j)$. Let $a_{\varphi(j)}$ be the image of a_i in $A_{\varphi(j)}$. By compatibility, $a_{\varphi(j)} \mapsto a$.

(2) Suppose $a \in A_{\varphi(j)}$ maps to $0 \in \varinjlim_{\mathcal{I}} A$. Then there is some $i \in \mathcal{I}$ such that $\varphi(j) \leq i$ and $a \mapsto 0$ in A_i. But then there is $j' \in \mathcal{J}$ such that $i \leq \varphi(j')$, and $a \mapsto 0 \in A_{\varphi(j')}$ as well. $\qquad\square$

Proof of Theorem 35.2. Let (K, L) be a closed pair in the space X. We have

$$\check{H}^p(K) = \varinjlim_{U \in \mathcal{U}_K} H^p(U), \quad \check{H}^p(L) = \varinjlim_{V \in \mathcal{U}_L} H^p(V),$$

and

$$\check{H}^p(K, L) = \varinjlim_{(U,V) \in \mathcal{U}_{K,L}} H^p(U, V).$$

We can rewrite the entire sequence as the direct limit of a directed system of exact sequences indexed by $\mathcal{U}_{K,L}$, since the order-preserving maps

$$\mathcal{U}_K \leftarrow \mathcal{U}_{K,L} \to \mathcal{U}_L$$

$$U \leftarrow\!\shortmid (U, V) \mapsto V$$

are both surjective and hence cofinal. So the long exact sequence of a pair in Čech cohomology is the direct limit of the system of long exact sequences of the neighborhood pairs (U, V) and so is exact. $\qquad\square$

The proof of the excision theorem depends upon another pair of cofinalities.

Lemma 35.7. *Assume that X is a normal space and A, B closed subsets, or that X is a Hausdorff space and A, B compact subsets. Then the order-preserving maps*

$$\mathcal{U}_{(A \cup B, B)} \leftarrow \mathcal{U}_A \times \mathcal{U}_B \to \mathcal{U}_{(A, A \cap B)}$$

given by

$$(W \cup Y, Y) \leftarrow\!\shortmid (W, Y) \mapsto (W, W \cap Y)$$

are both cofinal.

Proof. The left map is surjective, because if $(U, V) \in \mathcal{U}_{A \cup B, B}$ then $U \in \mathcal{U}_A$, $V \in \mathcal{U}_B$, and $(U, V) = (U \cup V, V)$.

To see that the right map is cofinal, start with $(U, V) \in \mathcal{U}_{A, A \cap B}$.

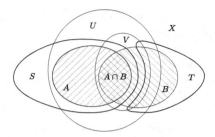

Note that A is disjoint from $B \cap (X - V)$, so by normality, or compactness in a Hausdorff space, there exist non-intersecting open sets S and T with $A \subseteq S$ and $B \cap (X-V) \subseteq T$. Then take $W = U \cap S \in \mathcal{U}_A$ and $Y = V \cup T \in \mathcal{U}_B$, and observe that $W \cap Y = V \cap S$ and so $(W, W \cap Y) \subseteq (U, V)$.   □

Proof of Theorem 35.3. Combine Lemma 35.7 with excision for singular cohomology:

$$
\begin{array}{ccc}
\varinjlim_{(W,Y)\in\mathcal{U}_A\times\mathcal{U}_B} H^p(W \cup Y, Y) & \xrightarrow{\;\cong\;} & \varinjlim_{(W,Y)\in\mathcal{U}_A\times\mathcal{U}_B} H^p(W, W \cap Y) \\
\Big\downarrow{\scriptstyle\cong} & & \Big\downarrow{\scriptstyle\cong} \\
\varinjlim_{(U,V)\in\mathcal{U}_{A\cup B,B}} H^p(U,V) & \longrightarrow & \varinjlim_{(U,V)\in\mathcal{U}_{A,A\cap B}} H^p(U,V) \\
\Big\| & & \Big\| \\
\check{H}^p(A \cup B, B) & \longrightarrow & \check{H}^p(A, A \cap B).
\end{array}
$$

The diagram commutes, and completes the proof.   □

The Mayer-Vietoris long exact sequence is a consequence of these two results.

Corollary 35.8 (Mayer-Vietoris). *Suppose A and B are closed subsets of a normal space, or compact subsets of a Hausdorff space. There is a natural long exact sequence:*

$$\to \check{H}^{p-1}(A \cup B) \to \check{H}^{p-1}(A) \oplus \check{H}^{p-1}(B) \to \check{H}^{p-1}(A \cap B) \to \check{H}^p(A \cup B) \to \,.$$

Proof. Apply Lemma 11.6 to the ladder below.   □

$$
\begin{array}{ccccccccc}
\to & \check{H}^{p-1}(A \cup B) & \to & \check{H}^{p-1}(B) & \to & \check{H}^p(A \cup B, B) & \to & \check{H}^p(A \cup B) & \to & \check{H}^p(B) & \to \\
& \downarrow & & \downarrow & & \downarrow{\scriptstyle\cong} & & \downarrow & & \downarrow & \\
\to & \check{H}^{p-1}(A) & \to & \check{H}^{p-1}(A \cap B) & \to & \check{H}^p(A, A \cap B) & \to & \check{H}^p(A) & \to & \check{H}^p(A \cap B) & \to
\end{array}
$$

36   Fully relative cap product

Čech cohomology appeared as the natural algebra acting on $H^*(X, X - K)$, where K is a closed subspace of X:

$$\cap : \check{H}^p(K) \otimes H_n(X, X - K) \to H_q(X, X - K), \quad p + q = n.$$

If we fix $x_K \in H_n(X, X - K)$, then capping with x_K gives a map

$$- \cap x_K : \check{H}^p(K) \to H_q(X, X - K), \quad p + q = n.$$

We will be very interested in showing that this map is an isomorphism under certain conditions. This is a kind of duality result, comparing cohomology and relative homology in complementary dimensions. We'll try to show that such a map is an isomorphism by embedding it in a map of long exact sequences and using the five-lemma.

For a start, let's think about how these maps vary as we change K. So let L be a closed subset of K, so $X - K \subseteq X - L$ and we get a "restriction map"

$$i_* : H_n(X, X - K) \to H_n(X, X - L).$$

Define x_L as the image of x_K. The diagram

$$
\begin{array}{ccc}
\check{H}^p(K) & \longrightarrow & \check{H}^p(L) \\
{\scriptstyle -\cap x_K} \downarrow & & \downarrow {\scriptstyle -\cap x_L} \\
H_q(X, X - K) & \longrightarrow & H_q(X, X - L)
\end{array}
$$

commutes by the projection formula (Proposition 34.1). This embeds into a ladder shown in the theorem below. We will accompany this ladder with a second one, to complete the picture. We will employ Notation 32.1.

Theorem 36.1 (Fully relative cap product). *Let $L \subseteq K$ be closed subspaces of a space X. There is a "fully relative" cap product*

$$\cap : \check{H}^p(K, L) \otimes H_n(X, X - K) \to H_q(X - L, X - K), \quad p + q = n,$$

such that for any $x_K \in H_n(X, X - K)$ the ladder

$$
\begin{array}{ccccccccc}
\cdots \longrightarrow & \check{H}^p(K, L) & \longrightarrow & \check{H}^p(K) & \longrightarrow & \check{H}^p(L) & \xrightarrow{\delta} & \check{H}^{p+1}(K, L) & \longrightarrow \cdots \\
& \downarrow {\scriptstyle \cap x_K} & & \downarrow {\scriptstyle \cap x_K} & & \downarrow {\scriptstyle \cap x_L} & & \downarrow {\scriptstyle \cap x_K} & \\
\cdots \to & H_q(X - L | K) & \to & H_q(X | K) & \to & H_q(X | L) & \xrightarrow{\partial} & H_{q-1}(X - L | K) & \to \cdots
\end{array}
$$

commutes, where x_L is the restriction of x_K to $H_n(X, X - L)$; and for any $x \in H_n(X)$

$$\to \check{H}^p(X, K) \longrightarrow \check{H}^p(X, L) \longrightarrow \check{H}^p(K, L) \xrightarrow{\ \delta\ } \check{H}^{p+1}(X, K) \to$$

$$\downarrow \cap x \qquad\qquad \downarrow \cap x \qquad\qquad \downarrow \cap x_K \qquad\qquad \downarrow \cap x$$

$$\to H_q(X - K) \to H_q(X - L) \to H_q(X - L, X - K) \xrightarrow{\ \partial\ } H_{q-1}(X - K) \to$$

commutes, where x_K is the restriction of x to $H_n(X|K) = H_n(X, X - K)$.

Proof. What I have to do is define a cap product along the bottom row of the diagram (with $p + q = n$)

$$\check{H}^p(K) \otimes H_n(X, X - K) \xrightarrow{\ \cap\ } H_q(X, X - K)$$

$$\uparrow \qquad\qquad\qquad\qquad\qquad\qquad \uparrow$$

$$\check{H}^p(K, L) \otimes H_n(X, X - K) \xrightarrow{\ \cap\ } H_q(X - L, X - K).$$

This requires going back to the origin of the cap product. Our map $\check{H}^p(K) \otimes H_n(X, X - K) \to H_q(X, X - K)$ came (via excision) from compatible chain maps $S^p(U) \otimes S_n(U, U - K) \to S_q(U, U - K)$ where $U \supseteq K$, defined by $\beta \otimes \sigma \mapsto \beta(\sigma \circ \alpha_p) \cdot (\sigma \circ \omega_q)$. Now given inclusions

$$L \subseteq K$$
$$\cap \qquad \cap$$
$$V \subseteq U$$

we can excise $X - U$ from $X - L$ to see that

$$H_*(U - L, U - K) \xrightarrow{\ \cong\ } H_*(X - L, X - K).$$

So we try to construct a map

$$H^p(U, V) \otimes H_n(U, U - K) \to H_q(U - L, U - K)$$

compatible with $H^p(U) \otimes H_n(U, U - K) \to H_q(U, U - K)$. We can certainly fill in the bottom row of the diagram

$$S^p(U) \otimes S_n(U)/S_n(U - K) \longrightarrow S_q(U)/S_q(U - K)$$

$$\uparrow \qquad\qquad\qquad\qquad\qquad\qquad \uparrow$$

$$S^p(U) \otimes S_n(U - L)/S_n(U - K) \longrightarrow S_q(U - L)/S_q(U - K).$$

We now restrict to the subgroup $S^p(U, V) \subseteq S^p(U)$. Since cochains in $S^p(U, V)$ kill chains in V, we can form the diagram

$$
\begin{array}{ccc}
S^p(U) \otimes S_n(U)/S_n(U - K) & \longrightarrow & S_q(U)/S_q(U - K) \\
\uparrow & & \uparrow \\
S^p(U, V) \otimes (S_n(U - L) + S_n(V))/S_n(U - K) & \longrightarrow & S_q(U - L)/S_q(U - K) \\
\downarrow{\scriptstyle \simeq} & & \\
S^p(U, V) \otimes S_n(U)/S_n(U - K). & &
\end{array}
$$

But $L \subseteq V \subseteq U$, so $(U - L) \cup V = U$, and the locality principle (11.4) then guarantees that $S_n(U - L) + S_n(V) \to S_n(U)$ is a quasi-isomorphism. This gives us maps

$$
H_q(S^*(U, V) \otimes S_*(U, U - K)) \to H_q(U - L, U - K)
$$

and composing with our standard map $\mu : H_*(C) \otimes H_*(D) \to H_*(C \otimes D)$ gives the required maps. They are compatible as we change the neighborhood pair (U, V), and give us the fully relative cap product. We leave the checks of commutativity to the listener. $\qquad \square$

The diagram

$$
\begin{array}{ccc}
\check{H}^p(L) & \xrightarrow{\ \delta\ } & \check{H}^{p+1}(K, L) \\
\downarrow{\scriptstyle -\cap x_L} & & \downarrow{\scriptstyle -\cap x_K} \\
H_q(X, X - L) & \xrightarrow{\ \partial\ } & H_{q-1}(X - L, X - K)
\end{array}
$$

provides us with the memorable formula

$$
(\delta b) \cap x_K = \partial(b \cap x_L) .
$$

The construction of the Mayer-Vietoris sequences now gives:

Theorem 36.2. *Let A, B be closed in a normal space or compact in a Hausdorff space. The Čech cohomology and singular homology Mayer-Vietoris sequences are compatible: for any $x_{A \cup B} \in H_n(X, X - A \cup B)$, there is a commutative ladder (where again we use the notation $H_q(X|A) = H_q(X, X - A)$, and again $p + q = n$)*

$$
\begin{array}{ccccccccc}
\cdots \to & \check{H}^p(A \cup B) & \to & \check{H}^p(A) \oplus \check{H}^p(B) & \to & \check{H}^p(A \cap B) & \to & \check{H}^{p+1}(A \cup B) & \to \cdots \\
 & \downarrow{\scriptstyle \cap x_{A \cup B}} & & \downarrow{\scriptstyle (\cap x_A) \oplus (\cap x_B)} & & \downarrow{\scriptstyle \cap x_{A \cap B}} & & \downarrow{\scriptstyle \cap x_{A \cup B}} & \\
\cdots \to & H_q(X|A \cup B) & \to & H_q(X|A) \oplus H_q(X|B) & \to & H_q(X|A \cap B) & \to & H_{q-1}(X|A \cup B) & \to \cdots
\end{array}
$$

in which the homology classes $x_A, x_B, x_{A\cap B}$ *are restrictions of the class* $x_{A\cup B}$ *in the diagram*

$$
\begin{array}{ccc}
H_n(X, X - A \cup B) & \longrightarrow & H_n(X, X - A) \\
\downarrow & & \downarrow \\
H_n(X, X - B) & \longrightarrow & H_n(X, X - A \cap B).
\end{array}
$$

Exercises

Exercise 36.3. Check commutativity in Theorem 36.1.

37 Poincaré duality

Let M be a n-manifold and K a compact subset. By Theorem 32.2,
$$
j_K : H_n(M, M - K; R) \xrightarrow{\cong} \Gamma(K; o_M \otimes R).
$$
An *R-orientation along* K is a section of $o_M \otimes R$ over K that restricts to a generator of $H_n(M, M - x; R)$ for every $x \in K$. The corresponding class in $H_n(M, M - K; R)$ is a *fundamental class along* K, which we will denote by $[M]_K$. We recall also the fully relative cap product pairing (in which $p + q = n$ and L is a closed subset of K)
$$
\cap : \check{H}^p(K, L; R) \otimes_R H_n(M, M - K; R) \to H_q(M - L, M - K; R).
$$
If $[M]_K$ is a fundamental class along K, its restriction to L is a fundamental class along L, which we will denote by $[M]_L$.

We now combine all of this in the following climactic result.

Theorem 37.1 (Fully relative Poincaré duality). *Let* M *be an n-manifold and* $K \supseteq L$ *a pair of compact subsets. Assume given an R-orientation along* K, *with corresponding fundamental class* $[M]_K$. *With* $p + q = n$, *the map*
$$
- \cap [M]_K : \check{H}^p(K, L; R) \to H_q(M - L, M - K; R)
$$
is an isomorphism.

We have seen that these isomorphisms are compatible; they form the rungs of the ladder

$$
\begin{array}{ccccccccc}
\cdots \longrightarrow & \check{H}^{p-1}(L) & \longrightarrow & \check{H}^p(K, L) & \longrightarrow & \check{H}^p(K) & \longrightarrow & \check{H}^p(L) & \longrightarrow \cdots \\
& \downarrow{\scriptstyle \cap [M]_L} & & \downarrow{\scriptstyle \cap [M]_K} & & \downarrow{\scriptstyle \cap [M]_K} & & \downarrow{\scriptstyle \cap [M]_L} & \\
\cdots \rightarrow & H_{q+1}(M|L) & \rightarrow & H_q(M - L|K) & \rightarrow & H_q(M|K) & \rightarrow & H_q(M|L) & \rightarrow \cdots .
\end{array}
$$

Also, if M is compact and R-oriented with fundamental class $[M]$ restricting along K to $[M]_K$, we have the ladder of isomorphisms

$$\cdots \to \check{H}^p(M,L) \longrightarrow \check{H}^p(K,L) \longrightarrow \check{H}^{p+1}(M,K) \to \check{H}^{p+1}(M,L) \to \cdots$$

$$\downarrow{\scriptstyle\cap[M]} \qquad\qquad \downarrow{\scriptstyle\cap[M]_K} \qquad\qquad \downarrow{\scriptstyle\cap[M]} \qquad\qquad \downarrow{\scriptstyle\cap[M]}$$

$$\cdots \to H_q(M-L) \to H_q(M-L, M-K) \to H_{q-1}(M-K) \to H_{q-1}(M-L) \to \cdots$$

To prove this theorem, we will follow the same five-step process we used to prove the Orientation Theorem 32.2. We have already prepared the Mayer-Vietoris ladder 36.2 for this purpose. We will also need:

Lemma 37.2. *Let $A_1 \supseteq A_2 \supseteq \cdots$ be a decreasing sequence of compact subspaces of M, with intersection A. Then*

$$\varinjlim_k \check{H}^p(A_k) \to \check{H}^p(A)$$

is an isomorphism.

Proof. This follows from the observation that a direct limit of direct limits is a direct limit. □

Proof of Theorem 37.1. By the top ladder and the five-lemma, we may assume $L = \varnothing$; so we want to prove that

$$- \cap [M]_K : \check{H}^p(K;R) \to H_q(M, M-K;R)$$

is an isomorphism.

(1) $M = \mathbb{R}^n$, K a compact convex set. We claim that

$$\check{H}^*(K) \xrightarrow{\cong} H^*(K)$$

(which is of course isomorphic to $H^*(*)$). For any $\epsilon > 0$, let U_ϵ denote the ϵ-neighborhood of K,

$$U_\epsilon = \bigcup_{x \in K} B_\epsilon(x).$$

For any $y \in U_\epsilon$ there is a closest point in K, since the distance function to y is continuous and bounded below on the compact set K and so achieves its infimum. If $x', x'' \in K$ are the same distance from y, then the midpoint of the segment joining x' and x'' is closer, but lies in K since K is convex. So there is a unique closest point, $f(y)$. We let the listener check that $f : U_\epsilon \to K$ is continuous. It is also clear that if $i : K \to U_\epsilon$ is the inclusion then $i \circ f$ is homotopic to the identity on Y, by an affine homotopy.

Now let D^n be a disk centered at the origin and containing the compact set K, and consider the commutative diagram

$$
\begin{array}{ccc}
H^p(K) & \xrightarrow{\ \cap [\mathbb{R}^n]_K\ } & H_q(\mathbb{R}^n, \mathbb{R}^n - K) \\
\uparrow{\scriptstyle\cong} & & \uparrow{\scriptstyle\cong} \\
H^p(D^n) & \xrightarrow{\ \cap [\mathbb{R}^n]_{D^n}\ } & H_q(\mathbb{R}^n, \mathbb{R}^n - D^n) \\
\downarrow{\scriptstyle\cong} & & \downarrow{\scriptstyle\cong} \\
H^p(*) & \xrightarrow{\hspace{3em}} & H_q(\mathbb{R}^n, \mathbb{R}^n - *).
\end{array}
$$

The groups are zero unless $p = 0, q = n$. By naturality of the cap product, the bottom map is given by $1 \mapsto 1 \cap [\mathbb{R}^n]_*$, and this is $[\mathbb{R}^n]_*$ since capping with 1 is the identity. This fundamental class is a generator of $H_n(\mathbb{R}^n, \mathbb{R}^n - *)$, so the top map in the diagram is an isomorphism.

(2) K a finite union of compact convex subsets of $\mathbb{R}^n$. This follows by induction and the five lemma applied to the Mayer-Vietoris ladder 36.2.

(3) K is any compact subset of $\mathbb{R}^n$. This follows as before by a limit argument, using Lemmas 32.5 and 37.2.

(4) M arbitrary, K is a finite union of compact Euclidean subsets of M. This follows from (3) and Theorem 36.2.

(5) M arbitrary, K an arbitrary compact subset. This follows just as in the proof of Theorem 32.2. $\square$

Let me point out some special cases. With $K = M$, we get:

Corollary 37.3. *Suppose that M is a compact R-oriented n-manifold, and let L be a closed subset. Then (with $p + q = n$) we have the commuting ladder whose rungs are isomorphisms:*

$$
\begin{array}{ccccccccc}
\longrightarrow & \check{H}^{p-1}(L) & \longrightarrow & \check{H}^p(M, L) & \longrightarrow & H^p(M) & \longrightarrow & \check{H}^p(L) & \longrightarrow \\
& \downarrow{\scriptstyle\cap [M]_L} & & \downarrow{\scriptstyle\cap [M]} & & \downarrow{\scriptstyle\cap [M]} & & \downarrow{\scriptstyle\cap [M]_L} & \\
\longrightarrow & H_{q+1}(M, M-L) & \longrightarrow & H_q(M-L) & \longrightarrow & H_q(M) & \longrightarrow & H_q(M, M-L) & \longrightarrow
\end{array}
$$

With $L = \varnothing$, we get:

Corollary 37.4. *Suppose that M is an n-manifold, and let K be a compact subset. An R-orientation $[M]_K$ along K determines (with $p + q = n$) an isomorphism*

$$
-\cap [M]_K : \check{H}^p(K; R) \to H_q(M, M - K; R).
$$

The intersection of these two special cases is:

Corollary 37.5 (Poincaré duality). *Let M be a compact R-oriented n-manifold. Then (with $p + q = n$)*

$$- \cap [M] : H^p(M; R) \to H_q(M; R)$$

is an isomorphism.

Lefschetz duality

Continuing the theme that everything is more useful in a relative form, we end by stating a relative version of Poincaré duality. It will use the notion of a manifold-with-boundary. In the definition, a "Euclidean half-space" is a topological space of the form $\mathbb{R}^{n-1} \times [0, \infty)$.

Definition 37.6. A *manifold-with-boundary* is a Hausdorff space such that every point has a neighborhood that is homeomorphic to either a Euclidean space or a Euclidean half-space.

The *boundary* of a manifold-with-boundary M of dimension n is the set of points that do not admit a Euclidean neighborhood; in any Euclidean half-space neighborhood they lie on $\mathbb{R}^{n-1} \times \{0\} \subseteq \mathbb{R}^{n-1} \times [0, \infty)$. This subspace is written ∂M. It is an $(n-1)$-manifold, compact if M is compact. It may be empty, in which case M is a manifold. The complement of the boundary is the *interior*. It is an n-manifold, non-compact if $\partial M \neq \varnothing$.

An R-orientation of a manifold-with-boundary is an orientation of the interior.

Theorem 37.7 (Lefschetz duality, e.g. [72, Theorem 18.6.1] or [10, VI Theorem 9.2]). *Let M be an R-oriented compact manifold-with-boundary of dimension n. There is a unique class $[M] \in H_n(M, \partial M; R)$ that maps to the orientation class in $H_n(M, M - x)$ for every $x \in M - \partial M$. The boundary $\partial[M] \in H_{n-1}(\partial M; R)$ serves as a fundamental class $[\partial M]$ for the $(n-1)$-manifold ∂M. There are cap product maps making the following ladder commute (at least up to sign), with $p + q = n$.*

$$\cdots \to H^p(M, \partial M) \longrightarrow H^p(M) \longrightarrow H^p(\partial M) \to H^{p+1}(M, \partial M) \to \cdots$$
$$\downarrow \cap [M] \qquad \downarrow \cap [M] \qquad \downarrow \cap [\partial M] \qquad \downarrow \cap [M]$$
$$\cdots \longrightarrow H_q(M) \longrightarrow H_q(M, \partial M) \to H_{q-1}(\partial M) \longrightarrow H_{q-1}(M) \longrightarrow \cdots$$

Exercises

Exercise 37.8. Prove this theorem. Begin by attaching a "collar" to M: form $P = M \cup_{\partial M} (\partial M \times I)$ where ∂M is embedded into $\partial M \times I$ along $t = 0$.

38　Applications

Today we harvest consequences of Poincaré duality. We'll use the form

Theorem 38.1. *Let M be an n-manifold and K a compact subset. An R-orientation along K determines a fundamental class $[M]_K \in H_n(M, M - K)$, and capping gives an isomorphism:*

$$- \cap [M]_K : \check{H}^{n-q}(K; R) \xrightarrow{\cong} H_q(M, M - K; R).$$

Corollary 38.2. $\check{H}^p(K; R) = 0$ *for $p > n$.*

We can contrast this with singular (co)homology. Here's an example:

Example 38.3 (Barratt-Milnor [7]). Let K be a two-dimensional version of the "Hawaiian earring," i.e., nested spheres all tangent to a point whose radii are going to zero. What they proved is that $H_q(K; \mathbb{Q})$ is uncountable for every $q > 1$. But Čech cohomology is much more well-behaved:

Theorem 38.4 (Alexander duality). *For any compact subset K of $\mathbb{R}^n$, the composite*

$$\check{H}^{n-q}(K; R) \xrightarrow{\cap [\mathbb{R}^n]_K} H_q(\mathbb{R}^n, \mathbb{R}^n - K; R) \xrightarrow{\partial} \widetilde{H}_{q-1}(\mathbb{R}^n - K; R)$$

is an isomorphism.

Proof. $\widetilde{H}^*(\mathbb{R}^n; R) = 0.$ $\qquad\square$

This is extremely useful! For example

Corollary 38.5. *If K is a compact subset of $\mathbb{R}^n$ then $\check{H}^n(K; R) = 0$.*

Corollary 38.6. *The complement of a knot in S^3 is a homology circle.*

Example 38.7. Take the case $q = 1$:

$$\check{H}^{n-1}(K; R) \xrightarrow{\cong} \widetilde{H}_0(\mathbb{R}^n - K; R) = \ker(\varepsilon : R\pi_0(\mathbb{R}^n - K) \to R).$$

The augmentation is a split surjection, so this is a free R-module. This shows, for example, that $\mathbb{R}\mathbb{P}^2$ can't be embedded in $\mathbb{R}^3$ — at least not with a regular neighborhood.

If we take $n = 2$ and suppose that $\check{H}^*(K) = H^*(S^1)$, we find that the complement of K has *two* path components. This is the *Jordan Curve Theorem*.

Perfect pairings

There is a useful purely cohomological consequence of Poincaré duality, obtained by combining it with the Universal Coefficient Theorem 27.1

$$0 \to \mathrm{Ext}^1_{\mathbb{Z}}(H_{q-1}(X), \mathbb{Z}) \to H^q(X) \to \mathrm{Hom}(H_q(X), \mathbb{Z}) \to 0 \, .$$

First, note that $\mathrm{Hom}(H_q(X), \mathbb{Z})$ is always torsion-free. If we assume that $H_{q-1}(X)$ is finitely generated, then $\mathrm{Ext}^1_{\mathbb{Z}}(H_{q-1}(X), \mathbb{Z})$ is a finite abelian group. So the UCT implies that

$$H^q(X)/\mathrm{tors} \xrightarrow{\cong} \mathrm{Hom}(H_q(X)/\mathrm{tors}, \mathbb{Z}) \, .$$

That is to say, the Kronecker pairing descends to a perfect pairing

$$\frac{H^q(X)}{\mathrm{tors}} \otimes \frac{H_q(X)}{\mathrm{tors}} \to \mathbb{Z} \, .$$

Let's combine this with Poincaré duality. Let $X = M$ be a compact oriented n-manifold, so that

$$- \cap [M] : H^{n-q}(M) \xrightarrow{\cong} H_q(M) \, .$$

We get a perfect pairing

$$\frac{H^q(X)}{\mathrm{tors}} \otimes \frac{H^{n-q}(X)}{\mathrm{tors}} \to \mathbb{Z} \, .$$

And what is that pairing? It's given by the composite

$$
\begin{array}{ccc}
H^q(M) \otimes H^{n-q}(M) & \longrightarrow & \mathbb{Z} \\
{\scriptstyle 1 \otimes (-\cap [M])} \Big\downarrow & \nearrow {\scriptstyle \langle -, - \rangle} & \\
H^q(M) \otimes H_q(M) & &
\end{array}
$$

and we've seen (Proposition 34.1) what this is:

$$\langle a, b \cap [M] \rangle = \langle a \cup b, [M] \rangle \, .$$

We have used $R = \mathbb{Z}$, but the same argument works for any PID — in particular for any field, in which case $\mathrm{tors}V = 0$. We have proven:

Theorem 38.8 (Poincaré duality, cohomological form). *Let R be a PID and M a compact R-oriented n-manifold. Then*

$$a \otimes b \mapsto \langle a \cup b, [M] \rangle$$

induces a perfect pairing (with $p + q = n$)

$$\frac{H^p(M; R)}{\text{tors}} \otimes_R \frac{H^q(M; R)}{\text{tors}} \to R.$$

Corollary 38.9. *Let M be a compact manifold. Then $H_*(M; \mathbb{F}_2)$ is of finite type, and if M is oriented so is $H_*(M; k)$ for any field k.*

Proof. If M is k-oriented, then $H^p(M; k)$ is the dual of $H^q(M; k)$ and $H^q(M; k)$ is the dual of $H^p(M; k)$, so $H^p(M; k)$ is its own double dual. This implies that it is finite dimensional, and so the homology is too. □

Remark 38.10. In fact a compact manifold has the homotopy type of a finite CW complex [78].

Example 38.11. The complex projective plane $\mathbb{CP}^2$ is a compact 4-manifold, orientable since it is simply connected. It has a cell structure with cells in dimensions 0, 2, and 4, so its homology is $\mathbb{Z}$ in those dimensions and 0 elsewhere, and so the same is true of its cohomology. Up till now the cup product structure has been a mystery. But now we know that

$$H^2(\mathbb{CP}^2) \otimes H^2(\mathbb{CP}^2) \to H^4(\mathbb{CP}^2) \xrightarrow{\cong} \mathbb{Z}$$

is a perfect pairing. So if we write b for a generator of $H^2(\mathbb{CP}^2)$, then $b \cup b = b^2$ is a free generator for $H^4(\mathbb{CP}^2)$. We have discovered that

$$H^*(\mathbb{CP}^2) = \mathbb{Z}[b]/b^3.$$

By the way, notice that if we had chosen $-b$ as a generator, we would still produce the same generator for $H^4(\mathbb{CP}^2)$: so there is a preferred orientation, the one whose fundamental class pairs to 1 against b^2.

This calculation shows that while $\mathbb{CP}^2$ and $S^2 \vee S^4$ are both simply connected and have the same homology, they are not homotopy equivalent. This implies that the attaching map $S^3 \to S^2$ for the top cell in $\mathbb{CP}^2$ — the *Hopf map* — is essential.

How about $\mathbb{CP}^3$? It just adds a 6-cell, so now $H^6(\mathbb{CP}^3) \cong \mathbb{Z}$. The pairing $H^2(\mathbb{CP}^3) \otimes H^4(\mathbb{CP}^3) \to H^6(\mathbb{CP}^3) = \mathbb{Z}$ is perfect, so we find that b^3 generates $H^6(\mathbb{CP}^3)$. Continuing in this way, we have

$$H^*(\mathbb{CP}^n) = \mathbb{Z}[b]/(b^{n+1}).$$

Example 38.12. Exactly the same argument shows that

$$H^*(\mathbb{RP}^n; \mathbb{F}_2) = \mathbb{F}_2[a]/(a^{n+1})$$

where $|a| = 1$.

I'll end with the following application.

Theorem 38.13 (Borsuk-Ulam). *Think of S^n as the space of unit vectors in $\mathbb{R}^{n+1}$. For any continuous function $f : S^n \to \mathbb{R}^n$, there exists $x \in S^n$ such that $f(x) = f(-x)$.*

Proof. Suppose that no such x exists. Then we may define a continuous function $g : S^n \to S^{n-1}$ by

$$g : x \mapsto \frac{f(x) - f(-x)}{\|f(x) - f(-x)\|}.$$

Note that $g(-x) = -g(x)$: g is equivariant with respect to the antipodal action. It descends to a map $\overline{g} : \mathbb{RP}^n \to \mathbb{RP}^{n-1}$.

We claim that $\overline{g}_* : H_1(\mathbb{RP}^n) \to H_1(\mathbb{RP}^{n-1})$ is nontrivial. To see this, pick a basepoint $b \in S^n$ and choose a 1-simplex $\sigma : \Delta^1 \to S^n$ such that $\sigma(e_0) = b$ and $\sigma(e_1) = -b$. The group $H_1(\mathbb{RP}^n)$ is generated by the class of the cycle $p\sigma$, where $p : S^n \to \mathbb{RP}^n$ is the covering map. The image of this cycle in $H_1(\mathbb{RP}^{n-1})$ is represented by the loop $\overline{g}p\sigma$ at $\overline{b} = pb$, which is the image of the 1-simplex $g\sigma$ in S^{n-1} joining gb to $g(-b) = -g(b)$. The class of this 1-simplex thus generates $H_1(\mathbb{RP}^{n-1})$.

Therefore $\overline{g}$ is nontrivial in $H_1(-; \mathbb{F}_2)$, and hence also in $H^1(-; \mathbb{F}_2)$. Writing a_n for the generator of $H^1(\mathbb{RP}^n; \mathbb{F}_2)$, we must have $a_n = g^* a_{n-1}$, and consequently $a_n^n = (g^* a_{n-1})^n = g^*(a_{n-1}^n)$. But $H^n(\mathbb{RP}^{n-1}; \mathbb{F}_2) = 0$, so $a_{n-1}^n = 0$; while $a_n^n \neq 0$. This is a contradiction. $\square$

Exercises

Exercise 38.14. Let M be a compact connected n-manifold. Show that if M is orientable then $H_{n-1}(M)$ is free abelian, while if it is not orientable there is a unique nonzero element of finite order in $H_{n-1}(M)$.

Chapter 4

Basic homotopy theory

39 Limits, colimits, and adjunctions

Limits and colimits

I want to begin by developing a little more category theory. I still refer to the classic text *Categories for the Working Mathematician* by Saunders Mac Lane [34] for this material.

Definition 39.1. Suppose $\mathcal{I}$ is a small category (so that it has a *set* of objects), and let $\mathcal{C}$ be another category. Let $X : \mathcal{I} \to \mathcal{C}$ be a functor. A *cone under X* is a natural transformation e from X to a constant functor; to be explicit, this means that for every object i of $\mathcal{I}$ we have a map $e_i : X_i \to Y$, and these maps are compatible in the sense that for every $f : i \to j$ in $\mathcal{I}$ the following diagram commutes:

$$
\begin{array}{ccc}
X_i & \xrightarrow{\ e_i\ } & Y \\
{\scriptstyle f_*}\downarrow & & \downarrow{\scriptstyle =} \\
X_j & \xrightarrow{\ e_j\ } & Y
\end{array}
$$

A *colimit* of X is an initial cone (L, t_i) under X; to be explicit, this means that for any cone (Y, e_i) under X, there exists a unique map $h : L \to Y$ such that $h \circ t_i = e_i$ for all i.

Any two colimits are isomorphic by a unique isomorphism compatible with the structure maps; but existence is another matter. Also, as always for category theoretic concepts, some examples are in order.

Example 39.2. If $\mathcal{I}$ is a discrete category (that is, the only maps are identity maps; $\mathcal{I}$ is entirely determined by its set of objects), the colimit of a functor $\mathcal{I} \to \mathcal{C}$ is the coproduct in $\mathcal{C}$ (if this coproduct exists!).

Example 39.3. In Lecture 23 we discussed directed posets and the direct limit of a directed system $X : \mathcal{I} \to \mathcal{C}$. The colimit simply generalizes this to arbitrary indexing categories rather than restricting to directed partially ordered sets.

Example 39.4. Let G be a group; we can view this as a category with one object, where the morphisms are the elements of the group and composition is given by the group structure. If $\mathcal{C}$ is the category of topological spaces, a functor $G \to \mathcal{C}$ is simply a group action on a topological space X. The colimit of this functor is the orbit space of the G-action on X (together with the projection map to the orbit space).

Similarly, a functor from G into vector spaces over a field k is a representation of G on a vector space. Question for you: What is the colimit in this case?

Example 39.5. Let $\mathcal{I}$ be the category whose objects and non-identity morphisms are described by the following directed graph:

$$b \leftarrow a \to c.$$

The colimit of a diagram $\mathcal{I} \to \mathcal{C}$ is called a *pushout*. With $\mathcal{C} = \mathbf{Top}$, again, a functor $\mathcal{I} \to \mathcal{C}$ is determined by a diagram of spaces:

$$B \xleftarrow{f} A \xrightarrow{g} C.$$

The colimit of such a functor is just given by $B \cup_A C = B \sqcup C / \sim$, where $f(a) \sim g(a)$ for all $a \in A$. We have already seen this in action before: a special case of this construction appears in the process of attaching cells to build up a CW-complex.

If $\mathcal{C}$ is the category of groups, instead, the colimit of such a functor is the free product quotiented out by a certain relation; this is called the *amalgamated free product*.

Example 39.6. Suppose $\mathcal{I}$ is the category with two objects and two parallel morphisms:

$$a \xrightarrow{\quad\quad} b.$$

The colimit of a diagram $\mathcal{I} \to \mathcal{C}$ is called the *coequalizer* of the diagram. If $\mathcal{C} = \mathbf{Top}$, the coequalizer of $f, g : A \rightrightarrows B$ is the quotient of B by the equivalence relation generated by $f(a) \sim g(a)$ for $a \in A$.

One can also consider cones *over* a diagram $X : \mathcal{I} \to \mathcal{C}$: this is simply a cone in the opposite category.

Definition 39.7. The *limit* of a diagram $X : \mathcal{I} \to \mathcal{C}$ is a terminal object in cones over X.

Definition 39.8. A category $\mathcal{C}$ is *cocomplete* if all functors from small categories to $\mathcal{C}$ have colimits. Similarly, $\mathcal{C}$ is *complete* if all functors from small categories to $\mathcal{C}$ have limits.

All the large categories we typically deal with are both cocomplete and complete; in particular both **Set** and **Top** are, as well as algebraic categories like **Gp** and **Mod**$_R$.

Adjoint functors

The notion of a colimit is a special case of the more general concept of an adjoint functor, as long as we are dealing with a cocomplete category.

Let's write $\mathcal{C}^{\mathcal{I}}$ for the category of functors from $\mathcal{I}$ to $\mathcal{C}$, and natural transformations between them. There is a functor $c : \mathcal{C} \to \mathcal{C}^{\mathcal{I}}$, given by sending any object to the constant functor taking on that value. The process of taking the colimit of a diagram supplies us with a functor $\mathrm{colim}_{\mathcal{I}} : \mathcal{C}^{\mathcal{I}} \to \mathcal{C}$. (To be precise, we pick a specific colimit for each diagram, and then observe that a natural transformation of diagrams canonically defines a morphism between the corresponding colimits; and that these morphisms compose correctly.) We can characterize this functor via the formula

$$\mathcal{C}(\mathrm{colim}_{i \in \mathcal{I}} X_i, Y) = \mathcal{C}^{\mathcal{I}}(X, c_Y),$$

where X is any functor from $\mathcal{I}$ to $\mathcal{C}$, Y is any object of $\mathcal{C}$, and c_Y denotes the constant functor with value Y. This formula is reminiscent of the adjunction operation in linear algebra, and is in fact our first example of a category-theoretic adjunction.

Definition 39.9. Let $\mathcal{C}, \mathcal{D}$ be categories, and suppose given functors $F : \mathcal{C} \to \mathcal{D}$ and $G : \mathcal{D} \to \mathcal{C}$. An *adjunction between F and G* is an isomorphism

$$\mathcal{D}(FX, Y) = \mathcal{C}(X, GY)$$

that is natural in X and Y. In this situation, we say that F is a *left adjoint* of G and G is a *right adjoint* of F.

This notion was invented by the late MIT Professor Dan Kan, in 1958 [30]. The name was suggested by Kan's thesis advisor Sammy Eilenberg.

We've already seen one example of adjoint functors. Here is another one.

Example 39.10 (Free groups). There is a forgetful functor $u : \mathbf{Grp} \to \mathbf{Set}$. Any set X gives rise to a group FX, the free group on X. It is determined by a universal property: For any group Γ, set maps $X \to u\Gamma$ are the same as group homomorphisms $FX \to \Gamma$. This is exactly saying that the free group functor the left adjoint to the forgetful functor u.

In general, "free objects" come from left adjoints of forgetful functors.

As a general notational practice, try to write the left adjoint as the top arrow:

$$F : \mathcal{C} \rightleftarrows \mathcal{D} : G \qquad \text{or} \qquad G : \mathcal{D} \leftrightarrows \mathcal{C} : F .$$

These examples suggest that if a functor G has a left adjoint then any two left adjoints are canonically isomorphic. This is true and easily checked. We'll always speak of *the* left adjoint, or *the* right adjoint.

Lemma 39.11. *Suppose that*

$$\mathcal{C} \overset{F}{\underset{G}{\rightleftarrows}} \mathcal{D} \overset{F'}{\underset{G'}{\rightleftarrows}} \mathcal{E}$$

is a composable pair of adjoint functors. Then $F'F, GG'$ form an adjoint pair.

Proof. Compute:

$$\mathcal{E}(F'FX, Z) = \mathcal{D}(FX, G'Y) = \mathcal{C}(X, GG'Y) . \qquad \square$$

Proposition 39.12. *Let $F : \mathcal{C} \to \mathcal{D}$ be a functor. If F admits a right adjoint then it preserves colimits, in the sense that if $X : \mathcal{I} \to \mathcal{C}$ is a diagram in $\mathcal{C}$ with colimit cone $X \to c_L$, then $F \circ X \to F(c_L)$ is a colimit cone in $\mathcal{D}$. Dually, if F admits a left adjoint then it preserves limits.*

Proof. The key observation is that an adjoint pair $F : \mathcal{C} \rightleftarrows \mathcal{D} : G$ induces an adjoint pair on functor categories. So we can compute (using $Gc_Y = c_{GY}$):

$$\mathcal{D}^I(FX, c_Y) = \mathcal{C}^I(X, c_{GY}) = \mathcal{C}(L, GY) = \mathcal{D}(FL, Y) . \qquad \square$$

For example, the free group on a disjoint union of sets is the free product of the two groups (which is the coproduct in the category of groups). The dual statement says, for example, that the product (in the category of groups) of groups is a group structure on the product of their underlying sets.

The Yoneda lemma

An important and rather Wittgensteinian principle in category theory is that an object is determined by the collection of all maps out of it. The Yoneda lemma is a way of making this precise. Observe that for any $X \in \mathcal{C}$ the association $Y \mapsto \mathcal{C}(X, Y)$ gives us a functor $\mathcal{C} \to \mathbf{Set}$. This functor is said to be *corepresented* by X. Suppose that $G : \mathcal{C} \to \mathbf{Set}$ is any functor. An element $x \in G(X)$ determines a natural transformation

$$\theta_x : \mathcal{C}(X, -) \to G$$

in the following way. Let $Y \in \mathcal{C}$ and $f : X \to Y$, and define

$$\theta_x(f) = f_*(x) \in G(Y).$$

Lemma 39.13 (Yoneda Lemma). *The association $x \mapsto \theta_x$ provides a bijection*

$$G(X) \xrightarrow{\cong} \mathrm{nt}(\mathcal{C}(X, -), G).$$

Proof. The inverse sends a natural transformation $\theta : \mathcal{C}(X, -) \to G$ to $\theta_X(1_X) \in G(X)$. $\qquad\qquad\square$

In particular, if G is also corepresentable — $G = \mathcal{C}(Y, -)$, say — then

$$\mathrm{nt}(\mathcal{C}(X, -), \mathcal{C}(Y, -)) \cong \mathcal{C}(Y, X).$$

That is, each natural transformation $\mathcal{C}(X, -) \to \mathcal{C}(Y, -)$ is induced by a unique map $Y \to X$. Consequently any natural isomorphism $\mathcal{C}(X, -) \xrightarrow{\cong} \mathcal{C}(Y, -)$ is induced by a unique isomorphism $Y \xrightarrow{\cong} X$.

Exercises

Exercise 39.14. Revisit the examples provided above: what is the limit of each diagram? For instance, a product is a limit over a discrete category, and the limit of a group action is just the fixed points. If the indexing category is the opposite of a directed poset, the limit is called the *inverse limit* and may be denoted $\varprojlim$. A diagram indexed by the category $b \to a \leftarrow c$ is a diagram $B \xrightarrow{f} A \xleftarrow{g} C$, and its limit is the "pullback," denoted $B \times_A C$. In $\mathbf{Set}$, or $\mathbf{Top}$,

$$B \times_A C = \{(b, c) \in B \times C : f(b) = g(c) \in A\}.$$

Exercise 39.15. Show that any limit in a complete category can be expressed as an equalizer of two maps between products.

Exercise 39.16. Let $\mathcal{C}$ be a cocomplete category. Carry out the construction of a functor $\mathrm{colim}_{\mathcal{I}} : \mathcal{C}^{\mathcal{I}} \to \mathcal{C}$ suggested above, and show that any two choices of such a functor are canonically isomorphic.

Exercise 39.17. Let $\mathcal{C}$ and $\mathcal{D}$ be two categories and $F : \mathcal{C} \to \mathcal{D}$ and $G : \mathcal{D} \to \mathcal{C}$ two functors. An adjunction between F and G is an isomorphism

$$\mathcal{D}(FX, Y) \cong \mathcal{C}(X, GY)$$

that is natural in both variables. Show that this is equivalent to giving natural transformations

$$\alpha_X : X \to GFX, \quad \beta_Y : FGY \to Y,$$

such that the following two diagrams commute.

The map α is the "unit" of the adjunction, and β is the "counit." They are called "adjunction morphisms."

Exercise 39.18. Suppose that F and F' are both left adjoint to $G : \mathcal{D} \to \mathcal{C}$. Show that there is a unique natural isomorphism $F \to F'$ that is compatible with the adjunction morphisms.

40   Cartesian closure and compactly generated spaces

The category of topological spaces has a lot to recommend it, but it does not accommodate certain constructions from algebraic topology gracefully. For example, the product of two CW complexes may fail to have a CW structure. (This is a classic example due to Clifford Dowker, 1952, nicely explained in [24, Appendix]. The CW complexes involved are one-dimensional!) This is closely related to the observation that if $X \to Y$ is a quotient map, the induced map $W \times X \to W \times Y$ may fail to be a quotient map.

It turns out that these problems can be avoided by working in a carefully designed subcategory of **Top**, the category k**Top** of "compactly generated spaces." The key idea is that the unwanted behavior of **Top** is related to the fact that there isn't a well-behaved topology on the set of continuous maps between two spaces. The compact-open topology is available to us —

and we'll recall it later. But it suffers from some defects. To clarify how a mapping object should behave in an ideal world, I want to make another category-theoretic digression. Again, Mac Lane's book [34] is a good reference.

Cartesian closure

How *should* function objects behave? In the category **Set**, for example, the set of maps from X to Y can be characterized by the natural bijection
$$\mathbf{Set}(W \times X, Y) = \mathbf{Set}(W, \mathbf{Set}(X, Y))$$
under which $f : W \times X \to Y$ corresponds to $w \mapsto (x \mapsto f(w, x))$ and $g : W \to \mathbf{Set}(X, Y)$ corresponds to $(w, x) \mapsto g(w)(x)$. This suggests the following definition.

Definition 40.1. Let $\mathcal{C}$ be a category with finite products. It is *Cartesian closed* if for any object X in $\mathcal{C}$, the functor $- \times X$ has a right adjoint.

We'll write the right adjoint to $- \times X$ using exponential notation,
$$Y \mapsto Y^X,$$
so that there is a bijection natural in the pair (W, Y):
$$\mathcal{C}(W \times X, Y) = \mathcal{C}(W, Y^X).$$
In a Cartesian closed category, Y^X serves as a "mapping object" from X to Y. Let me convince you that this is reasonable. Take $Y = W \times X$: the identity map on $W \times X$ then corresponds to a map
$$\eta_W : W \to (W \times X)^X.$$
Take $W = Y^X$: the identity map $Y^X \to Y^X$ corresponds to a map
$$\varepsilon_Y : Y^X \times X \to Y.$$
These maps are natural transformations. In the example of **Set**, the first is given by
$$w \mapsto (x \mapsto (w, x)), \quad \text{inclusion of a slice},$$
and the second is given by
$$(f, x) \mapsto f(x), \quad \text{evaluation}.$$
Here are some direct consequences of Cartesian closure. Note: the assumption that finite products exist in $\mathcal{C}$ includes the case in which the indexing set is empty, in which case the universal property of the product characterizes the terminal object of $\mathcal{C}$, which thus exist in a Cartesian closed category. We'll denote it by $*$. You might call $\mathcal{C}(*, X)$ the "set of points" in X.

Proposition 40.2. *Let C be Cartesian closed.*
(1) $(X, Z) \mapsto Z^X$ *extends canonically to a functor* $C^{op} \times C \to C$, *and the bijection* $C(X \times Y, Z) = C(Y, Z^X)$ *is natural in all three variables.*
(2) $C(X, Z) = C(*, Z^X)$.
(3) $X \times -$ *preserves colimits: If* $Y : \mathcal{I} \to C$ *has a colimit, then the natural map* $X \times Y \to X \times \operatorname{colim} Y$ *is a colimit cone.*
(4) $-^X$ *preserves limits: if* $Z : \mathcal{I} \to C$ *has a limit, then the natural map* $(\lim Z)^X \to (Z^X)$ *is a limit cone.*

Many otherwise well-behaved categories are not Cartesian closed. A category is *pointed* if it has an initial object $\varnothing$ and a final object $*$, and the unique map $\varnothing \to *$ is an isomorphism. There are many pointed categories! — abelian groups **Ab** and groups **Gp**, for example. By (2), the only way a pointed category can be Cartesian closed is if there is exactly one map between any two objects.

There are deep connections between Cartesian closure and the type theory of computer science; see [52] for example.

k-spaces

The category **Top** is not Cartesian closed. We can see this using the observation (e.g. [51, p. 143]) that if $X \to Y$ is a quotient map, the induced map $W \times X \to W \times Y$ may fail to be a quotient map. We can characterize quotient maps in **Top** categorically using the following definition.

Definition 40.3. An *effective epimorphism* in a category C is a map $X \to Y$ in C such that the pullback $X \times_Y X$ exists and the map $X \to Y$ is the coequalizer of the two projection maps $X \times_Y X \to X$.

It's easy to check that in a Cartesian closed category, if $X \to Y$ is an effective epimorphism then so is $W \times X \to W \times Y$.

Lemma 40.4. *A map in* **Top** *is a quotient map if and only if it is an effective epimorphism.*

So, sadly, **Top** is not Cartesian closed.
On the other hand, Henry Whitehead showed that crossing with a locally compact Hausdorff space *does* preserve quotient maps. (See e.g. [51, pp. 186 and 289].) This will often suffice, but often not: for example CW complexes may fail to be locally compact. And the convenience of working in a Cartesian closed category is compelling.

Inspired by Whitehead's theorem, we agree to accept only properties of a space that can be observed by mapping compact Hausdorff spaces into it.

Definition 40.5. Let X be a space. A subspace $F \subseteq X$ is said to be *compactly closed*, or k-closed, if for any map $k : K \to X$ from a compact Hausdorff space K the preimage $k^{-1}(F) \subseteq K$ is closed.

It is clear that any closed subset is compactly closed, but there might be compactly closed sets that are not closed in the topology on X. This motivates the definition of a k-space:

Definition 40.6. A topological space X is *compactly generated* or is a *k-space* if every compactly closed set is closed.

The k comes from the German "*kompact*," though it might have referred to the general topologist John Kelley, who explored this condition.

A more categorical characterization of this property is: X is compactly generated if and only if a map $X \to Y$ is continuous precisely when for every compact Hausdorff space K and map $k : K \to X$ the composite $K \to X \to Y$ is continuous. For instance, compact Hausdorff spaces are k-spaces. First countable spaces (so for example metric spaces) and CW complexes are also k-spaces.

While not all topological spaces are k-spaces, any space can be "k-ified." The procedure is simple: endow the underlying set of a space X with an new topology, one for which the closed sets are precisely the sets that are compactly closed with respect to the original topology. You should check that this is indeed a topology on X. The resulting topological space is denoted kX. This construction immediately implies that the identity $kX \to X$ is continuous, and is the terminal map to X from a k-space.

Let $k\mathbf{Top}$ be the category of k-spaces, as a full subcategory of $\mathbf{Top}$. We will write $j : k\mathbf{Top} \to \mathbf{Top}$ for the inclusion functor. The process of k-ification gives a functor $k : \mathbf{Top} \to k\mathbf{Top}$ with the property that

$$k\mathbf{Top}(X, kY) = \mathbf{Top}(jX, Y).$$

This is another example of an adjunction! In this case the unit $\eta : X \to kjX$ is a homeomorphism.

We can conclude from this that limits in $k\mathbf{Top}$ may be computed by k-ifying limits in $\mathbf{Top}$: For any functor $X : \mathcal{I} \to k\mathbf{Top}$,

$$\lim{}^{k\mathbf{Top}} X \xrightarrow{\cong} \lim{}^{k\mathbf{Top}} kjX \xleftarrow{\cong} k \lim{}^{\mathbf{Top}} jX.$$

The second map is an isomorphism because k is a right adjoint. In particular, the product in $k\mathbf{Top}$ is formed by k-ifying the product in $\mathbf{Top}$.

Similarly, colimit (in $k\mathbf{Top}$) of any diagram of k-spaces can be computed by k-ifying the colimit in $\mathbf{Top}$:

$$\operatorname{colim}^{k\mathbf{Top}} X \xrightarrow{\cong} kj \operatorname{colim}^{k\mathbf{Top}} X \xleftarrow{\cong} k \operatorname{colim}^{\mathbf{Top}} jX \,.$$

The second map is an isomorphism because j is a left adjoint.

The category $k\mathbf{Top}$ has good categorical properties inherited from $\mathbf{Top}$: it is a complete and cocomplete category. In fact it has even better categorical properties than $\mathbf{Top}$ does:

Proposition 40.7. *The category $k\mathbf{Top}$ is Cartesian closed.*

Proof. See [21, 65]. □

I owe you a description of the mapping object Y^X. It consists of the set of continuous maps from X to Y endowed with a certain topology. For general topological spaces X and Y, the set $\mathbf{Top}(X, Y)$ can be given the "compact-open topology": a basis for open sets for the compact-open topology is given by

$$V(F, U) = \{f : X \to Y : f(F) \subseteq U\},$$

where F runs over compact subsets of X and U runs over open subsets of Y. This space is not generally compactly generated, however, and does not serve as a right adjoint to the product.

If X and Y are k-spaces, it's natural to make a slight modification: To start with, replace the compact subsets F in this definition by "k-compact" subsets, that is, subsets that are compact from the perspective of compact Hausdorff spaces: A subset $F \subseteq X$ is k-compact if there exists a compact Hausdorff space K and a map $k : K \to X$ such that $k(K) = F$. This is to overcome the sad fact that there are compact spaces that do not accept surjections from compact Hausdorff spaces.

The sets $V(F, U)$ where F runs over k-compact subsets of X and U runs over open subsets of Y form the basis of a new topology on $\mathbf{Top}(X, Y)$. Even if we assume that X and Y are k-spaces, this new topology may not be compactly generated. But we know what to do: k-ify it. This defines a k-space Y^X, and this turns out to witness the fact that $k\mathbf{Top}$ is Cartesian closed.

Exercises

Exercise 40.8. Verify that in the category of topological spaces effective epimorphisms and quotient maps coincide. Then check that in a Cartesian closed category if $X \to Y$ is an effective epimorphism then so is $W \times X \to W \times Y$.

Exercise 40.9. We have used the notation Z^X for a mapping object in a Cartesian closed category, and $\mathcal{C}^{\mathcal{I}}$ for the category of functors to $\mathcal{C}$ from a small category $\mathcal{I}$. Does this constitute a conflict of notation? Explain.

Exercise 40.10. Let $\mathcal{C}$ be a Cartesian closed category.

(a) Verify the exponential laws: construct natural isomorphisms

$$Z^{X \times Y} \cong (Z^X)^Y , \quad (Y \times Z)^X \cong Y^X \times Z^X .$$

The first of these shows that the adjunction bijection $\mathcal{C}(X \times Y, Z) \cong \mathcal{C}(Y, Z^X)$ "enriches" to an isomorphism in $\mathcal{C}$. The second says that the product in $\mathcal{C}$ is actually an "enriched" product.

(b) Construct a "composition" natural transformation

$$Y^X \times Z^Y \to Z^X$$

using the evaluation maps, and show that it is associative and unital.

Exercise 40.11. Construct left and right adjoints to the forgetful functor

$$u : \mathbf{Top} \to \mathbf{Set} ,$$

and conclude that for any small category I, the limit and the colimit of a functor $X : I \to \mathbf{Top}$ consists of the corresponding limit or colimit of underlying sets endowed with a suitable topology.

Exercise 40.12. Show that the colimit (in **Top**) of any diagram of k-spaces is again a k-space, and serves as the colimit in $k\mathbf{Top}$. (Suggestion: Show that in **Top** any coproduct of k-spaces is a k-space and that any quotient of a k-space is a k-space, and then use the dual of Exercise 39.15.)

41 Basepoints and the homotopy category

More on k-spaces

The ancients (mainly Felix Hausdorff, in 1914) came up with a good definition of a topology — but k-spaces are better!

Most spaces encountered in real life are k-spaces already, and many operations in **Top** preserve the subcategory $k\mathbf{Top}$.

Proposition 41.1 (see [21, 65]). (1) *Any locally compact Hausdorff space is compactly generated.*

(2) *Quotient spaces and closed subspaces of compactly generated spaces are compactly generated.*

(3) *If X is a locally compact Hausdorff space and Y is compactly generated then $X \times Y$ is again compactly generated.*

(4) *The colimit of any diagram of compactly generated spaces is compactly generated.*

As a result of (4), in the homeomorphism

$$k \operatorname{colim}^{\mathbf{Top}} jX_\bullet \to \operatorname{colim}^{k\mathbf{Top}} X_\bullet$$

that we considered in the last lecture, the space $\operatorname{colim}^{\mathbf{Top}} jX_\bullet$ is in fact already compactly generated; no k-ification is necessary — the colimit constructed in **Top** is the same as the colimit constructed in k**Top**.

When we say "space" in this course, we will always mean k-space, and the various constructions — products, mapping spaces, and so on — will take place in k**Top**.

I should add that there is a version of the Hausdorff condition that is well suited to the compactly generated setting. Check out the sources [21, 65] for this.

Here's a simple example of how useful the formation of mapping spaces can be. We already know 5.2 that a *homotopy* between maps $f, g : X \to Y$ is a map $h : I \times X \to Y$ such that $h(0, x) = f(x)$ and $h(1, x) = g(x)$. This gives us an equivalence relation on the set $\mathbf{Top}(X, Y)$, and we write $[X, Y] = \mathbf{Top}(X, Y)/\sim$ for the set of homotopy classes of maps from X to Y. The maps f and g are points in the space Y^X, and the homotopy h is the same thing as a path $\hat{h} : I \to Y^X$ from f to g. So

$$[X, Y] = \pi_0(Y^X).$$

There is another important reason why k-spaces are useful.

Theorem 41.2 (see [24, Theorem A.6]). *Let X and Y be CW complexes with skeleta $\mathrm{Sk}_i X$ and $\mathrm{Sk}_j Y$. Then the k-space product $X \times Y$ admits the structure of a CW complex in which*

$$\mathrm{Sk}_n(X \times Y) = \bigcup_{i+j=n} \mathrm{Sk}_i X \times \mathrm{Sk}_j Y.$$

Basepoints

To talk about the fundamental group and higher homotopy groups we have to get basepoints (Definition 10.4) into the picture. The term "basepoint" leads some people refer to "based spaces," but to my ear this makes it sound as if we are doing chemistry, or worse, and I prefer "pointed." We may put restrictions on the choice of basepoint; for example we may require that $\{*\}$ be a closed subset. We will put a further restriction on $\{*\} \hookrightarrow X$ in 44.2.

This gives a category $k\mathbf{Top}_*$ where the morphisms respect the basepoints. This category is complete and cocomplete. For example

$$(X, *) \times (Y, *) = (X \times Y, (*, *)).$$

The coproduct is not the disjoint union; which basepoint would you pick? So you identify the two basepoints; the coproduct in $k\mathbf{Top}_*$ is the *wedge*

$$X \vee Y = \frac{X \sqcup Y}{*_X \sim *_Y}.$$

The one-point space $*$ is the terminal object in $k\mathbf{Top}_*$, as in $k\mathbf{Top}$, but it is also *initial* in $k\mathbf{Top}_*$: $k\mathbf{Top}_*$ is a pointed category. As we saw, this precludes it from being Cartesian closed. But we still know what we would like to take as a "mapping object" in $k\mathbf{Top}_*$: Define Y_*^X to be the subspace of Y^X consisting of the pointed maps. In general we may have to k-ify this subspace, but if $\{*\}$ is closed in Y then Y_*^X is closed in Y^X and hence is already a k-space. As a replacement for Cartesian closure, let's ask: For fixed $X \in k\mathbf{Top}_*$, does the functor $Y \mapsto Y_*^X$ have a left adjoint? This would be an analogue in $k\mathbf{Top}_*$ of the functor $A \otimes -$ in $\mathbf{Ab}$. Compute:

$$k\mathbf{Top}(W, Y^X) \quad = \quad \{f : W \times X \to Y\}$$

$$\cup\mathsf{I} \qquad\qquad\qquad \cup\mathsf{I}$$

$$k\mathbf{Top}(W, Y_*^X) \quad = \quad \{f : f(w, *) = * \ \forall w \in W\}$$

$$\cup\mathsf{I} \qquad\qquad\qquad \cup\mathsf{I}$$

$$k\mathbf{Top}_*(W, Y_*^X) \quad = \quad \left\{f : \begin{array}{l} f(w, *) = * \ \forall w \in W \\ f(*, x) = * \ \forall x \in X \end{array}\right\}.$$

So the map $W \times X \to Y$ corresponding to $f : W \to Y_*^X$ sends the wedge $W \vee X \subseteq X \times W$ to the basepoint of Y, and hence factors (uniquely) through the *smash product*

$$W \wedge X = \frac{W \times X}{W \vee X}$$

obtained by pinching the "axes" in the product to a point. We have an adjoint pair

$$- \wedge X : k\mathbf{Top}_* \rightleftarrows k\mathbf{Top}_* : (-)^X_* .$$

A good way to produce a pointed space is to start with a pair (X, A) (with A a closed subspace of X) and collapse A to a point. Thus

$$k\mathbf{Top}_*(X/A, Y) = \{f : X \to Y : f(A) \subseteq \{*\}\} = \mathrm{map}((X, A), (Y, *)) .$$

We have another adjoint pair!

It's often useful to know that if $A \subseteq X$ and $B \subseteq Y$ then

$$(X/A) \wedge (Y/B) = \frac{X \times Y}{(A \times Y) \cup_{A \times B} (X \times B)} .$$

For example, if we think of $I^m/\partial I^m$ as our model of S^m as a pointed space,

$$S^m \wedge S^n = (I^m/\partial I^m) \wedge (I^n/\partial I^n)$$

$$= \frac{I^{m+n}}{(\partial I^m \times I^n) \cup (I^m \times \partial I^n)} = I^{m+n}/\partial I^{m+n} = S^{m+n} .$$

Smashing with S^1 is a critically important operation in homotopy theory, known as (reduced) *suspension*:

$$\Sigma X = S^1 \wedge X = \frac{I \times X}{(\partial I \times X) \cup (I \times *)} .$$

That is, the suspension is obtained from the cylinder by collapsing the top and the bottom to a point, as well as the line segment along a basepoint.

You are invited to check the various properties enjoyed by the smash product, analogous to properties of the tensor product. So it's functorial in both variables; the two-point pointed space serves as a unit; and it is associative and commutative. Associativity is a blessing bestowed by assuming compact generation; notice that in forming it we are mixing limits (the product) with colimits (the quotient by the axes), and indeed the smash product turns out *not* to be associative in the full category of spaces. By induction, the n-fold suspension is thus

$$\Sigma^n X = S^1 \wedge \Sigma^{n-1}X = S^1 \wedge (S^{n-1} \wedge X) = (S^1 \wedge S^{n-1}) \wedge X = S^n \wedge X .$$

The smash product and its adjoint render $k\mathbf{Top}_*$ a "closed symmetric monoidal category."

We can also think about the *loop space* of a pointed space,

$$\Omega X = X^{S^1}_* ,$$

or the iterated loop space $\Omega^n X$, which we claim equals $X^{S^n}_*$: by induction,

$$\Omega^n X = \Omega(\Omega^{n-1}X) = (X^{S^{n-1}}_*)^{S^1}_* = X^{S^{n-1} \wedge S^1}_* = X^{S^n}_* .$$

You may be alarmed at the prospect of trying to understand the algebraic topology of a function space like ΩX. Perhaps the following theorem of John Milnor will be of some solace.

Theorem 41.3 (Milnor; see [20]). *If X is a pointed countable CW complex, then ΩX has the homotopy type of a pointed countable CW complex.*

The homotopy category

From now on, **Top** will mean k**Top**.

In Lecture 5 we introduced the *homotopy category* (of spaces) Ho**Top**. The objects of Ho**Top** are the same as those of **Top**, but the set of morphisms from X to Y is given by $[X, Y]$. You should check that composition in **Top** descends to composition in Ho**Top**.

Be warned that the homotopy category has rather poor categorical properties. Products and coproducts in **Top** provide products and coproducts in Ho**Top**, but most other types of limits and colimits do not exist in Ho**Top**.

If we have basepoints around, we will naturally want our homotopies to respect them. A *pointed homotopy* between pointed maps is a function $h : I \times X \to Y$ such that $h(t, -)$ is pointed for all t. This means that it factors through the quotient of $I \times X$ obtained by pinching $I \times *$ to a point. This quotient space may be expressed in terms of the smash product:
$$\frac{I \times X}{I \times *} = I_+ \wedge X \,.$$
Pointed homotopy is again an equivalence relation, and we have the *pointed homotopy category*, or, more properly, the *homotopy category of pointed spaces* Ho**Top**$_*$. We'll write $[X, Y]_*$ for the set of maps in this category.

Definition 41.4. Let $(X, *)$ be a pointed space and n a positive integer. The nth *homotopy group* of X is
$$\pi_n(X) = [S^n, X]_* \,.$$

We could add that the set $\pi_0(X)$ of path-components of X acquires a basepoint if X is pointed. The next case, $\pi_1(X, *)$, is the *fundamental group* or *Poincaré group* of the pointed space X (Lecture 31). It is a group under concatenation of loops.

Note the long list of aliases for this set: for any k with $0 \leq k \leq n$,
$$\pi_n(X) = [S^n, X]_* = [S^k, \Omega^{n-k} X]_* = \pi_k(\Omega^{n-k} X) \,.$$
Since π_1 is group-valued, $\pi_n(X)$ is indeed a group for any $n \geq 1$. These groups look innocuous, but they turn out to hold the solutions to many important geometric problems, and are correspondingly difficult to compute. For example, if a simply connected finite complex is not contractible then infinitely many of its homotopy are nonzero [61], and only finitely many of them are known.

Exercises

Exercise 41.5. Show that the smash product is associative as a functor $k\mathbf{Top}_* \times k\mathbf{Top}_* \to k\mathbf{Top}_*$.

Exercise 41.6. Let W be a pointed k-space. Show that the functors

$$W \wedge - : k\mathbf{Top}_* \to k\mathbf{Top}_*$$

and

$$(-)_*^W : k\mathbf{Top}_* \to k\mathbf{Top}_*$$

are *homotopy functors*: they descend to well-defined functors

$$W \wedge - : \mathrm{ho}(k\mathbf{Top}_*) \to \mathrm{ho}(k\mathbf{Top}_*)$$

and

$$(-)_*^W : \mathrm{ho}(k\mathbf{Top}_*) \to \mathrm{ho}(k\mathbf{Top}_*) .$$

Hint: Construct a map $A \wedge (X_*^W) \to (A \wedge X)_*^W$. (Careful about the definition of X_*^W: if (god forbid) the basepoint of X is not closed, this may not be a closed subspace of X^W and so may not be a k-space. But we know how to fix that: k-ify it!)

42   Fiber bundles

Much of this course will revolve around variations on the following concept.

Definition 42.1. A *fiber bundle* is a map $p : E \to B$ with the property that for every $b \in B$ there exists an open subset $U \subseteq B$ containing b and a map $p^{-1}(U) \to p^{-1}(b)$ such that $p^{-1}(U) \to U \times p^{-1}(b)$ is a homeomorphism.

When $p : E \to B$ is a fiber bundle, E is called the *total space*, B the *base space*, and p the *projection*. The point pre-image $p^{-1}(b) \subseteq E$ for $b \in B$ is the the *fiber over b*. We may use the symbol ξ for the bundle, and write $\xi : E \downarrow B$.

An *isomorphism* from $p : E \to B$ to $p' : E' \to B$ is a homeomorphism $f : E \to E'$ such that $p' \circ f = p$. The map $p : E \to B$ is a fiber bundle if it is "locally trivial," i.e. locally (in the base) isomorphic to a "trivial" bundle $\mathrm{pr}_1 : U \times F \to U$.

Fiber bundles are naturally occurring objects. For instance, a covering space $E \to B$ is precisely a fiber bundle with discrete fibers.

Example 42.2. The "Hopf fibration" provides a beautiful example of a fiber bundle. Let $S^3 \subset \mathbb{C}^2$ be the unit 3-sphere. Write $p : S^3 \to \mathbb{CP}^1 \cong S^2$ for the map sending a vector v to the complex line through v and the origin. This is a fiber bundle whose fiber is S^1.

We said "the fiber" of p is S^1. It's not hard to see that any two fibers of a fiber bundle over a path connected base space are homeomorphic, so this language isn't too bad. If we envision S^3 as the one-point compactification of $\mathbb{R}^3$, we can visualize how the various fibers relate to each other. The fiber through the point at infinity is a line in $\mathbb{R}^3$; imagine it as the z-axis. All the other fibers are circles. It's a great exercise to envision [29] how they fill up Euclidean space.

This map $S^3 \to S^2$ is the attaching map for the 4-cell in the standard CW structure on $\mathbb{C}P^2$. The nontriviality of the cup-square in $H^*(\mathbb{C}P^2)$ shows that it is *essential*, that is, not null-homotopic. This example is due to Heinz Hopf (1894–1971), a German mathematician working mainly at ETH in Zürich. He discovered the Hopf fibration and its nontriviality during a visit to Princeton in 1927–28. This was the first indication that spheres might have interesting higher homotopy groups.

Example 42.3. The *Stiefel manifold* $V_k(\mathbb{R}^n)$ is the space of orthogonal "k-frames," that is, ordered k-element orthonormal sets of vectors in $\mathbb{R}^n$. Equivalently, it is the space of linear isometric embeddings of $\mathbb{R}^k$ into $\mathbb{R}^n$; or the set of $n \times k$ matrices A such that $AA^T = I_k$. It is a compact manifold. (Eduard Stiefel (1909–1978) was a Swiss mathematician at ETH Zürich.)

We also have the *Grassmannian* $\mathrm{Gr}_k(\mathbb{R}^n)$, the space of k-dimensional vector subspaces of $\mathbb{R}^n$. (Hermann Grassmann (1809–1877) discovered much of the theory of linear algebra, but his work was not appreciated during his lifetime. He taught at a Gymnasium in Stettin, Poland, and wrote on linguistics.) By forming the span, we get a map

$$V_k(\mathbb{R}^n) \to \mathrm{Gr}_k(\mathbb{R}^n)$$

generalizing the double cover $S^{n-1} \to \mathbb{R}P^{n-1}$ (which is the case $k = 1$). There is of course a complex analogue,

$$V_k(\mathbb{C}^n) \to \mathrm{Gr}_k(\mathbb{C}^n)$$

generalizing the Hopf bundle (which is the case $n = 2, k = 1$).

These maps are fiber bundles (with fiber over V given by the space of ordered orthonormal bases of V). We can regard fact this as a special case of the following general theorem about homogeneous spaces of compact Lie groups (such as $O(n)$, $U(n)$, or a finite group).

Proposition 42.4. *Let G be a compact Lie group and let $G \supseteq H \supseteq K$ a sequence of closed subgroups (also then compact Lie groups in their own right). Then the projection map between homogeneous spaces $G/K \to G/H$ is a fiber bundle. The fiber over $H/H \in G/H$ is the subspace H/K of G/K.*

The orthogonal group $O(n)$ acts on the Stiefel manifold $V_k(\mathbb{R}^n)$ from the left, by postcomposition. This action is transitive, and the isotropy group of the k-frame $\{e_{n-k+1}, \ldots, e_n\}$ is the subgroup $O(n-k) \times I_k \subseteq O(n)$. This means that

$$V_k(\mathbb{R}^n) = O(n)/(O(n-k) \times I_k),$$

and we have a fibration $O(n) \to V_k(\mathbb{R}^n)$ with fiber $O(n-k)$. For example, $V_1(\mathbb{R}^n)$ is the unit sphere S^{n-1} in $\mathbb{R}^n$, so we have a fibration $O(n) \to S^{n-1}$ with fiber $O(n-1)$. This will be useful in an analysis of this topological group.

Another interesting map occurs if we forget all but the first vector in a k-frame. This gives us a map $V_k(\mathbb{R}^n) \to S^{n-1}$. This is the bundle of tangent $(k-1)$-frames on the $(n-1)$-sphere. The problem of determining the maximal number of everywhere linearly independent vector fields on S^{n-1} was a touchstone challenge in algebraic topology, resolved in a famous paper [2] by Frank Adams.

The Grassmannian $\mathrm{Gr}_k(\mathbb{R}^n)$ is obtained by dividing by the larger subgroup $O(n-k) \times O(k)$, and Proposition 42.4 implies that the map $V_k(\mathbb{R}^n) \to \mathrm{Gr}_k(\mathbb{R}^n)$ is a fiber bundle.

Proposition 42.4 is a corollary of the following general criterion.

Theorem 42.5 (Ehresmann, 1951; see [17]). *Suppose E and B are smooth manifolds, and let $p : E \to B$ be a smooth (i.e., C^∞) map. If p is a proper (preimages of compact sets are compact) submersion (that is, $dp : T_e E \to T_{p(e)}B$ is a surjection for all $e \in E$), then it is a fiber bundle.*

Much of this course will consist of a study of fiber bundles such as these through various essentially algebraic lenses. To bring them into play, we will always demand a further condition of our bundles.

Definition 42.6. An open cover $\mathcal{U}$ of a space X is *numerable* if there exists subordinate partition of unity; that is, a family of functions $\phi_U : X \to [0, 1]$, indexed by the elements of $\mathcal{U}$, such that $\phi_U^{-1}((0, 1]) = U$ and any $x \in X$ belongs to only finitely many $U \in \mathcal{U}$. The space X is *paracompact* if any open cover admits a numerable refinement. A fiber bundle is *numerable* if it admits a numerable trivializing cover.

So any fiber bundle over a paracompact space is numerable. This isn't too restrictive for us:

Proposition 42.7 (Miyazaki; see Theorem 1.3.5 in [20]). *CW complexes are paracompact.*

Exercises

Exercise 42.8. Show that the fiber bundle $SO(n) \to S^{n-1}$ sending an orthogonal matrix with determinant 1 to its first column has a section if and only if S^{n-1} is parallelizable. What is the situation for $n = 3$? for $n = 4$?

43 Fibrations, fundamental groupoid

Fibrations and path liftings

During the 1940s, much effort was devoted to extracting homotopy-theoretic features of fiber bundles. It came to be understood that the desired consequences relied entirely on a "homotopy lifting property." One of the revolutions in topology around 1950 was the realization that it was advantageous to simply take that property as a *definition*. This extension of the notion of a fiber bundle included wonderful new examples, but still retained the homotopy theoretic consequences. Here is the definition.

Definition 43.1. A *fibration* is a map $p : E \to B$ that satisfies the *homotopy lifting property* ("HLP"): Given any $f : W \to E$ and any homotopy $h : I \times W \to B$ with $h(0, w) = pf(w)$, there is a map $\overline{h}$ that lifts h and extends f: that is, making the following diagram commute.

$$
\begin{array}{ccc}
W & \xrightarrow{\ f\ } & E \\
{\scriptstyle \mathrm{in}_0} \downarrow & {\overline{h}\ \nearrow} & \downarrow {\scriptstyle p} \\
I \times W & \xrightarrow{\ h\ } & B
\end{array}
$$

For example, for any space X (even the empty space!) the unique map $X \to *$ is a fibration — a lift is given by $\overline{h}(t, w) = f(w)$ — as is the unique map $\varnothing \to X$ (Why?). In general, though, this seems like an alarming definition, since the HLP has to be checked for *all* spaces W, *all* maps f, and *all* homotopies h!

On the other hand, an advantage of this type of definition, by means of a lifting condition, is that it enjoys various easily checked persistence properties.

- Base change: If $p : E \to B$ is a fibration and $X \to B$ is any map, then the induced map $E \times_B X \to X$ is again a fibration. In particular, any product projection is a fibration.
- Products: If $p_i : E_i \to B_i$ is a family of fibrations then the product map $\prod p_i$ is again a fibration.
- Exponentiation: If $p : E \to B$ is a fibration and A is any space, then $E^A \to B^A$ is again fibration.
- Composition: If $p : E \to B$ and $q : B \to X$ are both fibrations, then the composite $qp : E \to X$ is again a fibration.

Not all of these persistence properties are true for fiber bundles. Which ones fail?

There is a nice geometric interpretation of what it means for a map to be a fibration, in terms of "path liftings." We'll use Cartesian closure! The adjoint of the solid arrow part of the diagram in Definition 43.1 is

$$
\begin{array}{ccc}
W & \xrightarrow{\ f\ } & E \\
{\scriptstyle \widehat{h}}\big\downarrow & & \big\downarrow{\scriptstyle p} \\
B^I & \xrightarrow{\ \mathrm{ev}_0\ } & B.
\end{array}
$$

By the definition of the pullback, the data of this diagram is equivalent to a map $W \to B^I \times_B E$. Explicitly,

$$
B^I \times_B E = \{(\omega, e) \in B^I \times E : \omega(0) = p(e)\} \,.
$$

This space comes equipped with a map from E^I, given by sending a path $\omega : I \to E$ to

$$
\widetilde{p}(\omega) = (p\omega, \omega(0)) \in B^I \times_B E \,.
$$

In these terms, giving a lift $\overline{h}$ as in Definition 43.1 is equivalent to giving a lift

$$
\begin{array}{ccc}
 & & E^I \\
 & {\scriptstyle \widetilde{h}}\nearrow & \big\downarrow{\scriptstyle \widetilde{p}} \\
W & \longrightarrow & B^I \times_B E.
\end{array}
$$

This again needs to be checked for every W and every map to $B^I \times_B E$. But at least there is now a universal case to consider: $W = B^I \times_B E$ mapping by the identity map! So p is a fibration if and only if a lift λ exists in the following diagram; that is, a *section* of $\widetilde{p}$.

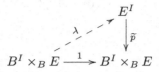

The section λ is called a *path lifting function*. To understand why, suppose $(\omega, e) \in B^I \times_B E$, so that ω is a path in B with $\omega(0) = p(e)$. Then $\lambda(\omega, e)$ is a path in E lying over ω and starting at e. The path lifting function provides a continuous lift of paths in B. The existence (or not) of a section of $\widetilde{p}$ provides a single condition that needs to be checked if you want to see that p is a fibration.

There is no mention of local triviality in this definition. However:

Theorem 43.2 (Albrecht Dold, 1963; see [72], Chapter 13). *Let $p : E \to B$ be a continuous map. Assume that there is a numerable cover of B, say $\mathcal{U}$, such that for every $U \in \mathcal{U}$ the restriction $p|_{p^{-1}(U)} : p^{-1}U \to U$ is a fibration. Then p itself is a fibration.*

Corollary 43.3. *Any numerable fiber bundle is a fibration.*

Comparing fibers over different points

If $p : E \to B$, let's write F_b for the fiber $p^{-1}(b)$ over b. If p is a covering space, then unique path lifting provides, for any path ω from a to b, a homeomorphism $F_a \to F_b$ depending only on the path homotopy class of ω. Our next goal is to construct an analogous map for a general fibration.

Consider the solid arrow diagram:

$$
\begin{array}{ccc}
F_a & \xrightarrow{\hspace{3cm}} & E \\
{\scriptstyle \text{in}_0} \downarrow & {\scriptstyle h} \nearrow & \downarrow p \\
I \times F_a & \xrightarrow{\text{pr}_1} I \xrightarrow{\omega} & B .
\end{array}
$$

This commutes since $\omega(0) = a$. By the homotopy lifting property, there is a dotted arrow that makes the entire diagram commute. If $x \in F_a$, the image $h(1, x)$ is in F_b. This supplies us with a map $f : F_a \to F_b$, given by $f(x) = h(1, x)$.

Since we are not working with a covering space, there will in general be many lifts h and so many choices of f. But we may at least hope that the homotopy class of f is determined by the path homotopy class of ω.

So suppose we have two paths ω_0, ω_1, with $\omega_0(0) = \omega_1(0) = a$ and $\omega_0(1) = \omega_1(1) = b$, and a homotopy $g : I \times I \to B$ between them (so that $g(0,t) = \omega_0(t)$, $g(1,t) = \omega_1(t)$, $g(s,0) = a$, $g(s,1) = b$). Here's a picture.

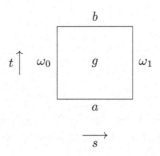

Choose lifts h_0 and h_1 as above. These data are captured by a diagram of the form

$$((\partial I \times I) \cup (I \times \{0\})) \times F_a \xrightarrow{\hspace{4cm}} E$$

with maps in_0, p, pr_1, and

$$I \times I \times F_a \xrightarrow{\mathrm{pr}_1} I \times I \xrightarrow{g} B.$$

The map along the top is given by h_0 and h_1 on $\partial I \times I \times F_a$ and by $\mathrm{pr}_2 : I \times F_a \to F_a$ followed by the inclusion on the other summand.

If the dotted lift exists, it would restrict on $I \times \{1\} \times F_a$ to a homotopy between f_0 and f_1. Well, the subspace $(\partial I \times I) \cup (I \times \{0\})$ of $I \times I$ wraps around three edges of the square. It's easy enough to create a homeomorphism with the pair $(I \times I, \{0\} \times I)$, so the HLP (with $W = I \times F_a$) gives us the dotted lift.

So the map $F_a \to F_b$ is well-defined up to homotopy by the path homotopy class of the path ω from a to b. We'll denote that homotopy class by f_ω.

The fundamental groupoid

We can set this up in categorical terms. The space B defines a category whose objects are the points of B and in which a morphism from a to b is a homotopy class of paths from a to b. Composition is given by the juxtaposition rule

$$(\sigma \cdot \omega)(t) = \begin{cases} \omega(2t) & 0 \le t \le 1/2 \\ \sigma(2t - 1) & 1/2 \le t \le 1. \end{cases}$$

The constant path c_a serves as an identity up to homotopy: Here are pictures of the homotopy between $c_b \cdot \omega$ and ω, and between $\sigma \cdot c_a$ and σ.

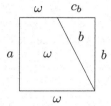 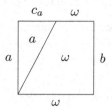

Similar pictures show that $(\alpha \cdot \sigma) \cdot \omega \simeq \alpha \cdot (\sigma \cdot \omega)$ and that every morphism has an inverse, given by $\overline{\omega}(t) = \omega(1 - t)$.

This gives us a *groupoid* — a small category in which every morphism is an isomorphism — called the *fundamental groupoid* of B, and written with a capital π: $\Pi_1(X)$.

Our work can be succinctly summarized as follows.

Proposition 43.4. *Formation of fibers of a fibration $p : E \to B$ determines a functor $\Pi_1(B) \to \mathrm{HoTop}$.*

Proof. We should check functoriality: if $\omega : a \sim b$ and $\sigma : b \sim c$, then hopefully the induced homotopy classes compose:

$$f_{\sigma\omega} = f_\sigma \circ f_\omega .$$

To see this, pick lifts h_ω and h_σ in

so that $f_\omega(e) = h_\omega(1, e)$ and $f_\sigma(e) = h_\sigma(1, e)$. Then construct a lifting in

by using h_ω in the left half of the interval and $h_\sigma \circ f_\omega$ in the right half. The resulting map $F_a \to F_b$ is then precisely $f_\sigma \circ f_\omega$. $\qquad\square$

Remark 43.5. In Lecture 31 we defined the product of loops as juxtaposition but in the reverse order. That convention would have produced a contravariant functor $\Pi_1(X) \to \mathbf{HoTop}$.

Remark 43.6. Since any functor carries isomorphisms to isomorphisms, Proposition 43.4 implies that a path from a to b determines a homotopy class of homotopy equivalences from F_a to F_b.

Homotopy invariance

Fix a map $p : E \to Y$. The pullback of E along a map $f : X \to Y$ can vary wildly as f is deformed; it is far from being a homotopy invariant. Just think of the case $X = *$, for example, when the pullback along $f : * \to Y$ is the point preimage $p^{-1}(f(*))$. One of the great features of fibrations is this:

Proposition 43.7. Let $p : E \to Y$ be a fibration and $f_0, f_1 : X \to Y$ two maps. Write E_0 and E_1 for pullbacks of E along f_0 and f_1. If f_0 and f_1 are homotopic then E_0 and E_1 are homotopy equivalent.

Proof. We construct a fibration over Y^X whose fiber over f is f^*E, the pullback of $E \to Y$ along f. It occurs as the middle vertical composite in the following diagram of pullbacks.

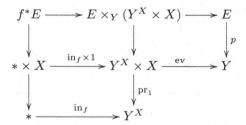

The middle horizontal composite is the map f, so the pullback is f^*E as shown. Now a homotopy between f_0 and f_1 is a path in Y^X from f_0 to f_1, and so by Lemma 43.4 the fibers over them are homotopy equivalent. $\square$

Remark 43.8. We could ask for more: We could ask that E_0 and E_1 are homotopy equivalent by maps and homotopies respecting the projections to X: that there is a *fiber homotopy equivalence* between them. This is in fact the case, as you will show for homework.

Corollary 43.9. Let $p : E \to B$ be a fibration. If B is contractible to $* \in B$, then the inclusion of the fiber $p^{-1}(*) \hookrightarrow E$ is a homotopy equivalence.

Proof. The identity map 1_B and the constant map $c : B \to B$ with value $*$ are homotopic, so pulling back $E \downarrow B$ along them produce homotopy equivalent spaces. One gives E, the other $B \times p^{-1}(*)$. The projection $\mathrm{pr}_2 : B \times p^{-1}(*) \to p^{-1}(*)$ is a homotopy equivalence since B is contractible. We leave you to check that the resulting equivalence is the inclusion. $\square$

Exercises

Exercise 43.10. Which of the properties of fibrations listed above are true for fiber bundles? Which fail?

Exercise 43.11. Let $p : E \to B$ and $p' : E' \to B$ be fibrations, and let $f : E \to E'$ be a homotopy equivalence such that $p' \circ f = p$. Show that f is in fact a fiber-homotopy equivalence.

In particular, suppose that B is contractible. Then the identity map 1_B is homotopic to a map factoring through $*$. The pullback of E along this trivial map is $B \times p^{-1}(*)$, while the pullback along the identity is E. So any fibration over a contractible space is fiber homotopical equivalent to a product projection.

Hint: First show that it suffices to find a map $g : E' \to E$ such that $p \circ g = p'$ and $f \circ g$ is fiber-homotopic to $1_{E'}$.

Then reduce this to the following (where E will be what used to be E', and f is something else again): Suppose that $p : E \to B$ is a fibration and that $f : E \to E$ is such that $pf = p$ and $f \sim 1_E$. Then there is a map $g : E \to E$ such that $pg = p$ and fg is fiber homotopic to the identity.

Further hint: Dualize the proof in [36, §6.5].

44 Cofibrations

Let $i : A \to X$ be a map of spaces, and Y some other space. When is the induced map $Y^X \to Y^A$ a fibration? For example, if $a \in X$, does evaluation at a produce a fibration $Y^X \to Y$?

By the definition of a fibration, we want a lifting in the solid-arrow diagram

$$
\begin{array}{ccc}
W & \longrightarrow & Y^X \\
\downarrow{\scriptstyle in_0} & \nearrow & \downarrow \\
I \times W & \longrightarrow & Y^A .
\end{array}
$$

Adjointing over, we get:

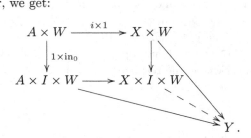

Adjointing over again, this diagram transforms to:

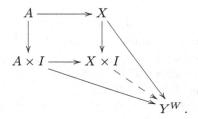

This discussion motivates the following definition of a cofibration, "dual" to the notion of fibration.

Definition 44.1. A *cofibration* is a map $i : A \to X$ that satisfies *homotopy extension property* (sometimes abbreviated as "HEP"): for any solid-arrow commutative diagram as below, a dotted arrow exists making the whole diagram commutative.

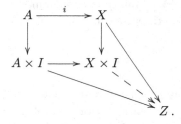

How shall we check that a map is a cofibration? By the universal property of a pushout, $A \to X$ is a cofibration if and only if there is an extension in

$$(X \times 0) \cup_A (A \times I) \xrightarrow{\ j\ } X \times I$$

with f to Z.

for every map f. Now there is a universal example, namely $Z = (X \times 0) \cup_A (A \times I)$, $f = \mathrm{id}$. So a map i is a cofibration if and only if the map $j : (X \times 0) \cup_A (A \times I) \to X \times I$ admits a retraction: a map $r : X \times I \to (X \times 0) \cup_A (A \times I)$ such that $rj = 1$.

The space involved here is called the *mapping cylinder*. It's not hard to check (using the mapping cylinder) that any cofibration is a subspace embedding. But the map j may not be an embedding; the map $(X \times 0) \cup_A (A \times I) \to \mathrm{im}(j) \subseteq X \times I$ is a continuous bijection but it may not be a homeomorphism. If $A \subseteq X$ is a *closed* subset then $A \times I \subseteq X \times I$ is a closed map, and $X \times 0 \subseteq X \times I$ is also, so the map from the pushout is a closed map and hence is then a homeomorphism to its image. In fact the image of a cofibration $A \to X$ is automatically closed if X is Hausdorff [67].

So the inclusion of a closed subspace $A \subseteq X$ is a cofibration if and only if there is a retraction from $X \times I$ onto its subspace $(X \times 0) \cup (A \times I)$.

Definition 44.2. A basepoint $*$ in X is *nondegenerate* if $\{*\} \hookrightarrow X$ is a closed cofibration. One also says that $(X, *)$ is *well-pointed*.

Any point in a CW complex, for example, will serve as a nondegenerate basepoint. If $*$ is a nondegenerate basepoint of A, the evaluation map $\mathrm{ev} : X^A \to X$ is a fibration. The fiber of ev over the basepoint of X is then exactly the space of pointed maps X^A_*.

Whenever it's convenient we will assume our basepoints are nondegenerate.

Example 44.3. $i : S^{n-1} \hookrightarrow D^n$ is a cofibration: The map

$$j : D^n \cup_{S^{n-1}} (S^{n-1} \times I) \hookrightarrow D^n \times I$$

is the inclusion of the open tin can into the closed can full of soup —

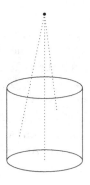

The illustrated retraction of $D^n \times I$ onto the open can sends a point in the soup to its shadow on the open tin can.

In particular, setting $n = 1$ in this example, $\{0,1\} \hookrightarrow I$ is a cofibration, so evaluation at the pair of endpoints,

$$\mathrm{ev}_{0,1} : Y^I \to Y \times Y,$$

is a fibration. Every point in I is nondegenerate, so $\mathrm{ev}_a : Y^I \to Y$ is a fibration for any $a \in I$.

The class of cofibrations is closed under the following operations.

- Cobase change: if $A \to X$ is a cofibration and $A \to B$ is any map, the pushout $B \to X \cup_A B$ is again a cofibration.
- Coproducts: if $A_j \to X_j$ is a cofibration for every j, then the coproduct map $\coprod A_j \to \coprod X_j$ is again a cofibration.
- Product: If $A \to X$ is a cofibration and B is any space, then $A \times B \to X \times B$ is again a cofibration.
- Composition: If $A \to B$ and $B \to X$ are both cofibrations, then the composite $A \to X$ is again a cofibration.

It follows from these inheritance properties and the single example $S^{n-1} \hookrightarrow D^n$ that if X is a CW complex and A is a subcomplex then $A \to X$ is a cofibration.

Cofibrance provides a natural condition under which a contractible subspace can be collapsed without damage.

Proposition 44.4. *Let $A \to X$ be a cofibration, and write X/A for the pushout of $* \leftarrow A \to X$. If A is contractible then $X \to X/A$ is a homotopy equivalence.*

Proof. Pick a contracting homotopy $h : A \times I \to A$, so that $h(a,0) = a$ and $h(a,1) = * \in A$ for all $a \in A$. By cofibrance there is an extension of $f \circ h$ to a homotopy $g : X \times I \to X$ such that $g(x,0) = x$. The map $g(-,1)$ then factors through the projection $p : X \to X/A$: there is a map $r : X/A \to X$ such that $r \circ p$ is homotopic the identity.

To construct a homotopy from $p \circ r$ to $1 : X/A \to X/A$, note that the homotopy g sends $A \times I$ into A, so its composite with $p : X \to X/A$ factors through a map $\overline{g} : (X/A) \times I \to X/A$. At $t = 0$ this is the identity; at $t = 1$ it is just $p \circ r$. $\qquad\square$

Exercises

Exercise 44.5. Let $C \subset [0,1]$ be the Cantor set. This is a closed subset; but show that the inclusion is not a cofibration. (Hint: the Hahn-Mazurkiewicz Theorem (e.g. [26, Theorem 3-30]) may be useful.)

45   Cofibration sequences and co-exactness

There is a pointed version of the cofibration condition: but you only ask to extend *pointed* homotopies; so the condition is weaker than the unpointed version. (It's true that we seek an extension to a *pointed homotopy*, but since the basepoint is in the source space this is automatic.) A pointed homotopy can be thought of as a pointed map

$$X \wedge I_+ = \frac{X \times I}{* \times I} \to Y.$$

The condition that the embedding of a closed subspace $i : A \subseteq X$ is a pointed cofibration can again be expressed as requiring that the inclusion of the (now "reduced") mapping cylinder

$$M(i) = (X \times 0) \cup_{A \times 0} (A \wedge I_+)$$

into $X \wedge I_+$ admits a retraction. Today we'll work entirely in the pointed context, and I'll tend to omit the adjectives "reduced" and "pointed."

Any pointed map $f : X \to Y$ admits a canonical factorization as a closed pointed cofibration followed by a pointed homotopy equivalence

$$\begin{array}{ccc}
& & X \\
& {}^{f}\swarrow & \downarrow{}^{i} \\
Y & \xleftarrow{\;\simeq\;} & M(f)
\end{array}$$

where i embeds X along $t = 1$. For example, the *cone* on a space X is a mapping cylinder:

$$CX = M(X \to *) = X \wedge I.$$

The map $X \to *$ factors as the cofibration $X \to CX$ followed by the homotopy equivalence $CX \to *$.

Since i is a cofibration, we should feel entitled to collapse X to a point; that is, form the pushout in

$$\begin{array}{ccccc}
& & X & \longrightarrow & * \\
& {}^{f}\swarrow & \downarrow{}^{i} & & \downarrow \\
Y & \xleftarrow{\;\simeq\;} & M(f) & \longrightarrow & C(f).
\end{array}$$

$C(f)$ is the *mapping cone* of f. If the mapping cylinder is a top hat, the mapping cone is a witch's hat. One example: the suspension functor is given by

$$\Sigma X = C(X \to *).$$

Since i is a cofibration, the pushout $* \to C(f)$ is again a cofibration; the cone point is always nondegenerate.

This pushout can be expressed differently: Instead of replacing $f : X \to Y$ with a cofibration, let's replace $X \to *$ with a cofibration, namely, the inclusion $X \hookrightarrow CX$. So we have a pushout diagram

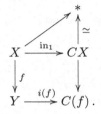

This pushout is homeomorphic to the earlier one; but notice that the homeomorphism uses the automorphism of the unit interval sending t to $1 - t$.

If f is already a cofibration, the cobase change property implies that $CX \to C(f)$ is again cofibration. CX is contractible, so by Proposition 44.4, collapsing it to a point is a homotopy equivalence. But collapsing CX in $C(f)$ is the same as collapsing Y in X:

Lemma 45.1. *If $f : X \to Y$ is a cofibration then the collapse map $C(f) \to Y/X$ is a homotopy equivalence.*

Co-exactness

Definition 45.2. A *cofibration sequence* is a diagram that is homotopy equivalent to

$$X \xrightarrow{f} Y \xrightarrow{i(f)} C(f)$$

for some map f.

The composite $X \to C(f)$ is *null-homotopic*; that is, it's homotopic to the constant map (with value the basepoint). The homotopy is given by $h : (x, t) \mapsto [x, t]$: When $t = 0$ we can use $[x, 0] \sim f(x)$ to see the composite, while when $t = 1$ we get the constant map.

The pair $(i(f), h)$ is *universal* with this property: giving a map $g : Y \to Z$ along with a null-homotopy of the composite $g \circ f$ is the same thing as giving a map $C(f) \to Z$ that extends g.

An implication of this is the following:

Lemma 45.3. *For any pointed map $f : X \to Y$ and any pointed space Z, the sequence of pointed sets*

$$[X, Z]_* \xleftarrow{f^*} [Y, Z]_* \xleftarrow{i(f)^*} [C(f), Z]_*$$

is exact, *in the sense that*

$$\mathrm{im}(i(f)^*) = \{g : Y \to Z : g \circ f \simeq *\}.$$

A sequence of composable arrows with this property is "*co-exact*": so cofibration sequences are coexact.

The map $f : X \to Y$ functorially determines the map $i(f) : Y \to C(f)$, and we may form *its* mapping cone; and continue:

$$X \xrightarrow{f} Y \xrightarrow{i(f)} C(f) \xrightarrow{i^2(f)} C(i(f)) \xrightarrow{i^3(f)} C(i^2(f)) \xrightarrow{i^4(f)} \cdots .$$

This looks like it will lead off into the wilderness, but luckily there is a kind of periodicity at work. Here's a picture of $C(i(f))$:

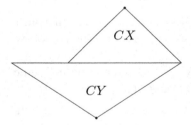

The map $i(f)$ is the pushout of the cofibration $X \to CX$ along $X \to Y$, so it is a cofibration. Therefore, by Lemma 45.1 the collapse map $C(i(f)) \to C(f)/Y$ is a homotopy equivalence. But

$$C(f)/Y = \Sigma X ,$$

the suspension of X. So we have the commutative diagram

$$X \xrightarrow{f} Y \xrightarrow{i(f)} C(f) \xrightarrow{i^2(f)} C(i(f))$$
$$\searrow^{\pi(f)} \quad \downarrow{\simeq}$$
$$\Sigma X .$$

Now we have two ways to continue! I combine them in the homotopy commutative diagram

$$X \xrightarrow{f} Y \xrightarrow{i(f)} C(f) \xrightarrow{i^2(f)} C(i(f)) \xrightarrow{i^3(f)} C(i^2(f)) \xrightarrow{i^4(f)} \cdots$$

with $\pi(f)$, $\simeq$, $\pi(i(f))$, $\simeq$, $\pi(i^2(f))$ and

$$\Sigma X \xrightarrow{-\Sigma f} \Sigma Y \xrightarrow{-\Sigma i(f)} \cdots$$

Notice the minus sign! It means that instead of $[t, x] \mapsto [t, f(x)]$, we have to use $[t, x] \mapsto [1 - t, f(x)]$. This is needed to make the triangle commute, even up to homotopy, as you can see by being careful with the parametrization of the cones.

The resulting long sequence of maps
$$X \to Y \to C(f) \to \Sigma X \to \Sigma Y \to \Sigma C(f) \to \Sigma^2 X \to \cdots$$
is the *Barratt-Puppe sequence* associated to the map f. Each two-term subsequence is a cofibration sequence and is co-exact. (Michael Barratt (1927–2015), professor at Manchester and then Northwestern, and Dieter Puppe (1930–2005), professor at Heidelberg, were two visionary topologists.)

The Barratt-Puppe sequence is a "homotopy theoretic" version of the long exact homology sequence of a pair. Suppose that A is a subspace of X. Then I claim that
$$\overline{H}_*(X \cup CA) \cong H_*(X, A).$$
If you combine that with the suspension isomorphism in reduced homology, the Barratt-Puppe sequence gives you the homology long exact sequence of the pair.

To see the equality, just use homotopy invariance and excision:
$$\overline{H}_*(X \cup CA) = H_*(X \cup CA, *) = H_*(X \cup CA, CA)$$
$$= H_*(X \cup C_{\leq(1/2)}A, C_{\leq(1/2)}A) = H_*(X \cup A \times I, A \times I) = H_*(X, A).$$

Since $X \cup CA \simeq X/A$ if $A \to X$ is a cofibration, this is a good condition to guarantee that
$$H_*(X, A) = \overline{H}_*(X/A).$$

Exercises

Exercise 45.4. (a) Use a homotopy $h : A \times I \to Y$ between the branches of the diagram

$$\begin{array}{ccc} A & \xrightarrow{i} & X \\ \downarrow{f} & & \downarrow{g} \\ B & \xrightarrow{j} & Y \end{array}$$

to construct a map $C(f) \to C(g)$ such that in the diagram

$$
\begin{array}{ccccc}
X & \longrightarrow & C(i) & \longrightarrow & \Sigma A \\
\downarrow{\scriptstyle g} & & \downarrow & & \downarrow{\scriptstyle \Sigma f} \\
Y & \longrightarrow & C(j) & \longrightarrow & \Sigma B
\end{array}
$$

the left square commutes and the right one commutes up to homotopy.

(b) Use a homotopy $f \simeq g : X \to Y$ to construct a homotopy equivalence $C(f) \simeq C(g)$. [See e.g. [4, Proposition 3.2.15].]

46 Weak equivalences and Whitehead's theorems

We now have defined the homotopy groups of a pointed space,

$$
\pi_n(X) = [S^n, X]_* \, .
$$

So $\pi_0(X)$ is the pointed set of path components. For $n > 0$, π_n only sees the path component of the basepoint. It's a group for $n = 1$, and hence also for $n \geq 1$ since $\pi_n(X) = \pi_1(\Omega^{n-1}X)$.

Here's another very useful way to represent an element of $\pi_n(X, *)$. Recall our description of the n-sphere as a pointed space:

$$
S^n = I^n / \partial I^n \, .
$$

So an element of $\pi_n(X, *)$ is a homotopy class of maps of pairs

$$
(I^n, \partial I^n) \to (X, *) \, .
$$

The Eckmann-Hilton argument

Lemma 46.1. *For $n \geq 2$, $\pi_n(X)$ is abelian.*

Proof. I'll give you two proofs of this fact; both variants of the "Eckmann-Hilton argument." Since $\pi_n(X) = \pi_2(\Omega^{n-2}X)$, it suffices to consider $n = 2$.

First, geometric: Given $f, g : I^2 \to X$, both sending ∂I^2 to $*$, we can form another one by putting the two side by side (and compressing the horizontal coordinate by a factor of 2 in each). This is the sum in $\pi_n(X)$. This is homotopic to the map that does f and g in much smaller rectangles and fills in the rest of the square with maps to the basepoint. Now I'm free to move these two smaller rectangles around one another, exchanging positions. Then I can re-expand, to get the addition $g + f$.

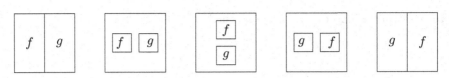

Now, algebraic: An *H-space* is a pointed space Y together with map $\mu : Y \times Y \to Y$ such that

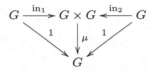

commutes in $\mathrm{Ho}(\mathbf{Top}_*)$. The relevant example here is $Y = \Omega X$. Then $\pi_1(Y, *)$ has extra structure: Since $\pi_1(Y \times Y, *) = \pi_1(Y, *) \times \pi_1(Y, *)$ (as groups) we get a group G together with a group homomorphism $\mu : G \times G \to G$ such that

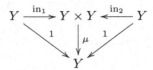

commutes. That is to say, $\mu(a, 1) = a$, $\mu(1, d) = d$, and, since $(a, b) \cdot (c, d) = (ac, bd)$ in $G \times G$,

$$\mu(ac, bd) = \mu(a, b) \cdot \mu(c, d).$$

Take $b = 1 = c$ so $\mu(a, d) = ad$: that is, the "multiplication" μ is none other than the group multiplication. Then take $a = 1 = d$ so $\mu(c, b) = bc$: that is, the group structure is commutative.   □

Change of basepoint

We can trace what happens when we move the basepoint. Let $\omega : I \to X$ be a path from a to b. It induces a map

$$\omega_\# : \pi_n(X, a) \to \pi_n(X, b)$$

in the following way. Given $f : I^n \to X$ representing $\alpha \in \pi_n(X, *)$, define a map

$$(I^n \times 0) \cup (\partial I^n \times I) \to X$$

by

$$(v, t) \mapsto \begin{cases} f(v) & \text{for } v \in I^n, t = 0 \\ \omega(t) & \text{for } v \in \partial I^n . \end{cases}$$

Precompose this map with the map from the face $I^n \times 1$ given by projecting from the point $(b, 2)$, where b is the center of I^n. The result is a new map $I^n \to X$; it sends the middle part of the cube by f, and the peripheral part by ω.

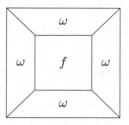

It's easy to check that this gives rise to a functor $\Pi_1(X) \to \mathbf{Set}$, and hence to an action of $\pi_1(X, *)$ on $\pi_n(X, *)$. For $n = 1$, this is the conjugation action,

$$\omega \cdot \alpha = \omega \alpha \omega^{-1}.$$

For all $n \geq 1$ it is an action by group homomorphisms; for $n \geq 2$, $\pi_n(X, *)$ is a $\mathbb{Z}[\pi_1(X, *)]$-module.

Definition 46.2. A space is *simple* if this action is trivial for every choice of basepoint.

Example 46.3. If all path components are simply connected, the space is simple. A topological group is a simple space.

This action can be used to explain how homotopic maps act on homotopy groups.

Proposition 46.4. *Let $h : f_0 \sim f_1$ be a ("free," as opposed to pointed) homotopy of pointed maps $X \to Y$. Let $* \in X$, and let $\omega : I \to X$ by $\omega(t) = h(*, t)$. Then*

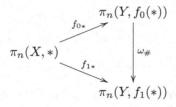

commutes.

Proof. The homotopy h fills in the cube $I^n \times I$, and provides a pointed homotopy from $\omega \cdot f_0$ to f_1. $\qquad \square$

Weak homotopy equivalence

While it may be hard to compute homotopy groups, we can think about what sort of maps induce isomorphisms in them.

Definition 46.5. A map $f : X \to Y$ is a *weak equivalence* or *weak homotopy equivalence* if it induces an isomorphism in π_0 and in π_n for all $n \geq 1$ and every choice of basepoint in X.

A space that is weakly equivalent to a point is said to be *weakly contractible*.

Of course it suffices to pick one point in each path component.

Weak equivalences may not have any kind of map going in the opposite direction. The definition seems very base-point focused, but in fact it is not. For example:

Proposition 46.6. *Any homotopy equivalence is a weak equivalence.*

Proof. Let $f : X \to Y$ be a homotopy equivalence with homotopy inverse $g : Y \to X$, and pick a homotopy $h : 1_X \sim gf$. Define $\omega : I \to X$ by $\omega(t) = h(*, t)$. Then by Proposition 46.4 we have a commutative diagram

$$\pi_n(X, *) \xrightarrow{\ f_* \ } \pi_n(Y, f(*))$$

with $\omega_\#$ on the diagonal and g_* down to $\pi_n(X; gf(*))$

in which the diagonal is an isomorphism. Picking a homotopy $1_Y \sim fg$ gives the rest of the diagram

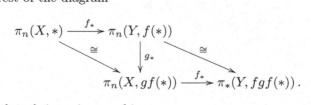

$$\pi_n(X, *) \xrightarrow{\ f_* \ } \pi_n(Y, f(*))$$
$$\pi_n(X, gf(*)) \xrightarrow{\ f_* \ } \pi_*(Y, fgf(*)) \,.$$

It follows that f_* is an isomorphism. $\qquad\square$

Here are three fundamental theorems about weak equivalences, all due more or less to J.H.C. Whitehead.

Theorem 46.7. *Any weak equivalence induces an isomorphism in singular homology.*

Since $H_0(X)$ is the free abelian group generated by $\pi_0(X)$, this is obvious in dimension 0, and on each path component Poincaré's theorem implies it in dimension 1. See Exercise 64.6 for the general case.

Theorem 46.8. *Let X and Y be simple spaces. Any map from X to Y that induces an isomorphism in homology is a weak equivalence.*

Theorem 46.9. *Let X and Y be CW complexes. Any weak equivalence from X to Y is in fact a homotopy equivalence.*

Theorem 46.8 clearly provides a powerful way to construct weak equivalences, and, when combined with Theorem 46.9, homotopy equivalences. We will prove a vast generalization of Theorem 46.8 in Lecture 69.

Here is a useful strengthening of Theorem 46.9:

Theorem 46.10 ("Whitehead's little theorem"). *A map $f : X \to Y$ is a weak equivalence if and only if $f \circ - : [W, X] \to [W, Y]$ is bijective for all CW complexes W.*

We will prove this in Lecture 48.

Proof of 46.10⇒46.9. We assume that

$$f \circ - : [W, X] \to [W, Y]$$

is bijective for every CW complex W. Taking $W = Y$, we find that there is a map $g : Y \to X$ such that $f \circ g = 1_Y$. We claim that $g \circ f = 1_X$ as well. To see this we take $W = X$: so

$$f \circ - : [X, X] \to [X, Y]$$

is a monomorphism. Under it $1 \mapsto f$, but $g \circ f$ does as well:

$$g \circ f \mapsto f \circ (g \circ f) = (f \circ g) \circ f = 1_Y \circ g = f \,.$$

So $g \circ f = 1_X$. $\qquad\square$

Remark 46.11. There is a deep shift of focus involved here. In the beginning, homotopy theory dealt with what happens when you define an equivalence relation ("homotopy") on maps. Focusing on weak equivalences is an entirely different perspective: we are picking out a collection of maps that will be regarded as "equivalences." They are to become the isomorphism in the homotopy category. The fact that they satisfy 2-out-of-3 makes the collection of weak equivalences an appropriate choice.

This change in perspective may be attributed to Daniel Quillen, who, in *Homotopical Algebra* [54] (written while Quillen (1940–2011, Fields Medal

1978) was a professor at MIT, in collaboration with his colleague Dan Kan), set out an axiomatization of homotopy theory using three classes of maps, which he termed "weak equivalences," "cofibrations," and "fibrations." They are assumed to be related to each other through appropriate factorization and lifting properties that axiomatize what we have just been doing. The resulting theory of "model categories" dominated the underlying framework of homotopy theory for thirty years, and is still a critically important tool.

Exercises

Exercise 46.12. Verify the claim that a topological group is a simple space.

Exercise 46.13. Show that weak equivalences satisfy "2 out of 3": in

$$X \underset{gf}{\overset{f}{\rightrightarrows}} Y \overset{g}{\Longrightarrow} Z$$

if two of f, g, and gf are weak equivalences then so is the third.

Exercise 46.14. Let $\omega \in \pi_1(S^1 \vee S^2)$ and $\alpha \in \pi_2(S^1 \vee S^2)$ be represented by the inclusion of the two spheres into the wedge. Form a new CW complex X by attaching a 3-cell by means of a map representing the homotopy class $2\alpha - \omega \cdot \alpha \in \pi_2(S^1 \vee S^2)$. Show that the inclusion of S^1 into X induces isomorphisms in π_1 and in homology, but that X is not weakly equivalent to the circle.

So no simple adjustment to the Whitehead theorem will work. Notice however that the map on universal covers is not an isomorphism in homology.

47   Homotopy long exact sequence and homotopy fibers

Relative homotopy groups

We'll continue to think of $\pi_n(X, *)$ as a set of homotopy classes of maps of pairs:

$$\pi_n(X, *) = [(I^n, \partial I^n), (X, *)].$$

As usual in algebraic topology, there is much to be gained from establishing a "relative" version. We will use the sequence of subspaces

$$I^n \supseteq \partial I^n \supseteq \partial I^{n-1} \times I \cup I^{n-1} \times 0$$

in this definition. We will write J_n for the last subspace, so for example $J_1 = \{0\} \subset I$. In general it's an "open box":

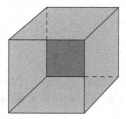

Definition 47.1. Let $(X, A, *)$ be a pointed pair. For $n \geq 1$, define a pointed set

$$\pi_n(X, A, *) = [(I^n, \partial I^n, J_n), (X, A, *)].$$

This definition is set up in such a way that

$$\pi_n(X, \{*\}, *) = \pi_n(X, *)$$

so that the inclusion $\{*\} \hookrightarrow A$ induces a map

$$\pi_n(X, *) \to \pi_n(X, A, *).$$

Also, restricting to the "back face" $I^{n-1} \times 0$ provides a map

$$\partial : \pi_n(X, A, *) \to \pi_{n-1}(A, *)$$

and the composite of these two is obviously "trivial," meaning that its image is the basepoint $* \in \pi_{n-1}(A, *)$. We get a sequence of pointed sets

$$\cdots \longrightarrow \pi_2(X, *) \longrightarrow \pi_2(X, A, *)$$
$$\partial$$
$$\pi_1(A, *) \longrightarrow \pi_1(X, *) \longrightarrow \pi_1(A, X, *)$$
$$\partial$$
$$\pi_0(A, *) \longrightarrow \pi_0(X, *).$$

We claim that this is an exact sequence of pointed sets: the *long exact homotopy sequence of a pair*. For example, an element of $\pi_1(X, A, *)$ is represented by a path starting at the basepoint and ending in A. Its boundary is the component of that point in A. Saying that the component of $a \in A$ maps to the base point component of X is exactly saying that $[a] \in \pi_0(A)$ is in the image of $\partial : \pi_1(X, A, *) \to \pi_0(A, *)$.

We will investigate the structure of these relative homotopy groups, and explain why the sequence is exact, by developing an analogue of the Barratt-Puppe sequence that will turn out to give rise to the homotopy long exact sequence of a pair.

Fiber sequences

In the pointed category, we could redefine "fibration" slightly (as is done in [36], for example) so that $p : E \to B$ is a fibration if every pointed solid arrow diagram

$$
\begin{array}{ccc}
W & \longrightarrow & E \\
\downarrow{\scriptstyle \mathrm{in}_0} & \nearrow & \downarrow{\scriptstyle p} \\
I_+ \wedge W & \longrightarrow & B
\end{array}
$$

admits a lift. There are fewer diagrams, but more is demanded of the lift.

Instead we'll leave the fibrations as they are, but in compensation insist that our basepoints should be nondegenerate. Lifting is then contained in the following lemma. See [67] for the proof, which we forgo, preferring to give the proof of similar result 48.6 later.

Lemma 47.2 (Relative homotopy lifting property). *Let $A \subseteq X$ be a closed cofibration and $E \to B$ a fibration. Then a lifting exists in any solid arrow diagram*

$$
\begin{array}{ccc}
(X \times 0) \cup (A \times I) & \longrightarrow & E \\
\downarrow & \nearrow & \downarrow \\
X \times I & \longrightarrow & B .
\end{array}
$$

Exactly the same proof we did before shows that if $A \to B$ is a pointed cofibration and the basepoints are nondegenerate then $X_*^B \to X_*^A$ is a fibration. For example we can take $(B, A, *) = (I, \partial I, 0)$ to see that the map from the path space

$$
P(X) = X_*^I = \{\omega : I \to X : \omega(0) = *\}
$$

to X by evaluation at 1 is a fibration.

Taking A to be a singleton in Lemma 47.2:

Corollary 47.3. *Let $p : E \to B$ be a fibration and suppose given $g : W \to E$ and $f : W \to B$ such that $pg \simeq f$: so g is a lift of f up to homotopy. Then g is homotopic to a lift "on the nose," that is, a function $\overline{g} : W \to E$ such that $p\overline{g} = f$.*

So if $g : W \to E$ is such that $pg \simeq *$, then g is homotopic to a map that lands in the fiber $p^{-1}(*) = F$ of p over $*$. This shows that the sequence — the "fiber sequence" — of pointed spaces

$$F \to E \to B$$

is "exact," in the sense that for any well-pointed space W the sequence

$$[W, F]_* \to [W, E]_* \to [W, B]_*$$

is exact.

Not every map is a fibration, but every map factors as

$$X \overset{\simeq}{\longrightarrow} T(f) \quad = \quad \{(x, \omega) \in X \times Y^I : \omega(1) = f(x)\}$$

where $X \to T(f)$ is a homotopy equivalence and p is a fibration.

The fiber of p is the *homotopy fiber* of f, written $F(f)$:

$$F(f) = \{(x, \omega) \in X \times Y_*^I : \omega(1) = f(x)\}.$$

Here we take $0 \in I$ as the basepoint, so ω is a path in Y from $*$ to $f(x)$.

As in our discussion of the Barratt-Puppe cofibration sequence, there is an equivalent way of constructing $F(f)$, by replacing $* \to Y$ with a fibration, namely the path space Y_*^I, and forming the pullback over X:

$$
\begin{array}{ccc}
F(f) & \longrightarrow P(Y) & = \quad Y_*^I \\
\downarrow & \downarrow & \\
X & \longrightarrow Y &
\end{array}
$$

Lemma 47.4. *Let* $p : E \to B$ *be a fibration and* $* \in B$. *The natural map* $p^{-1}(*) \to F(p)$ *is a homotopy equivalence.*

Proof. Regard $F(p)$ as the pullback of p along $PB \downarrow B$. The induced map on fibers is a homeomorphism; but PB is contractible so the inclusion of the fiber of $F(p) \downarrow PB$ into $F(p)$ is a homotopy equivalence, by Corollary 43.9. $\square$

Continuing with the analogy with cofibrations, the map $p(f) : F(f) \to X$ is a fibration, with fiber ΩX, and we have the *Barratt-Puppe fibration sequence*

$$Y \overset{f}{\leftarrow} X \overset{p}{\leftarrow} F(f) \overset{i}{\leftarrow} \Omega Y \overset{\Omega f}{\leftarrow} \Omega X \overset{\Omega p}{\leftarrow} \Omega F(f) \overset{\Omega i}{\leftarrow} \cdots .$$

It is exact, and it gives rise to the long exact homotopy sequence by virtue of the following lemma.

Lemma 47.5. *Let $(X, A, *)$ be a pointed pair, and let F denote the homotopy fiber of the inclusion $A \to X$. For each $n \geq 1$ there is a natural isomorphism*

$$\pi_n(X, A) \xrightarrow{\cong} \pi_{n-1}(F, *)$$

such that

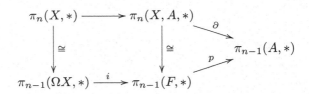

commutes.

Corollary 47.6. *The sequence homotopy long exact sequence of a pair is in fact exact; for $n \geq 2$ the set $\pi_n(X, A, *)$ is a group, abelian for $n \geq 3$; and all the maps between groups in the sequence are homomorphisms.*

Proof of Lemma 47.5. To begin with, notice that $\pi_1(X, A, *)$ is the set of path components of the space of maps

$$(I, \partial I, J_1) \to (X, A, *).$$

This is the space of paths in X from $*$ to some element of a: that is, it's precisely $F(A \to X)$.

In fact, for any $n \geq 1$, the space of maps

$$(I^n, \partial I^n, J_n) \to (X, A, *)$$

is precisely $\Omega^{n-1} F(A \to X)$. For example, when $n = 2$, an element in the given space is given by a map $I^2 \to X$ that is the basepoint along the bottom and takes values in A along the top — so a path in $F(A \to X)$ — and also is the basepoint along the left and right edges — so it's a loop in $F(A \to X)$.

The diagram is easily seen to commute. $\qquad \square$

There is another perspective on the homotopy long exact sequence, arising from Lemma 47.4.

Lemma 47.7. *Let $p : E \to B$ be a fibration and $* \in E$. Write $*$ also for the image of $*$ in B, and let F be the fiber over $*$. Then*

$$p_* : \pi_*(E, F, *) \to \pi_*(B, *)$$

is an isomorphism.

Proof. $F(p) \to E$ is a fibration, so by Proposition 43.7 and Lemma 47.4,

$$\text{hofib}(F \to E) \simeq \text{hofib}(F(p) \to E) \simeq \text{fib}(F(p) \to E) = \Omega B \,.$$

We leave to you the check that the resulting composite isomorphism

$$\pi_n(E, F) \to \pi_{n-1}(\text{hofib}(F \to E)) \to \pi_{n-1}(\Omega B) \to \pi_n(B)$$

is indeed the map induced by p_*. $\qquad\qquad\qquad\qquad\qquad\qquad\square$

We saw that $\pi_1(A, *)$ acts on $\pi_n(A, *)$. The map $\pi_n(A, *) \to \pi_n(X, *)$ is equivariant, if we let $\pi_1(A, *)$ act on $\pi_n(X, *)$ via the group homomorphism $\pi_1(A, *) \to \pi_1(X, *)$. The group $\pi_1(A, *)$ also acts on $\pi_n(X, A, *)$, compatibly, as illustrated by the following diagram.

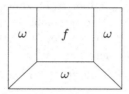

It's clear from the picture that the maps in the homotopy long exact sequence are equivariant.

Exercises

Exercise 47.8. Let $f : \Sigma X \to Y$ be a pointed map, and let $\hat{f} : X \to \Omega Y$ be its adjoint. Construct a map $g : CX \to PY$ from the cone to the path space $PY = Y_*^I$ such that the diagram

$$
\begin{array}{ccc}
X & \longrightarrow CX \longrightarrow & \Sigma X \\
\downarrow{\scriptstyle \hat{f}} & \downarrow{\scriptstyle g} & \downarrow{\scriptstyle f} \\
\Omega Y & \longrightarrow PY \longrightarrow & Y
\end{array}
$$

commutes.

Exercise 47.9. Let $p : E \to B$ be a fibration and $A \subseteq B$ a subset, and write E_A for the pullback of E to a fibration over A. Show that

$$\pi_n(E, E_A) \to \pi_n(B, A)$$

is an isomorphism for any $n \geq 1$.

Exercise 47.10. By passing to π_0, the action described in Exercise 47.11 provides a right action of the group $\pi_1(Y)$ on $\pi_0(F(f))$.

(a) Show that two elements in $\pi_0(F(f))$ map to the same element of $\pi_0(X)$ if and only if they are in the same orbit under this action.

(b) Suppose ω is a path in Y from $*$ to y. Write $\omega_\# : \pi_1(Y, *) \to \pi_1(Y, y)$ for the group isomorphism sending σ to $\omega\sigma\omega^{-1}$. Show that the isotropy group of the component of (x, ω) in $F(f)$ is

$$\omega_\#^{-1} \operatorname{im}(\pi_1(X, x) \to \pi_1(Y, f(x))) \subseteq \pi_1(Y, *).$$

(c) Suppose that X is path connected, and pick $* \in X$. Conclude from **(a)** that the evident surjection $\pi_n(X) \to [S^n, X]$ can be identified with the orbit projection for the action of $\pi_1(X)$ on $\pi_n(X)$.

Exercise 47.11. Let $f : X \to Y$ and fix $* \in Y$. Assume that Y is path-connected. The homotopy fiber $F(f)$ comes equipped with a fibration $p : F(f) \to X$ sending (x, ω) to x. The loop space ΩY "acts" on the homotopy fiber $F(f)$ from the right: let $\omega \in \Omega Y$ and $(x, \sigma) \in F(f)$, and define $(x, \sigma) \cdot \omega = (x, \sigma \cdot \omega)$ where

$$(\sigma \cdot \omega)(t) = \begin{cases} \omega(2t) & 0 \le t \le 1/2 \\ \sigma(2t - 1) & 1/2 \le t \le 1. \end{cases}$$

With $X = *$ we get the usual "multiplication" $\Omega Y \times \Omega Y \to \Omega Y$, which is known to be associative and unital up to homotopy (and to admit a homotopy inverse, sending ω to $\overline{\omega} : t \mapsto \omega(1 - t)$). The same proof shows that the action of ΩY on $F(f)$ is associative and unital up to homotopy.

Suppose a group G acts on a set S (from the right) with orbit space X. The fiber product $S \times_X S$ consists of pairs of elements in the same orbit. The action is free exactly when the map $S \times G \to S \times_X S$, sending (s, g) to (s, sg), is bijective.

Returning to the story of the homotopy fiber, note that $p((x, \sigma) \cdot \omega) = x = p(x, \sigma)$. We get a map

$$F(f) \times \Omega Y \to F(f) \times_X F(f)$$

to the fiber product by sending $((x, \sigma), \omega)$ to $((x, \sigma), (x, \sigma) \cdot \omega)$.

Finally, the problem: Show that this map is a homotopy equivalence.

So in a sense there is a "free" (a better word would be "principal") action of ΩY on $F(f)$ with orbit space X. In particular, $F(1_X)$ is the contractible path space PX; so X is the "orbit space" of an action of ΩX on a contractible space. This entitles us to regard X as the "classifying space" of ΩX.

Chapter 5

The homotopy theory of CW complexes

48   Serre fibrations and relative lifting

Relative CW complexes

We will do many proofs by induction over cells in a CW complex. We might as well base the induction arbitrarily. This suggests the following definition.

Definition 48.1. A *relative CW complex* is a pair (X, A) together with a filtration

$$A = X_{-1} \subseteq X_0 \subseteq X_1 \subseteq \cdots \subseteq X,$$

such that (1) for all n the space X_n sits in a pushout square:

$$
\begin{array}{ccc}
\coprod_{\alpha \in I_n} S_\alpha^{n-1} & \longrightarrow & \coprod_{\alpha \in I_n} D_\alpha^n \\
\downarrow & & \downarrow \\
X_{n-1} & \longrightarrow & X_n \, ,
\end{array}
$$

and (2) $X = \varinjlim X_n$ topologically.

The maps $S^{n-1} \to X_{n-1}$ are "attaching maps" and the maps $D^n \to X_n$ are "characteristic maps."

If $A = \varnothing$, this is just the definition of a CW complex. Often X will be a CW complex and A a subcomplex.

Our inductive strategy will involve constructing lifts inductively.

Definition 48.2. A map $p : E \to B$ is said to satisfy the *relative homotopy* *lifting property* with respect to $i : A \to X$ if every solid arrow diagram

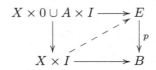

admits a "filler" as shown.

The absolute case is $A = \varnothing$.

This is in turn a special case of the following very general language, due to Quillen.

Definition 48.3. Fix maps $p : E \to B$ and $j : V \to W$. The map p satisfies the *right lifting property with respect to j*, and j satisfies the *left lifting property with respect to p*, if every solid arrow diagram

$$
\begin{array}{ccc}
V & \longrightarrow & E \\
\downarrow{\scriptstyle j} & \nearrow & \downarrow{\scriptstyle p} \\
W & \longrightarrow & B
\end{array}
$$

admits a filler as shown.

Serre fibrations

If we're going to restrict our attenton to CW complexes, we might as well weaken the lifting condition defining fibrations.

Definition 48.4. A map $p : E \to B$ is a *Serre fibration* if it has the homotopy lifting property with respect to all CW complexes. That is, for every CW complex X and every solid arrow diagram

$$
\begin{array}{ccc}
X & \longrightarrow & E \\
\downarrow{\scriptstyle \mathrm{in}_0} & \nearrow & \downarrow{\scriptstyle p} \\
X \times I & \longrightarrow & B
\end{array}
$$

there is a lift as indicated.

For contrast, what we called a fibration is also known as a *Hurewicz fibration*. (Witold Hurewicz (1904–1956) was a faculty member at MIT from 1945 till his death from a fall from the top of the Uxmal Pyramid in Mexico.)

Clearly things like the homotopy long exact sequence of a fibration extend to the context of Serre fibrations. So for example:

Lemma 48.5. *Suppose that* $p : E \to B$ *is both a Serre fibration and a weak equivalence. Then each fiber is weakly contractible; i.e. the map to* $*$ *is a weak equivalence.*

Proof. Since $\pi_0(E) \to \pi_0(B)$ is bijective, we may assume that both E and B are path connected. The long exact homotopy sequence shows that $\partial : \pi_1(B) \to \pi_0(F)$ is surjective with kernel given by the image of the surjection $\pi_1(E) \to \pi_1(B)$: so $\pi_0(F) = *$. Moving up the sequence then shows that all the higher homotopy groups of F are also trivial. $\quad\square$

No new ideas are required to prove the following two facts.

Proposition 48.6. *Let* $p : E \to B$. *The following are equivalent.*

(1) p is a Serre fibration.
(2) p has the HLP with respect to D^n for all $n \geq 0$.
(3) p has the relative HLP with respect to $S^{n-1} \hookrightarrow D^n$ for all $n \geq 0$.
(4) p has the relative HLP with respect to $A \hookrightarrow X$ for all relative CW complexes (X, A).

Proposition 48.7 (Relative straightening). *Assume that (X, A) is a relative CW complex and that $p : E \to B$ is a Serre fibration, and that the diagram*

$$
\begin{array}{ccc}
A & \longrightarrow & E \\
\downarrow{\scriptstyle j} & & \downarrow{\scriptstyle p} \\
X & \stackrel{g}{\longrightarrow} & B
\end{array}
$$

commutes. If g is homotopic to a map g' still making the diagram commute and for which there is a filler, then there is a filler for g.

Proof of "Whitehead's little theorem"

We are moving towards a proof of this theorem of J.H.C. Whitehead.

Theorem 48.8. *Let $f : X \to Y$ be a weak equivalence and W any CW complex. The induced map $[W, X] \to [W, Y]$ is bijective.*

The key fact is this:

Proposition 48.9. *Suppose that $j : A \hookrightarrow X$ is a relative CW complex and $p : E \to B$ is both a Serre fibration and a weak equivalence. Then a filler*

exists in any diagram

In the language of Quillen's *Homotopical Algebra*, this says that j satisfies the left lifting property with respect to "acyclic" Serre fibrations, and acyclic Serre fibrations satisfy the right lifting property with respect to relative CW complex inclusions.

Proof (following [48]). The proof will of course go by induction. The inductive step is this: Assuming that $p : E \to B$ is a Serre fibration and a weak equivalence, any diagram

$$
\begin{array}{ccc}
S^{n-1} & \longrightarrow & E \\
\Big\downarrow{\scriptstyle j} & \nearrow & \Big\downarrow{\scriptstyle p} \\
D^n & \longrightarrow & B
\end{array}
$$

admits a filler.

First let's think about the special case in which $B = *$. Then E is path connected, and for any choice of basepoint $* \in E$, $\pi_n(E, *) = 0$, so the extension exists.

For the general case, we begin by using Lemma 48.7 replacing the map g by a homotopic map g' with properties that will let us construct a filler. To define g', let $\varphi : D^n \to D^n$ by

$$
\varphi : v \mapsto \begin{cases} 0 & \text{if } |v| \leq 1/2 \\ (2|v| - 1)v & \text{if } |v| \geq 1/2. \end{cases}
$$

This map is homotopic to the identity (by a piecewise linear homotopy that fixes S^{n-1}), so $g' = g \circ \varphi \simeq g$.

The virtue of g' is that we can treat the two parts of D^n separately. The annulus $\{v \in D^n : |v| \geq 1/2\}$ is homeomorphic to $I \times S^{n-1}$, so a lifting exists on it since p is a Serre fibration. On the other hand g' is constant on the inner disk $D^n_{1/2}$, with value $g(0)$. We just constructed a lift on $S^{n-1}_{1/2}$, but it actually lands in the fiber of p over $g(0)$. We can fill in that map with a map $D^n_{1/2} \to p^{-1}(g(0))$ since the fiber is weakly contractible.   □

Proof of Theorem 48.8. Begin by factoring $f : X \to Y$ as a homotopy equivalence followed by a fibration; so as a weak equivalence followed by a

Serre fibration p. Weak equivalences satisfy "2 out of 3" (Exercise 46.13) so p is again a weak equivalence. Thus we may assume that f is a Serre fibration (as well as being a weak equivalence).

To see that the map is onto, apply Proposition 48.9 to

To see that the map is one-to-one, apply Proposition 48.9 to

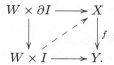

□

This style of proof — using lifting conditions and factorizations — is very much in the spirit of Daniel Quillen's formalization of homotopy theory in his development of "model categories" [54].

49 Connectivity and approximation

The language of connectivity

An analysis of the proof of "Whitehead's little theorem" shows that if the CW complex we are using as a source has dimension at most n, then we only needed to know that the map $X \to Y$ was an "n-equivalence" in the following sense.

Definition 49.1. Let n be a positive integer. A map $f : X \to Y$ is an *n-equivalence* provided that $f_* : \pi_0(X) \to \pi_0(Y)$ is an isomorphism, and for every choice of basepoint $a \in X$ the map $f_* : \pi_q(X, a) \to \pi_q(Y, f(a))$ is an isomorphism for $q < n$ and an epimorphism for $q = n$. It is a 0-*equivalence* if $f_* : \pi_0(X) \to \pi_0(Y)$ is an epimorphism.

So a map is a weak equivalence if it is an n-equivalence for all n. We restate:

Theorem 49.2. *Let n be a nonnegative integer and W a CW complex. If $f : X \to Y$ is an n-equivalence then the map $f_* : [W, X] \to [W, Y]$ is bijective if $\dim W < n$ and surjective if $\dim W = n$.*

The odd edge condition in the definition of n-equivalence might be made more palatable by noticing that the long exact homotopy sequence shows that (for $n > 0$) f is an n-equivalence if and only if $\pi_0(X) \to \pi_0(Y)$ is bijective and for any $b \in Y$ the group $\pi_q(F(f, b))$ is trivial for $q < n$.

This suggests some further language.

Definition 49.3. Let n be a positive integer. A space X is n-*connected* if it is path connected and for any choice of basepoint a the set $\pi_q(X, a)$ is trivial for all $q \leq n$. A space X is 0-*connected* if it is path connected.

So "1-connected" and "simply connected" are synonymous. The homotopy long exact sequence shows that for $n > 0$ a map $X \to Y$ is an n-equivalence if it is bijective on connected components and for every $b \in Y$ the homotopy fiber $F(f, b)$ is n-connected.

The language of connectivity extends to pairs:

Definition 49.4. Let n be a non-negative integer. A pair (X, A) is n-*connected* if $\pi_0(A) \to \pi_0(X)$ is surjective and for every basepoint $a \in A$ the set $\pi_q(X, A, a)$ is trivial for $q \leq n$.

That is, (X, A) is n-connected if the inclusion map $A \to X$ is an n-equivalence.

Skeletal approximation

Theorem 49.5 (Skeletal approximation theorem). *Let (X, A) and (Y, B) be relative CW complexes. Any map $f : (X, A) \to (Y, B)$ is homotopic rel A to a* skeletal *map — a map sending X_n into Y_n for all n. Any homotopy between skeletal maps can be deformed rel A to one sending X_n into Y_{n+1} for all n.*

I will not give a proof of this theorem. You have to inductively push maps off of cells, using smooth or simplicial approximation techniques. See for example [10, p. 208]. I am following Norman Steenrod in calling such a map "skeletal" rather than the more common "cellular," since it is after all not required to send cells to cells. With $A = \varnothing$:

Corollary 49.6. *Any map $X \to Y$ of CW complexes is homotopic to a skeletal map, and any homotopy between skeletal maps can be deformed to one sending X_n to Y_{n+1}.*

For example, the n-sphere $I^n/\partial I^n$ has a CW structure in which $\mathrm{Sk}_{n-1} S^n = *$ and $\mathrm{Sk}_n S^n = S^n$. The characteristic map is given by a choice of homeomorphism $D^n \to I^n$. So if $q < n$, then any map $S^q \to S^n$ factors through the basepoint up to homotopy. This shows that

$$\pi_q(S^n) = 0 \text{ for } q < n$$

— the n-sphere is $(n-1)$-connected. So also is any CW complex with one 0-cell and no other q-cells for $q < n$.

As a special case:

Proposition 49.7. *Let (X, A) be a relative CW complex in which all the cells of X are in dimension greater than n. Then (X, A) is n-connected.*

For example (with $A = \varnothing$) $\pi_0(X_0) \to \pi_0(X)$ is surjective: every path component of X contains a vertex. And $\pi_1(X_1) \to \pi_1(X)$ is surjective: any path between vertices can be deformed onto the 1-skeleton. Moreover, any homotopy between paths in the 1-skeleton can be deformed to lie in the 2-skeleton; $\pi_1(X_2) \to \pi_1(X)$ is an isomorphism.

For $n > 0$, this is saying that for any choice of basepoint in X, $\pi_q(X, X_n)$ is trivial for $q \leq n$.

CW approximation

Any space is weakly equivalent to a CW complex. In fact:

Theorem 49.8. *Any map $f : A \to Y$ admits a factorization as*

$$A \xrightarrow{\ i\ } X \xrightarrow{\ j\ } Y,$$
$$\underset{f}{}$$

where i is a relative CW inclusion and j is a weak equivalence.

This is analogous to the factorization as a cofibration followed by a homotopy equivalence. This factorization is part of the "Quillen model structure" [54] on spaces, while the earlier one is part of the "Strøm model structure" [68]. An important special case: $A = \varnothing$: so any space admits a weak equivalence from a CW complex. For a functorial way to do this, see Lecture 58.

Proof. To begin with, pick a point in each path component of Y not meeting A and adjoin to A a discrete set mapping surjectively to those points. This gives us a factorization $A \to X_0 \to Y$ in which X_0 is obtained from A by attaching 0-simplices and $X_0 \to Y$ is a 0-equivalence.

Next, for each pair of distinct path components of A that map to the same path component in Y pick points a, b in them and a path in Y from $f(a)$ to $f(b)$. These data determine a map to Y from the pushout

$$
\begin{array}{ccc}
\coprod S^0 & \longrightarrow & X_0 \\
\downarrow & & \downarrow \\
\coprod D^1 & \longrightarrow & X_1'
\end{array}
$$

that is bijective on π_0.

These constructions let us assume that both A and Y are path connected, and we do so henceforth. Pick a point in A to use as a basepoint, and use its image in Y as a basepoint there.

We want to add 1-cells to A to obtain a path-connected space X, along with an extension of f to a 1-equivalence $X \to Y$. This just means a surjection in π_1. So pick a subset of $\pi_1(Y)$ that together with $\operatorname{im}(\pi_1(A) \to \pi_1(Y))$ generate $\pi_1(Y)$, and pick a representative loop for each element of that set. This defines a map $X = A \vee \bigvee S^1 \to Y$ that is surjective on π_1.

Now suppose that $f : A \to Y$ is a 1-equivalence. We will adjoin 2-cells to A to produce a space X, together with an extension of f to a 2-equivalence.

As a convenience, we first factor f as $A \hookrightarrow Y' \to Y$ in which the first map is a closed cofibration and the second is a homotopy equivalence. This lets us assume that A is in fact a subspace of Y.

We want to adjoin 2-cells to produce an extension of f to a 2-equivalence $X \to Y$. The group $\pi_2(Y, A)$ measures the failure of f itself to be a 2-equivalence. It is a group with an action of $\pi_1(A)$. Pick generators of it as such, and for each pick a representative map

$$(D^2, S^1, *) \to (Y, A, *).$$

Together they determine a map to Y from the pushout in

$$
\begin{array}{ccc}
\coprod S^1 & \longrightarrow & A \\
\downarrow & & \downarrow \\
\coprod D^2 & \longrightarrow & X.
\end{array}
$$

We want to see that $\pi_1(X) \to \pi_1(Y)$ is an isomorphism and $\pi_2(X) \to \pi_2(Y)$ is an epimorphism. The factorization $A \to X \to Y$ determines a map of homotopy long exact sequences of groups:

$$
\begin{array}{ccccccccccc}
\pi_2(A) & \longrightarrow & \pi_2(X) & \longrightarrow & \pi_2(X, A) & \overset{\partial}{\longrightarrow} & \pi_1(A) & \longrightarrow & \pi_1(X) & \longrightarrow & * \\
\downarrow & & \downarrow & & \downarrow & & \parallel & & \downarrow & & \\
\pi_2(A) & \longrightarrow & \pi_2(Y) & \longrightarrow & \pi_2(Y, A) & \overset{\partial}{\longrightarrow} & \pi_1(A) & \longrightarrow & \pi_1(Y) & \longrightarrow & *.
\end{array}
$$

By construction, the middle arrow is surjective. The usual diagram chases show that $\pi_1(X) \to \pi_1(Y)$ is an isomorphism and that $\pi_2(X) \to \pi_2(Y)$ is an epimorphism.

An identical argument continues the induction. We carried out this case because it's slightly nonstandard, involving nonabelian groups.

At the end, we have to observe that the direct limit of a sequence of cell attachments enjoys the property that

$$\lim_{\to} \pi_q(X_n) \to \pi_q(\lim_{\to} X_n)$$

is an isomorphism. $\qquad\qquad\qquad\qquad\qquad\qquad\qquad\qquad\qquad\qquad\qquad\square$

Notice that if we only want to get to an n-equivalence, we need only add cells up to dimension n: Any space is n-equivalent to a CW complex of dimension at most n.

This construction is of course very ineffective: at each stage you have to compute some relative homotopy group! And since finite complexes have infinitely much homotopy, it seems that this process might go on for ever even for very simple spaces. In the simply connected case, though, you can do much better; indeed, as well as can be hoped for (Exercise 51.8).

Exercises

Exercise 49.9. Construct a CW approximation for ΩS^1.

50 Postnikov towers

Postnikov sections

The cell attaching method used in the proof of CW approximation has other applications.

Theorem 50.1 (Postnikov sections). *For any space X and any nonnegative integer n, there is a map $X \to P_n(X)$, the nth Postnikov section of X, with the following properties.*
(1) For every basepoint $* \in X$, $\pi_q(X, *) \to \pi_q(P_n(X), *)$ is an isomorphism for $q \leq n$.
(2) For every basepoint $* \in P_n(X)$, $\pi_q(P_n(X), *) = 0$ for $q > n$.
(3) $(P_n(X), X)$ is a relative CW complex with cells of dimension not less than $(n + 2)$.

When $n = 0$, the space $P_0(X)$ is "weakly discrete"; a CW approximation to it is given by a map $\pi_0(X) \to P_0(X)$.

When X is path connected and $n = 1$, this is asserting the existence of a path connected space $P_1(X)$ with $\pi_1(P_1(X)) = \pi_1(X)$ and *no higher homotopy groups*, and a map $X \to P_1(X)$ inducing an isomorphism on π_1. Assuming $P_1(X)$ is nice enough to have a universal cover, its universal cover will be weakly contractible. Such a space is said to be "aspherical." Thus any group π is the fundamental group of an aspherical space, because it occurs as $\pi_1(X)$ for a suitable 2-dimensional CW complex (remark following Proposition 18.6).

Proof. Work one connected component at a time. We'll progressively clean out the higher homotopy of the space X, constructing a sequence of spaces

$$X = X(n) \to X(n + 1) \to X(n + 2) \to \cdots$$

all sharing the same π_q for $q \le n$ but with

$$\pi_q(X(t)) = 0 \quad \text{for } n < q \le t.$$

We can take $X(n) = X$. Thereafter $X(t)$ will be built from $X(t - 1)$ by attaching $(t + 1)$-cells, so by Corollary 49.7 the pair $(X(t), X(t - 1))$ is t-connected: the inclusion induces isomorphisms in π_q for $q < t$ and $\pi_t(X(t), X(t - 1)) = 0$.

So we just want to be sure to kill $\pi_t(X(t - 1))$, while not introducing anything new in $\pi_t(X(t))$. Pick a set of generators for $\pi_t(X(t-1))$, and pick representatives $S^t \to X(t-1)$ for them. Attach $(t+1)$-cells to $X(t-1)$ using these maps as attaching maps, to form a space $X(t)$. Here's a fragment of the homotopy long exact sequence.

$$\pi_{t+1}(X(t), X(t-1)) \xrightarrow{\partial} \pi_t(X(t-1)) \to \pi_t(X(t)) \to \pi_t(X(t), X(t-1)) = 0.$$

By construction, the boundary map is surjective, so $\pi_t(X(t)) = 0$.

Now pass to the limit;

$$P_n(X) = \lim_{\to} X(t). \qquad \square$$

If X was a CW complex, we can use skeletal approximation to make all the attaching maps skeletal. They then join any cells of the same dimension in X, and the resulting space $P_n(X)$ admits the structure of a CW complex in which X is a subcomplex.

What's this about passing to the limit?

Lemma 50.2. *Any compact subspace of a CW complex lies in a finite subcomplex.*

Proof. The "interior" of D^n is $D^n - S^{n-1}$ (so for example the interior of D^0 is D^0 itself). A CW complex X is, as a set, the disjoint union of the interiors of its cells. These subspaces are sometimes called "open cells," but since they are rarely open in X I prefer "cell interiors." Any subset of X that meets each cell interior in a finite set is a discrete subspace of X. So any compact subset of X meets only finitely many cell interiors. In particular a CW complex is compact if and only if it is finite.

The boundary of an n-cell (i.e. the image of the corresponding attaching map) is a compact subspace of the $(n-1)$-skeleton. It meets only finitely many of the cell interiors in that $(n-1)$-dimensional CW complex. By induction on dimension, all of those cells lie in finite complexes, so the n-cell we began with lies in a finite subcomplex.

Now let K be a compact subspace of X. It lies in the union of the finite subcomplexes containing the finite number of cell interiors meeting K. This union is a finite subcomplex of X. $\qquad\square$

If (X, A) is a relative CW complex, the quotient X/A is a CW complex, where we can apply this lemma.

Corollary 50.3. *Let* $X(0) \subseteq X(1) \subseteq \cdots$ *be a sequence of relative CW inclusions. Then for each* q
$$\varinjlim \pi_q(X(n)) \xrightarrow{\cong} \pi_q(\varinjlim X(n)).$$

Proof. Both S^q and D^{q+1} are compact. $\qquad\square$

Now we have really gotten into homotopy theory! The space $P_n(X)$ is called the *nth Postnikov section* of X. (Mikhail Postnikov (1927–2004) worked at Steklov Institute in Moscow.) Most of the time they are infinite dimensional, and you usually can't compute their cohomology, even if you know the cohomology of X.

The Postnikov tower

How unique is the map $X \to P_n(X)$? How natural is this construction? To answer these questions, observe:

Proposition 50.4. *Let* n *be a nonnegative integer, and let* Y *be a space such that* $\pi_q(Y, *) = 0$ *for every choice of basepoint and all* $q > n$. *Let* (X, A) *be a relative CW complex. If all the cells in* $X - A$ *are of dimension at least* $n + 2$ *then the map*
$$[X, Y] \to [A, Y]$$
is bijective. If there are also $(n+1)$-*cells, the map is still injective.*

Proof. This uses the fact that if $\pi_q(Y, *) = 0$ then any map $S^q \to Y$ landing in the path component containing $*$ extends to a map from D^{q+1}.

Surjectivity: We extend a map $A \to Y$ to a map from X. For each attaching map $g : S^{q-1} \to \mathrm{Sk}_{q-1}X$ (where $q \geq n + 2$) the composite $f \circ g : S^{q-1} \to Y$ extends over the disk D^q since $q - 1 > n$.

Injectivity: Regard $(X \times I, X \times \partial I \cup A \times I)$ as a relative CW complex, in which the cells are of dimension one larger than those of X. $\square$

Corollary 50.5. *Let X be an n-connected CW complex and Y a space with homotopy concentrated in dimension at most n. Then every map from X to Y is homotopic to a constant map.*

Proof. By CW approximation, we may assume that X has a 0-cell and no other cells of dimension less than $n + 1$. The pair $(X, *)$ satisfies the requirement necessary to conclude that $[X, Y] \to [*, Y]$ is injective. $\square$

Now let $f : X \to Y$ be any map. Construct $X \to P_m(X)$ and $Y \to P_n(Y)$, so that $P_m(X)$ is attached using cells of dimension at least $m + 2$ and $\pi_q(P_n(Y)) = 0$ for $q > n$. If $m \geq n$, then by Proposition 50.4 there is a unique homotopy class of maps $P_m(X) \to P_n(X)$ making

$$
\begin{array}{ccc}
X & \xrightarrow{\ f\ } & Y \\
\downarrow & & \downarrow \\
P_m(X) & \dashrightarrow & P_n(Y)
\end{array}
$$

commute.

For example we could take $X = Y$ and use the identity map: For $m \geq n$ there is a unique homotopy class $P_m(X) \to P_n(X)$ making

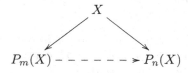

commute. When $m = n$, this shows that the map $X \to P_m(X)$ is unique up to a unique weak equivalence. When $m = n + 1$, it gives us a tower of spaces, the *Postnikov tower*:

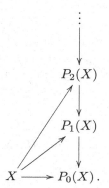

$$X \longrightarrow P_0(X).$$

As you go up in the tower you capture more and more of the homotopy groups of X. The Postnikov tower is functorial on the level of the homotopy category. We have a lot of control over how each space $P_n(X)$ is constructed, but very little control over what the resulting space looks like — e.g. what its homology is in high dimensions. There is likely to be a lot, even if X is a finite complex.

In a weak sense this tower is Eckmann-Hilton dual to a skeleton filtration: instead of building up a space as a direct limit of a sequence of spaces approximating the homology dimension by dimension, we are building it as the inverse limit of a sequence approximating the homotopy dimension by dimension.

More generally, Proposition 50.4 shows that $X \to P_n(X)$ is the *initial* map (in Ho**Top**) to a space with nontrivial homotopy only in dimension at most n.

Another common notation for $P_n(X)$ is $\tau_{\leq n}X$: the "truncation" of X at dimension n.

51   Hurewicz, Eilenberg, Mac Lane, and Whitehead

Hurewicz theorem

I have claimed that homotopy groups carry a lot of geometric information, but are correspondingly hard to compute. Homology groups are much easier; they are "local," in the sense that you can compute the homology of pieces of a space and glue the results together using Mayer-Vietoris. A cell structure quickly determines the homology (as we'll recall in the next lecture).

So it would be great if we had a way to compare homotopy and homology, maybe by means of a map

$$h : \pi_n(X) \to H_n(X).$$

First we have to fix an orientation for the sphere $S^n = I^n/\partial I^n$ (for $n > 0$). Do this by declaring the standard ordered basis to be positively ordered. This gives us a preferred generator $\sigma_n \in H_n(S^n)$.

Now let $\alpha \in \pi_n(X)$. This homotopy class of maps $S^n \to X$ determines a map $H_n(S^n) \to H_n(X)$. Define

$$h(\alpha) = \alpha_*(\sigma_n).$$

This is a well-defined map $h : \pi_n(X) \to H_n(X)$, the *Hurewicz map*.

Lemma 51.1. *h is a homomorphism.*

Proof. The product in $\pi_n(X)$ (or sum if $n > 1$) is given by the composite

$$
\begin{array}{ccc}
S^n & \xrightarrow{\;\alpha\beta\;} & X \\
\downarrow{\scriptstyle \delta} & & \uparrow{\scriptstyle \nabla} \\
S^n \vee S^n & \xrightarrow{\;\alpha \vee \beta\;} & X \vee X,
\end{array}
$$

where δ pinches an equator and ∇ is the fold map. Apply $\overline{H}_n$ and trace where σ_n goes:

$$
\begin{array}{ccc}
\sigma_n & & h(\alpha) + h(\beta) \\
\big\downarrow & & \big\uparrow \\
(\sigma_n, \sigma_n) & \longmapsto & (h(\alpha), h(\beta)).
\end{array}
$$
$\qquad\square$

When $n = 1$, the Hurewicz homomorphism factors through the abelianization of $\pi_1(X)$.

Theorem 51.2 (Hurewicz). *If X is path-connected, $\pi_1(X)^{ab} \to H_1(X)$ is an isomorphism. If X is $(n-1)$-connected for $n > 1$, $\pi_n(X) \to H_n(X)$ is an isomorphism.*

This can be proved by "elementary means," but we'll prove an improved form of this theorem later and I'd prefer to defer the proof. The $n = 1$ case is due to Poincaré; see [72, Theorem 9.2.1] for example.

This lowest dimension in which homotopy can occur is the "Hurewicz dimension." If X is an $(n-1)$-connected CW complex, it has a CW approximation that begins in dimension n, and the reduced homology (being isomorphic to the cellular homology) vanishes below dimension n.

In the simply connected case there is a converse.

Corollary 51.3. *Let X be a simply connected space. If $\overline{H}_q(X) = 0$ for $q < n$ then X is $(n-1)$-connected.*

Proof. If $n > 2$, the Hurewicz theorem says that $\pi_2(X) = H_2(X) = 0$, so X is 2-connected. And so on. $\square$

Simple connectivity is required here. A good example is provided by the "Poincaré sphere." Let I be the group of orientation-preserving symmetries of the regular icosahedron. It is a subgroup of $SO(3)$ of order 60. Its preimage $\tilde{I}$ in the double cover S^3 of $SO(3)$ is a perfect group (of order 120). The quotient space $S^3/\tilde{I}$ thus has $H_1 = 0$. The group acts freely by oriented diffeomorphisms, so the quotient is an oriented 3-manifold, and by Poincaré duality it has the same homology as S^3. But its fundamental group is $\tilde{I}$, so it is not homotopy equivalent to S^3. You can't decide whether or not you need 1-cells or 2-cells by looking at homology alone, in this non-simply connected example. In fact $\tilde{I}$ can be presented with two generator and two relations, so $S^3/\tilde{I}$ has a CW structure with two 1-cells and two 2-cells. The boundary map $C_2 \to C_1$ is an isomorphism.

Eilenberg Mac Lane spaces

Now let M be a Moore space for π, n, as described in Lecture 18. Our construction of it began with n-cells, so by skeletal approximation it has no homotopy below dimension n. (We don't need to appeal to Corollary 51.3 for this.) It probably has lots of homotopy above dimension n, but we can kill all that by forming the Postnikov stage or truncation

$$P_n(M) = \tau_{\leq n} M.$$

This is now a space with just one homotopy group, in dimension n. The Hurewicz theorem tells us that this single homotopy group is canonically isomorphic to π.

If $n = 1$ we can start with any group π, abelian or not, build a 2-dimensional complex with $\pi_1 = \pi$ as in the remark below Proposition 18.6, and form its Postnikov 1-section.

So we have now constructed a space with a single nonzero homotopy group, in dimension n. This is an *Eilenberg Mac Lane space*, denoted

$$K(\pi, n).$$

You know some examples of Eilenberg Mac Lane spaces already.

- $K(\mathbb{Z}, 1) = S^1$. $K(\mathbb{Z}^n, 1) = (S^1)^n$.

- Any closed surface other than S^2 and $\mathbb{R}P^2$ has contractible universal cover and so is aspherical. There are many other examples of aspherical compact manifolds. But as soon as there is torsion in a group, the Eilenberg Mac Lane space is infinite dimensional.
- The space $\mathbb{R}P^n$ has S^n as universal cover, and as $n \to \infty$ the space S^n loses all its homotopy groups. So

$$K(\mathbb{Z}/2\mathbb{Z}, 1) = \mathbb{R}P^\infty .$$

Similarly,

$$K(\mathbb{Z}, 2) = \mathbb{C}P^\infty .$$

In contrast to the Moore space $M(\pi, n)$, the Eilenberg Mac Lane space $K(\pi, n)$ can be constructed functorially in π.

The Whitehead tower

One further thing we can do at this point: Endow X with a basepoint $*$ and form the homotopy fiber of the map $X \to \tau_{\leq n} X$. By the homotopy long exact sequence, the map from the homotopy fiber will induce isomorphisms in π_q for $q > n$, while the homotopy groups of the homotopy fiber will be trivial for $q \leq n$: it is n-connected. Let's write $\tau_{>n} X$ for this space. For example, $\tau_{>0} X$ is the basepoint component of X (assuming $X \to \pi_0(X)$ is continuous). $\tau_{>2} X$ is the universal cover of X (assuming that X is path connected and is nice enough to admit a universal cover).

The example of covering spaces shows that $\tau_{>n} X \to X$ is not unique in quite the same sense that $X \to \tau_{>n} X$ is; you need a basepoint condition. In the pointed homotopy category, $\tau_{>n} X \to X$ is the terminal map from an n-connected space.

These spaces fit into a tower also, this time with X at the bottom:

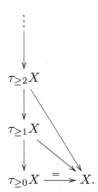

This is the *Whitehead tower*. (George Whitehead, 1918–2004, MIT faculty member, was a transitional figure in the development of algebraic topology [38], bridging the gap between an earlier "geometric" era and the more algebraic methods that appeared in the 1950's. He was apparently related neither to Alfred North Whitehead nor to J.H.C. Whitehead.)

Moore-Postnikov factorization

The Postnikov and Whitehead towers are extreme cases of a general factorization, which may be constructed just as we built the Postnikov system.

Theorem 51.4. *Let* $f : X \to Y$ *be any map between pointed path connected spaces. There is a diagram*

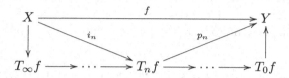

in which $\pi_q(i_n)$ *is an isomorphism for all* $q < n$ *and* $\pi_q(p_n)$ *is an isomorphism for all* $q \geq n$.

As one progresses through the Moore-Postnikov system, the space X morphs into Y, through a series of intermediate "gryphons," with lower part from X and upper part from Y.

Exercises

Exercise 51.5. Verify the claims above about the Poincaré sphere.

Exercise 51.6. Identify each of the spaces $\tau_{>2}\mathbb{C}P^n$ and $\tau_{\leq 2}S^2$ with known CW complexes (up to homotopy type, of course).

Exercise 51.7. (a) Let $N < G$ be a normal subgroup, with quotient group H. Show that there is a fibration $K(G,1) \to K(H,1)$ with fiber weakly equivalent to — a phrase we'll neglect below! — $K(N,1)$.

(b) Suppose that G is abelian. Then the same argument gives us a fibration $K(G,n) \to K(H,n)$ with fiber $K(N,n)$. But show also that there is a fibration $K(N,n) \to K(G,n)$ with fiber $K(H,n-1)$, and a fibration $K(H,n) \to K(N,n+1)$ with fiber $K(G,n)$. For example, what is the homotopy fiber of the map $\mathbb{C}P^\infty \to \mathbb{C}P^\infty$ represented by twice a generator of $H^2(\mathbb{C}P^\infty)$?

Exercise 51.8. Prove the following theorem of C.T.C. Wall [75], known as "homology approximation." Let Y be a simply connected space such that $H_n(Y)$ is finitely generated for all n. Let β_n be the nth Betti number (the rank of $H_n(Y)$) and let τ_n be the nth torsion number (the number of finite cyclic summands in $H_n(Y)$). Then there is a CW complex with $(\beta_n + \tau_n + \tau_{n-1})$ n-cells for each n that admits a weak equivalence to Y. This is clearly optimal, since in order to produce a finite cyclic summand in the nth homology of a chain complex of finitely generated abelian groups you need generators in dimension n and $n+1$.

52 Representability of cohomology

I want to think a little more about the significance of Eilenberg Mac Lane spaces. First, how unique are they?

Let π be an abelian group and n a positive integer. Pick a free resolution

$$0 \to F_1 \to F_0 \to \pi \to 0\,,$$

pick generators for F_0 and F_1, and build the corresponding cofiber sequence

$$\bigvee_j S^n \to \bigvee_i S^n \to M\,.$$

Our first model for $K(\pi, n)$ is the Postnikov section $\tau_{\leq n} M$.

Lemma 52.1. *Let n be a positive integer and let Y be any pointed space such that $\pi_q(Y, *) = 0$ for $q \neq n$, and write G for $\pi_n(Y, *)$. Then*

$$\pi_n : [\tau_{\leq n} M, Y]_* \to \mathrm{Hom}(\pi, G)$$

is an isomorphism.

Proof. Since $M \to \tau_{\leq n} M$ is universal among maps to spaces with homotopy concentrated in dimensions at most n, it's enough to show that

$$\pi_n : [M, Y]_* \to \mathrm{Hom}(\pi, G)$$

is an isomorphism. Since the sequence defining M is co-exact, we have an exact sequence

$$[\bigvee_j S^n, Y]_* \leftarrow [\bigvee_i S^n, Y]_* \leftarrow [M, Y]_* \leftarrow [\bigvee_j S^{n+1}, Y]_*\,.$$

Our assumptions on Y imply that this sequence reads

$$\mathrm{Hom}(F_1, G) \leftarrow \mathrm{Hom}(F_0, G) \leftarrow [M, Y]_* \leftarrow 0\,.$$

But a homomorphism $F_0 \to G$ that restricts to zero on F_1 is exactly a homomorphism $\pi \to G$. $\qquad\square$

In this proof π and G could be non-abelian if $n = 1$.

In particular, we could take $G = \pi$, and discover that there is a unique homotopy class of maps $\tau_{\leq n} M \to Y$ inducing the identity in π_n. This map is a weak equivalence. So if Y is also a CW complex, the map is a homotopy equivalence.

We learn from this that any two CW complexes of type $K(\pi, n)$ are homotopy equivalent by a homotopy equivalence inducing the identity on π_n, and that such a homotopy equivalence is unique up to homotopy. This leads to:

Corollary 52.2. *For any positive integer n there is a functor*

$$\mathbf{Ab} \to \mathrm{Ho}(\mathbf{CW}_*)$$

sending π to a space of type $K(\pi, n)$, unique up to isomorphism. When $n = 1$ this extends to a functor

$$\mathbf{Gp} \to \mathrm{Ho}(\mathbf{CW}_*)\,.$$

In fact it is possible to construct $K(\pi, n)$ as a functor from $\mathbf{Ab}$ to the category of topological abelian groups; see Lecture 58.

The case $n = 1$ is due to Heinz Hopf: There is, up to homotopy, a unique aspherical space with any prescribed fundamental group. The theory of covering spaces can be used in that case to check functoriality. This provides a collection of invariants of groups, $H_n(K(\pi, 1); G)$ and $H^n(K(\pi, 1); G)$. More generally, any π-module M determines a local coefficient system $\widetilde{M}$ over $K(\pi, 1)$, and one then has local homology and cohomology groups. It's not hard to show these are the homology and cohomology of the group with these coefficients:

$$H_n(K(\pi, 1); \widetilde{M}) = \mathrm{Tor}_n^{\mathbb{Z}[\pi]}(\mathbb{Z}, M)\,, \quad H^n(K(\pi, 1); \widetilde{M}) = \mathrm{Ext}_{\mathbb{Z}[\pi]}^n(\mathbb{Z}, M)\,.$$

Fundamental classes

Let n be a positive integer and Y an $(n-1)$-connected space. Then $\widetilde{H}_q(Y) = 0$ for $q < n$. Let π be an abelian group. The universal coefficient theorem asserts the existence of a short exact sequence

$$0 \to \mathrm{Ext}^1(H_{q-1}(Y), \pi) \to H^q(Y; \pi) \to \mathrm{Hom}(H_q(Y), \pi) \to 0$$

for any q. This shows that $H^q(Y; \pi) = 0$ for $q < n$. When $q = n$, the Ext term vanishes so the second map is an isomorphism. If we take $\pi = \pi_n(Y)$, for example, the inverse of the Hurewicz isomorphism is an element in Hom, and so delivers to us a canonical cohomology class in $H^n(Y; \pi_n(Y))$.

In particular, with $Y = K(\pi, n)$ we obtain a canonical class

$$\iota_n \in H^n(K(\pi, n); \pi)$$

called the *fundamental cohomology class*. Using it, we get a canonical natural transformation

$$[X, K(\pi, n)] \to H^n(X; \pi)$$

sending f to $f^*(\iota_n)$.

Theorem 52.3. *If X is a CW complex, this map is an isomorphism.*

That is: On CW complexes, cohomology is a *representable functor*; the representing object is the appropriate Eilenberg Mac Lane space; and ι_n is the universal n-dimensional cohomology class with coefficients in π.

Test cases: We decided that $K(\mathbb{Z}/2\mathbb{Z}, 1) = \mathbb{R}P^\infty$. So the claim is that $H^1(X; \mathbb{Z}/2\mathbb{Z}) = [X, \mathbb{R}P^\infty]$. We'll discuss this in more detail later, but $\mathbb{R}P^\infty$ carries the universal real line bundle, so the set of homotopy classes of maps into it (from a CW complex X) is in bijection with the set of isomorphism classes of real line bundles over X.

Similar story for $H^2(X; \mathbb{Z}) = [X, \mathbb{C}P^\infty]$.

One other case is of interest:

$$H^1(X, \mathbb{Z}) = [X, S^1] \,.$$

Other cases are less geometric!

Proof of Theorem 52.3. We'll prove a pointed version of the statement:

$$[X, K(\pi, n)]_* \xrightarrow{\cong} \overline{H}^n(X; \pi) \,.$$

Fix π, and pick any sequence of Eilenberg Mac Lane CW complexes, $K(\pi, n)$, $n \geq 0$. Thus for example $K(\pi, 0)$ is a CW complex that is homotopy equivalent to the discrete group π: we can take it to *be* π as a discrete group if we want.

The space $\Omega K(\pi, n+1)$ accepts a map from $K(\pi, n)$ that is an isomorphism on π_n; a CW replacement for $\Omega K(\pi, n+1)$ thus serves as another model for $K(\pi, n)$. Thus $K(\pi, n)$ has the structure of an H-group. In fact one can use $\Omega^2 K(\pi, n+2)$, by the same argument; so this H-group structure is abelian, and the functor $[-, K(\pi, n)]_*$ takes values in abelian groups.

The map $[X, K(\pi, n)]_* \to \overline{H}^n(X; \pi)$ is a homomorphism. To see this, use the pinch map $\Sigma X \to \Sigma X \vee \Sigma X$ to produce a homomorphism

$$\overline{H}^{n+1}(\Sigma X; \pi) \times \overline{H}^{n+1}(\Sigma X; \pi) \to \overline{H}^{n+1}(\Sigma X \vee \Sigma X; \pi) \to \overline{H}^{n+1}(\Sigma X; \pi) \,.$$

The argument proving that π_2 is abelian shows that this map coincides with the addition in the group $\overline{H}^{n+1}(\Sigma X; \pi) = \overline{H}^n(X; \pi)$.

The group structure in

$$[X, K(\pi, n)]_* = [X, \Omega K(\pi, n+1)]_* = [\Sigma X, K(\pi, n+1)]_*$$

has the same source; so the map is a homomorphism by naturality.

Now I will try to prove that the map is an isomorphism by induction on skelata.

When $X = X_0$, we can agree that

$$\mathrm{map}_*(X_0, \pi) = \overline{H}^0(X_0, \pi), \quad [X_0, K(\pi, n)]_* = 0 = \overline{H}^n(X_0; \pi) \quad \text{for } n > 0.$$

We may henceforth assume that X is connected. In general we have a cofiber sequence $\bigvee S^{q-1} \to X_{q-1} \to X_q$. It is co-exact and hence induces an exact sequence in $[-, K(\pi, n)]_*$. It also induces an exact sequence in reduced cohomology, one that can be regarded as coming from the same geometric source. Since both S^{q-1} and X_{q-1} are of dimension less than q, the map is an isomorphism for them. So by the 5-lemma it's an isomorphism on X_q.

There is still a limiting argument to worry about, if X is infinite dimensional, but I will not address that here. $\qquad\square$

Remark 52.4. One can also prove directly that cohomology is a representable functor on CW complexes, and then define Eilenberg Mac Lane spaces as the representing objects. The relevant theorem is "Brown representability" [12]. The fact that contravariant functors satisfying the kind of "descent" embodied by the Mayer-Vietoris theorem are representable gives homotopy theory a special character. Most of the time you can just work with spaces, which are much more concrete than functors!

Remark 52.5. Note that the suspension isomorphism in reduced cohomology is represented by the weak equivalence

$$K(\pi, n) \to \Omega K(\pi, n+1)$$

adjoint to the map representing the suspension of the fundamental class. A sequence of pointed spaces $\ldots, E_0, E_1, \ldots$ equipped with maps $E_n \to \Omega E_{n+1}$ (or equivalently $\Sigma E_n \to E_{n+1}$) is a *(topological) spectrum*. It's an Ω-*spectrum* if the maps $E_n \to \Omega E_{n+1}$ are all weak equivalences. Much of what we just did above carries over to Ω-spectra in general; the (abelian!) groups

$$\overline{E}^n(X) := [X, E_n]_*$$

form the groups in a (reduced generalized) cohomology theory. There are many examples. Any generalized cohomology theory is representable on CW complexes by an Ω spectrum.

Remark 52.6. One asset of representability is the Yoneda Lemma 39.13, which implies:

$$\text{n.t.}(H^m(-, A), H^n(-, B)) = [K(A, m), K(B, n)] = H^n(K(A, m); B).$$

Understanding the natural transformations acting between different dimensions of $H^*(-; \mathbb{F}_2)$, for example, is addressing the optimal value category for mod 2 cohomology. It's a graded $\mathbb{F}_2$ algebra, yes, but much more as well. This is the story of Steenrod operations (Lecture 75) and it's addressed in full by computing $H^*(K(\mathbb{F}_2, n); \mathbb{F}_2)$.

53 Obstruction theory

Cellular homology

We replay some of Lecture 16 in the relative case. Let (X, A) be a relative CW complex with skelata

$$A = X_{-1} \subseteq X_0 \subseteq X_1 \subseteq \cdots \subseteq X.$$

The inclusion $X_{n-1} \hookrightarrow X_n$ is a cofibration, so $H_*(X_n, X_{n-1}) \cong \overline{H}_*(X_n/X_{n-1})$. A choice of cell structure establishes a homeomorphism

$$X_n/X_{n-1} = \bigvee_{i \in I_n} S_i^n,$$

where I_n is the set of n-cells, so

$$H_*(X_n, X_{n-1}) \cong \mathbb{Z}[I_n].$$

This group is the *cellular chain group* $C_n = C_n(X, A)$.

There is a boundary map $d : C_{n+1} \to C_n$, defined by

$$d : C_{n+1} = H_{n+1}(X_{n+1}, X_n) \xrightarrow{\partial} H_n(X_n) \to H_n(X_n, X_{n-1}) = C_n.$$

This gives us the *cellular chain complex*. In terms of the basis given by a choice of cell structure, the differential $d : C_{n+1} \to C_n$ is giving exactly the data of the *relative attaching maps*

$$S^n \xrightarrow{\alpha_i} X_n \to X_n/X_{n-1},$$

where α_i runs through the attaching maps of the $(n+1)$-cells. Passage to the relative attaching maps forgets a great deal of information about the homotopy type of X; homology is a rather weak invariant in this sense.

Theorem 16.3 asserts (at least when $A = \varnothing$) that

$$H_n(X, A) \cong H_n(C_*(X, A)).$$

Of course, the same story runs for cohomology: one gets a chain complex which, in dimension n, is given by

$$C^n(X, A; \pi) = \mathrm{Hom}(C_n(X, A), \pi) = \mathrm{Map}(I_n, \pi),$$

where π is any abelian group, and

$$H^n(X, A; \pi) = H^n(C^*(X, A; \pi)).$$

Obstruction theory

We've seen that when the dimension of the CW complex X is less than the connectivity of the space Y, any map from X to Y is null-homotopic. What if there is some overlap? Here's a more general type of question we can try to answer.

Question 53.1. Let $f : A \to Y$ be a map from a space A to Y. Suppose (X, A) is a relative CW complex. When can we find an extension in the diagram below?

We've seen that answering this kind of question can also lead to results about the uniqueness of an extension, by considering $X \times \partial I \cup A \times I \subseteq X \times I$.

Let's try to make this extension skeleton by skeleton, and find what obstructions occur. We can start easily enough! If Y is empty then A is too, and there's an extension if and only if X is empty as well.

More realistically, as long as Y is nonempty we can certainly extend to X_0 by sending the new points anywhere you like in Y.

So make such a choice: $f : X_0 \to Y$. Can we extend f further over X_1? Well, we can extend if and only if for every pair a and b of 0-cells in X_0 that are in the same path component of X_1, the images $f(a)$ and $f(b)$ are in the same path component in Y. Note that we might do better at this stage if we could go back and choose f better. This simple observation serves as a model for the whole process.

Let's now assume we have constructed $f : X_n \to Y$, for $n \geq 1$, and hope to extend it over X_{n+1}. Pick attaching maps for the $(n+1)$-cells, so we have the diagram

$$
\begin{array}{ccccc}
\coprod_{i \in I_{n+1}} S^n & \xrightarrow{\ \alpha\ } & X_n & \xrightarrow{\ f\ } & Y \\
\downarrow & & \downarrow & \nearrow & \\
\coprod_{i \in I_{n+1}} D^{n+1} & \longrightarrow & X_{n+1}. & &
\end{array}
$$

The desired extension exists exactly when the composite $S^n \xrightarrow{\alpha_i} X_n \to Y$ is nullhomotopic for each $i \in I_{n+1}$.

Now is the moment to assume that Y is path connected and simple, so that

$$[S^n, Y] = \pi_n(Y, *)$$

canonically for any choice of basepoint. We will therefore omit basepoints from the notation.

This procedure produces a map $\theta_f : I_{n+1} \to \pi_n(Y)$, that is, an $(n+1)$-cochain, $\theta_f \in C^{n+1}(X, A; \pi_n(Y))$, and $\theta_f = 0$ if and only if f extends to a map $X_{n+1} \to Y$.

Proposition 53.2. *θ_f is a cocycle in $C^{n+1}(X, A; \pi_n(Y))$.*

Proof. θ_f is a homomorphism $H_{n+1}(X_{n+1}, X_n) \to \pi_n(Y)$. We would like to show that the composite

$$H_{n+2}(X_{n+2}, X_{n+1}) \xrightarrow{\partial} H_{n+1}(X_{n+1}) \to H_{n+1}(X_{n+1}, X_n) \xrightarrow{\theta_f} \pi_n(Y)$$

is trivial.

We'll see this by relating the homotopy long exact sequence to the homology long exact sequence. A relative homotopy class is represented by a map

$$(I^q, \partial I^q, J_q) \to (X, A, *).$$

Our choice of orientation for $I^q/\partial I^q$ specifies a generator for $H_q(I^q, \partial I^q)$. Evaluation of H_n then determines a map

$$h : \pi_q(X, A, *) \to H_q(X, A),$$

the *relative Hurewicz homomorphism.* It is again a homomorphism, extending the definition of the absolute Hurewicz homomorphism, and gives us a map of long exact sequences.

The characteristic maps in the cell structure for X give us elements of $\pi_{n+1}(X_{n+1}, X_n)$ that map to the generators of $H_{n+1}(X_{n+1}, X_n)$.

These observations lead to part of the commutative diagram below.

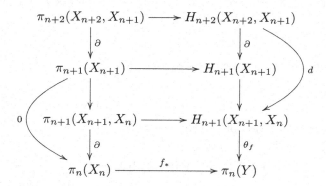

The bottom square commutes by definition of θ_f. Tracing around the left side goes through two successive maps in the homotopy long exact sequence, and so sends these elements to zero. $\qquad\square$

This cochain θ_f is the *obstruction cocycle* associated to $f : X_n \to Y$. It obstructs the extension of f over the $(n+1)$-skeleton. This theorem gives a way of extending a map $A \to Y$ skeleton by skeleton all the way to a map $X \to Y$.

But it could happen that the extension you made to X_n doesn't admit a further extension to X_{n+1}, while some other extension to X_n would. In order to maintain some control, let's fix the extension to X_{n-1}, but allow the extension to X_n to vary.

Theorem 53.3. *Let (X, A) be a relative CW complex and Y a path-connected simple space, and let $n \geq 1$. Let $f : X_n \to Y$ be a map from the n-skeleton of X, and let $\theta_f \in C^{n+1}(X, A; \pi_n(Y))$ be the associated obstruction cocycle. Then $f|_{X_{n-1}}$ extends to X_{n+1} if and only if $[\theta_f] \in H^{n+1}(X, A; \pi_n(Y))$ is zero.*

Proof. The proof begins with the construction a *difference cochain* $\delta_{f',f''}$ associated to maps $f', f'' : X_n \to Y$ together with a homotopy from $f'|_{X_{n-1}}$ to $f''|_{X_{n-1}}$ rel A. It will not be a cocycle. Instead, it will provide a homology between the obstruction cocycles associated to f' and f''. .

We'll lighten notation by dropping indication of the subspace A. Fix a cell structure on X. This is about homotopies, so let's begin by giving

$X \times I$ the CW structure in which

$$(X \times I)_n = (X_n \times \partial I) \cup (X_{n-1} \times I).$$

Each n-cell e in X produces in $X \times I$ an $(n+1)$-cell $e \times I$ and two n-cells $e \times 0$ and $e \times 1$. Thus there is a map

$$- \times I : C_n(X) \to C_{n+1}(X \times I),$$

given by linearly extending the assignment on cells. This is not a chain map; rather

$$d(e \times I) = (de) \times I + (-1)^n(e \times 1 - e \times 0)$$

(by choice of orientation of the unit interval).

This construction defines a map

$$C^{n+1}(X; \pi_n(Y)) \to C^n(X; \pi_n(Y))$$

by sending a cochain c to $e \mapsto c(e \times I)$.

Define a map $g : (X \times I)_n \to Y$ as follows. Send $X_n \times 0$ by f_0, $X_n \times 1$ by f_1, and $X_{n-1} \times I$ by a homotopy between the restrictions of f_0 and f_1 to X_{n-1}. We then have the obstruction cocycle $\theta_g \in C^{n+1}(X \times I; \pi_n(Y))$ associated to the map g.

Our difference cochain $\delta \in C^n(X; \pi_n(Y))$ is defined by

$$\delta(e) = \theta_g(e \times I).$$

For any n-cell e in X, calculate as follows, using the definition of the differential in the cellular cochain complex:

$$0 = (d\theta_g)(e \times I) = \theta_g(d(e \times I)) = \theta_g((de) \times I) \pm (\theta_g(e \times 0) - \theta_g(e \times 1)).$$

The three terms can be re-expressed as follows.

$$\theta_g((de) \times I) = \delta(de) = (d\delta)(e),$$

$$\theta_g(e \times 0) = \theta_{f'}(e), \quad \theta_g(e \times 1) = \theta_{f''}(e).$$

This verifies that

$$d\delta = \pm(\theta_{f'} - \theta_{f''}).$$

So for a map $f : X_n \to Y$, the cohomology class of the obstruction cocycle θ_f depends only on $f|_{X_{n-1}}$. In particular if $f|_{X_{n-1}}$ does extend to a map from X_{n+1}, then this cohomology class vanishes.

For the converse, we observe that for any $f' : X_n \to Y$ and $\delta \in C^n(X; \pi_n(Y))$ there exists an extension f'' of $f'|_{X_{n-1}}$ such that δ is precisely the difference cochain associated to the pair (f', f'') and the constant homotopy between their restrictions to X_{n-1}. We leave this to you; it uses the homotopy extension property.

We can now argue as follows. Suppose that $[\theta_{f'}] = 0 \in H^{n+1}(X; \pi_n(Y))$. Pick a null-homology δ of $\theta_{f'}$, and pick f'' in such a way that δ is the difference cocycle between f' and f''. Then (adjusting the sign if necessary)

$$\theta_{f''} = \theta_{f'} - d\delta = 0,$$

so f'' extends to X_{n+1}. $\qquad\qquad\square$

The easiest way to check that an obstruction class vanishes is to know that it lies in a zero group.

Corollary 53.4. *Let Y be a path connected simple space and (X, A) a relative CW complex. If $H^{n+1}(X, A; \pi_n(Y)) = 0$ for all $n \geq 1$ then any map $A \to Y$ extends to a map $X \to Y$. If moreover $H^n(X, A; \pi_n(Y)) = 0$ for all $n \geq 1$ then such an extension is unique up to homotopy rel A.*

Proof. The second assertion follows from the isomorphism

$$H^{n+1}(X \times I, A \times I \cup X \times \partial I; \pi) = H^n(X, A; \pi).\qquad\square$$

This raises important questions. The reduced cohomology of a space may well be trivial with coefficients in a finite p-group, for a fixed prime p, for example. Are there homological conditions on Y guaranteeing that each homotopy group is a finite p-group? The power to prove results of that sort is part of the revolution in homotopy theory engineered by Jean-Pierre Serre, developments we will get to later in this course.

Two extensions

First, the difference cochain has other uses. Let X be an n-dimensional CW complex, Y any space, and consider maps $f : X \to Y$. We have seen that if Y is n-connected then any map from X to Y is null-homotopic. If Y is only $(n-1)$ connected (with $n \geq 2$), we obtain a difference cochain $\delta_{*,f} \in C^n(X; \pi_n(Y))$, comparing the obstruction cocycles of f and the null map. Since $C^{n+1}(X; \pi_n(Y)) = 0$, this difference cochain is a cocyle.

Theorem 53.5 ("Hopf classification theorem"; [73]). *Fix $n \geq 2$. Let X be an n-dimensional CW complex and Y an $(n-1)$-connected space. Then $[f] \mapsto [\delta_{*,f}]$ provides an isomorphism*

$$[X, Y] \to H^n(X; \pi_n(Y)).$$

Second, there is a relative version of obstruction theory. Let $p : E \to B$ be a fibration. Assume that the base is path connected, and that the fiber (which is well-defined up to homotopy type) is path connected and simple. The homotopy groups $\pi_n(p^{-1}(-))$ then form local coefficient systems over B, and we will call the projection map *simple* if these conditions hold and these local systems are trivial.

Theorem 53.6 (e.g. [79, §VI.5]). *Let $p : E \to B$ be a simple fibration. Suppose we are given a relative CW complex (X, A) and a commutative diagram*

$$
\begin{array}{ccc}
A & \longrightarrow & E \\
\downarrow & & \downarrow{\scriptstyle p} \\
X & \longrightarrow & B
\end{array}
$$

and assume given a map $g : \mathrm{Sk}_n X \to E$ making the diagram commute when X is replaced by X_n. There is a cohomology class $\theta_g \in H^{n+1}(X, A; \pi_n(F))$ that vanishes if and only if $g|_{X_{n-1}}$ extends to a filler on X_{n+1}.

If the fiber is simple but the coefficient system is nontrivial, you still get a version of this using cohomology with local coefficients.

Exercises

Exercise 53.7. Verify the existence of f'' asserted in the proof of Theorem 53.3.

Exercise 53.8. Let Y be a simple space and N an integer, and suppose that $N\pi_*(Y) = 0$. Let (X, A) be a relative CW complex and assume that $H_*(X, A; \mathbb{F}_p) = 0$ whenever the prime p divides N. Show that the restriction map $[X, Y] \to [A, Y]$ is bijective.

Exercise 53.9. (a) A path connected space X is *even* if both $\pi_*(X)$ and $H_*(X)$ vanish in odd dimensions. Show that an even space admits the structure of an H-space.

(b) Let n be a positive integer. Construct a $(2n - 1)$-connected even space $F(2n)$ with $\pi_{2n}(F(2n)) = \mathbb{Z}$.

This is the beginning of an extensive theory of such spaces. $\mathbb{CP}^\infty$, BU, and BSU are classical examples.

Exercise 53.10. Prove Theorem 53.5.

Chapter 6

Vector bundles and principal bundles

54 Vector bundles

Each point in a smooth manifold M has a "tangent space." This is a real vector space, whose elements are equivalence classes of smooth paths $\sigma : \mathbb{R} \to M$ such that $\sigma(0) = x$. The equivalence relation retains only the velocity vector at $t = 0$. These vector spaces "vary smoothly" over the manifold. The notion of a vector bundle is a topological extrapolation of this idea.

Let B be a topological space. To begin with, let's define the "category of spaces over B," **Top**$/B$. An object is just a map $E \to B$. To emphasize that this is single object, and that it is an object "over B," we may give it a symbol and display the arrow vertically: $\xi : E \downarrow B$. A morphism from $p' : E' \to B$ to $p : E \to B$ is a map $E' \to E$ making

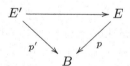

commute.

This category has products, given by the fiber product over B:

$$E' \times_B E = \{(e', e) : p'e' = pe\} \subseteq E' \times E.$$

Using it we can define an "abelian group over B": an object $E \downarrow B$ together with a "zero section" $0 : B \to E$ (that is, a map from the terminal object of **Top**$/B$) and an "addition" $E \times_B E \to E$ (of spaces over B) satisfying the usual properties.

As an example, any topological abelian group A determines an abelian group over B, namely $\mathrm{pr}_1 : B \times A \to B$ with its evident structure maps. If

A is a ring, then $\mathrm{pr}_1 : B \times A \to B$ is a "ring over B." For example, we have the "reals over B," and hence can define a "vector space over B." Each fiber has the structure of a vector space, and this structure varies continuously as you move around in the base.

Vector spaces over B form a category in which the morphisms are maps covering the identity map of B that are linear on each fiber.

Example 54.1. Let S be the subspace of $\mathbb{R}^2$ consisting of the x and y axes, and consider $\mathrm{pr}_1 : S \to \mathbb{R}$. Then $\mathrm{pr}_1^{-1}(0) = \mathbb{R}$ and $\mathrm{pr}_1^{-1}(s) = 0$ for $s \neq 0$. With the evident structure maps, this is a perfectly good ("skyscraper") vector space over $\mathbb{R}$. This example is peculiar, however; it is not locally constant. Our definition of vector bundles will exclude it and similar oddities. Sheaf theory is the proper home for examples like this.

But this example occurs naturally even if you restrict to trivial bundles and maps between them. The trivial bundle $\mathrm{pr}_1 : \mathbb{R} \times \mathbb{R} \to \mathbb{R}$ has as an endomorphism the map

$$(s, t) \mapsto (s, st).$$

This map is an isomorphism on almost all fibers, but is zero over $s = 0$. So if you want to form a kernel or the cokernel, you will get the skyscraper vector space over $\mathbb{R}$. The image will be a vector space over X with a complementary peculiarity.

Definition 54.2. A *vector bundle* over B is a vector space E over B that is locally trivial — that is, every point $b \in B$ has a neighborhood over which E is isomorphic to a trivial bundle — and whose fiber vector spaces are all of finite dimension.

Remark 54.3. As in our definition of fiber bundles, we will always assume that a vector bundle admits a numerable trivializing cover. On the other hand, there is nothing to stop us from replacing $\mathbb{R}$ with $\mathbb{C}$ or even with the quaternions $\mathbb{H}$, and talking about complex or quaternionic vector bundles.

If $\xi : E \downarrow B$ is a vector bundle, then E is called the *total space*, the map $p : E \to B$ is called the *projection map*, and B is called the *base space*. We may write $E(\xi), B(\xi)$ for the total space and base space, and ξ_b for the fiber of ξ over $b \in B$.

If all the fibers are of dimension n, we have an n-dimensional vector bundle or an "n-plane bundle."

Example 54.4. The "trivial" n-dimensional vector bundle over B is the projection $\mathrm{pr}_1 : B \times \mathbb{R}^n \to B$. We may write $n\epsilon$ for it.

Example 54.5. At the other extreme, Grassmannians support highly non-trivial vector bundles. We can form Grassmannians over any one of the three (skew)fields $\mathbb{R}, \mathbb{C}, \mathbb{H}$. Write K for one of them, and consider the (left) K-vector space K^n. The *Grassmannian* (or *Grassmann manifold*) $\mathrm{Gr}_k(K^n)$ is the space of k-dimensional K-subspaces of K^n. As we saw last term, this is a topologized as a quotient space of a Stiefel variety $V_k(K^n)$ of k-frames in K^n. To each point in $\mathrm{Gr}_k(K^n)$ is associated a k-dimensional subspace of K^n. This provides us with a k-dimensional K-vector bundle $\xi_{n,k}$ over $\mathrm{Gr}_k(K^n)$, with total space

$$E(\xi_{n,k}) = \{(V, x) \in \mathrm{Gr}_k(K^n) \times K^n : x \in V\}.$$

This is the *canonical* or *tautologous* vector bundle over $\mathrm{Gr}_k(K^n)$. It occurs as a subbundle of $n\epsilon$.

For instance, when $k = 1$, we have $\mathrm{Gr}_1(\mathbb{R}^n) = \mathbb{RP}^{n-1}$. The tautologous bundle $\xi_{n,1}$ is 1-dimensional; it is a *line bundle*, the canonical line bundle over $\mathbb{RP}^{n-1}$. We may write λ for this or any line bundle.

Example 54.6. Let M be a smooth manifold. Define τ_M to be the tangent bundle $TM \downarrow M$ over M. For example, if $M = S^{n-1}$, then

$$TS^{n-1} = \{(x, v) \in S^{n-1} \times \mathbb{R}^n : v \cdot x = 0\}.$$

Constructions with vector bundles

Just about anything that can be done for vector spaces can also be done for vector bundles:

(1) The pullback of a vector bundle is again a vector bundle: If $p : E \to B$ is a vector bundle then the map p' in the pullback diagram below is also a vector bundle.

$$
\begin{array}{ccc}
E' & \xrightarrow{\ \overline{f}\ } & E \\
\downarrow{\scriptstyle p'} & & \downarrow{\scriptstyle p} \\
B' & \xrightarrow[\ f\ ]{} & B.
\end{array}
$$

The pullback of $\xi : E \downarrow B$ may be denoted $f^*\xi$.

There's a convenient way characterize a pullback: the top map $\overline{f}$ in the pullback diagram has two key properties: It covers f, and it is a linear isomorphism on fibers. These conditions suffice to present p' as the pullback of p along f.

(2) If $p : E \to B$ and $p' : E' \to B'$ are two vector bundles, then the product map $p \times p' : E \times E' \to B \times B'$ is a vector bundle whose fiber over (x, y) is the vector space $p^{-1}(x) \times p'^{-1}(y)$.

(3) If $B = B'$, we can form the pullback:

$$
\begin{array}{ccc}
E \oplus E' & \longrightarrow & E \times E' \\
\downarrow & & \downarrow \\
B & \xrightarrow{\ \Delta\ } & B \times B.
\end{array}
$$

The bundle $\xi \oplus \xi' : E \oplus E' \downarrow B$ is called the *Whitney sum* of $\xi : E \downarrow B$ and $\xi' : E' \downarrow B$. (Hassler Whitney (1907–1989) working mainly at the Institute for Advanced Study in Princeton, is responsible for many early ideas in geometric topology.) For instance,

$$n\epsilon = \epsilon \oplus \cdots \oplus \epsilon.$$

(4) If $\xi : E \downarrow B$ and $\xi' : E' \downarrow B$ are two vector bundles over B, we can form another vector bundle $\xi \otimes \xi'$ over B by taking the fiberwise tensor product. Likewise, taking the fiberwise Hom produces a vector bundle $\mathrm{Hom}(\xi, \xi')$ over B.

Example 54.7. Recall from Example 54.5 the tautological bundle λ over $\mathbb{RP}^{n-1}$. The tangent bundle $\tau_{\mathbb{RP}^{n-1}}$ also lives over $\mathbb{RP}^{n-1}$. It is natural to wonder what is the relationship between these two bundles. We claim that

$$\tau_{\mathbb{RP}^{n-1}} = \mathrm{Hom}(\lambda, \lambda^{\perp}),$$

where $\lambda^{\perp}$ denotes the fiberwise orthogonal complement of λ in $n\epsilon$. To see this, make use of the double cover $S^{n-1} \downarrow \mathbb{RP}^{n-1}$. The projection map is smooth, and covered by a fiberwise isomorphism of tangent bundles. The fibers $T_x S^{n-1}$ and $T_{-x} S^{n-1}$ are both identified with the orthogonal complement of $\mathbb{R}x$ in $\mathbb{R}^n$, and the differential of the antipodal map sends v to $-v$. So the tangent vector to $\pm x \in \mathbb{RP}^{n-1}$ represented by (x, v) is the same as the tangent vector represented by $(-x, -v)$. This tangent vector determines a homomorphism $\lambda_x \to \lambda_x^{\perp}$ sending tx to tv.

Metrics and splitting exact sequences

A map of vector bundles, $\xi \to \eta$, over a fixed base can be identified with a section of $\mathrm{Hom}(\xi, \eta)$. We have seen that the kernel and cokernel of a homomorphism will be vector bundles only if the rank is locally constant.

In particular, we can form kernels of surjections and cokernels of injections; and consider short exact sequences of vector bundles. It is a characteristic of topology, as opposed to analytic or algebraic geometry, that short exact sequences of vector bundles always split. To see this we use a "metric."

Definition 54.8. A *metric* on a vector bundle is a continuous choice of inner products on the fibers.

Lemma 54.9. *Any (numerable) vector bundle ξ admits a metric.*

Proof. This will use the fact that if g, g' are both inner products on a vector space then $tg + (1 - t)g'$ (for t between 0 and 1) is another. So the space of metrics on a vector bundle $E \downarrow B$ forms a convex subset of the vector space of continuous functions $E \times_B E \to \mathbb{R}$.

Pick a trivializing open cover $\mathcal{U}$ for ξ, and for each $U \in \mathcal{U}$ an isomorphism $\xi|_U \cong U \times V_U$. Pick an inner product g_U on each of the vector spaces V_U. Pick a partition of unity subordinate to $\mathcal{U}$; that is, functions $\phi_U : U \to [0, 1]$ such that the preimage of $(0, 1]$ is U and

$$\sum_{x \in U} \phi_U(x) = 1 \,.$$

Now the sum

$$g = \sum_U \phi_U g_U$$

is a metric on ξ. $\qquad\square$

Corollary 54.10. *Any exact sequence $0 \to \xi' \to \xi \to \xi'' \to 0$ of vector bundles (over the same base) splits.*

Proof. Pick a metric for ξ. Using it, form the orthogonal complement $\xi'^\perp$. The composite

$$\xi'^\perp \hookrightarrow \xi \to \xi''$$

is an isomorphism. This provides a splitting of the surjection $\xi \to \xi''$ and hence of the short exact sequence. $\qquad\square$

Exercises

Exercise 54.11. Give an example of a trivial subbundle of a trivial vector bundle with nontrivial quotient bundle.

Exercise 54.12. Prove that $\xi_{n,k}$, as defined above, is locally trivial, so is a vector bundle over $\mathrm{Gr}_k(K^n)$.

Exercise 54.13. Prove that $\tau_{\mathrm{Gr}_k(\mathbb{R}^n)} = \mathrm{Hom}(\xi_{n,k}, \xi_{n,k}^\perp)$.

Exercise 54.14. Let $n > 0$ and suppose you have a bilinear map $\cdot :$ $\mathbb{R}^n \times \mathbb{R}^n \to \mathbb{R}^n$ such that $x \cdot y = 0$ only if either x or y is zero. Use this product to construct a parallelization of $\mathbb{R}P^{n-1}$ — a trivialization of its tangent bundle. (See Exercise 71.10 for some constraints on when this can happen. The definitive result [1] is that n must be $1, 2, 4$, or 8.)

55 Principal bundles, associated bundles

I-invariance

We will denote by $\mathrm{Vect}(B)$ the set of isomorphism classes of vector bundles over B, and $\mathrm{Vect}_n(B)$ the set of n-plane bundles.

Vector bundles pull back, and isomorphic vector bundles pull back to isomorphic vector bundles. This establishes Vect as a contravariant functor on **Top**:

$$\mathrm{Vect} : \mathbf{Top}^{op} \to \mathbf{Set}.$$

How computable is this functor? As a first step in answering this, we note that it satisfies the following characteristic property of bundle theories.

Theorem 55.1. *The functor* Vect *is I-invariant (where I denotes the unit interval): that is, the projection* $\mathrm{pr}_1 : X \times I \to X$ *induces an isomorphism* $\mathrm{Vect}(X) \to \mathrm{Vect}(X \times I)$.

The proof is deferred to Lecture 57. An important corollary of this result is:

Corollary 55.2. Vect *is a homotopy functor.*

Proof. Let $\xi : E \downarrow B$ be a vector bundle and suppose $H : B' \times I \to B$ a homotopy between two maps f_0 and f_1. We are claiming that $f_0^* \xi \cong f_1^* \xi$. This is far from obvious!

In the diagram

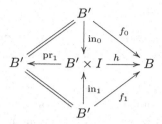

the map pr_1 induces a surjection in Vect by Theorem 55.1. It follows that $\mathrm{in}_0^* = \mathrm{in}_1^*$, so $f_0^* = \mathrm{in}_0^* \circ h^* = \mathrm{in}_1^* \circ h^* = f_1^*$. $\qquad\square$

Principal bundles

Definition 55.3. Let G be a topological group. A *principal G-bundle* is a right action of G on a space P such that the orbit projection $p : P \downarrow P/G = B$ is locally trivial, in the following sense. There exists an open cover $\mathcal{U}$ of B equipped with a section $s_U : U \to p^{-1}U$ for each $U \in \mathcal{U}$ such that the induced map $U \times G \to p^{-1}U$ sending (u,g) to $s(u)g$ is a homeomorphism.

In particular, the projection map is a fiber bundle. As in the case of fiber bundles, we will assume that this open cover is numerable (42.6) whenever needed.

There's a famous video of Jean-Pierre Serre talking about writing mathematics. In it he says you have to know the difference between "principle" and "principal". He contemplated what a "principle bundle" might be — principles varying over a moduli space of individuals, perhaps.

We will only care about Lie groups, among which are discrete groups. In that case we have already seen this concept, in Lecture 31. That story motivates constructions in the more general setting of principal G-bundles.

Construction 55.4. Let $P \downarrow B$ be a principal G-bundle. If F is a left G-space, we can define a new fiber bundle, "associated" to $P \downarrow B$, exactly as before:

Let's check that the fibers are homeomorphic to F. Let $x \in B$, and pick $y \in P$ over x. Map $F \to q^{-1}(x)$ by $z \mapsto [y, z]$. We claim that this is a homeomorphism. The inverse $q^{-1}(x) \to F$ is given by

$$[y', z'] = [y, gz'] \mapsto gz',$$

where $y' = yg$ for some g (which is necessarily unique since the G action is simply transitive on fibers of P). These two maps are inverse homeomorphisms.

If F is a finite dimensional vector space on which G acts linearly, then we get a vector bundle from this construction.

Let $\xi : E \downarrow B$ be an n-plane bundle. Construct a principal $\mathrm{GL}_n(\mathbb{R})$-bundle $P(\xi)$ by defining

$$P(\xi)_b = \{\text{ordered bases for } E(\xi)_b = \mathrm{Iso}(\mathbb{R}^n, E(\xi)_b)\}.$$

To define the topology, think of $P(\xi)$ as a quotient of the disjoint union of trivial bundles over the open sets in a trivializing cover for ξ; while for trivial bundles

$$P(B \times \mathbb{R}^n) = B \times \mathrm{Iso}(\mathbb{R}^n, \mathbb{R}^n)$$

topologically, where $\mathrm{Iso}(\mathbb{R}^n, \mathbb{R}^n) = \mathrm{GL}_n(\mathbb{R})$ is given the usual topology as a subspace of $\mathbb{R}^{n^2}$.

There is a right action of $\mathrm{GL}_n(\mathbb{R})$ on $P(\xi)$, given by precomposition. It is easy to see that this action is free and simply transitive on fibers. One therefore has a principal action of $\mathrm{GL}_n(\mathbb{R})$ on $P(\xi)$. The bundle $P(\xi)$ is called the *principalization* of ξ.

Given the principalization $P(\xi)$, we can recover the total space $E(\xi)$, using the defining linear action of $\mathrm{GL}_n(\mathbb{R})$ on $\mathbb{R}^n$:

$$E(\xi) \cong P(\xi) \times_{\mathrm{GL}_n(\mathbb{R})} \mathbb{R}^n.$$

These two constructions are inverses: the theories of n-plane bundles and of principal $GL_n(\mathbb{R})$-bundles are equivalent.

Remark 55.5. Suppose that we have a metric on ξ. Instead of looking at all ordered bases, we can use instead all ordered *orthonormal* bases in each fiber. This give the *frame bundle*

$$\mathrm{Fr}(\xi)_b = \{\text{ordered orthonormal bases of } E(\xi)_b\}$$
$$= \{\text{isometric isomorphisms } \mathbb{R}^n \to E(\xi)_b\}.$$

The orthogonal group $O(n)$ acts freely and fiberwise transitively on this space, endowing $\mathrm{Fr}(\xi)$ with the structure of a principal $O(n)$-bundle.

Providing a vector bundle with a metric, when viewed in terms of the associated principal bundles, is an example of "reduction of the structure group." We are giving a principal $O(n)$ bundle P together with an isomorphism of principal $GL_n(\mathbb{R})$ bundles from $P \times_{O(n)} GL_n(\mathbb{R})$ to the principalization of ξ. Many other geometric structures can be described in this way. An orientation of ξ, for example, consists of a principal $SL_n(\mathbb{R})$ bundle Q together with an isomorphism from $Q \times_{SL_n(\mathbb{R})} GL_n(\mathbb{R})$ to the principalization of ξ.

Suppose we have two principal G-bundles over a space X, P' and P. A "morphism" between them should obviously be a bundle map $f : P' \to P$, but we should also require that it respects the group action: it should be *equivariant*, $f(eg) = f(e)g$. But notice something now: Let A and B be G-spaces and $f : A \to B$ an equivariant map. If the G-action on B is transitive, and A is nonempty, then f is surjective. If the G-action on A is transitive and B is free, then f is injective. So (modulo some point-set issues that do not cause problems) any map of principal G-bundles is actually an isomorphism!

As a case of this observation, note that a section s of the principal G-bundle $P \downarrow X$ determines a principal G-bundle isomorphism $X \times G \to P$ by sending (x, g) to $s(x)g$:

Lemma 55.6. *A section of a principal bundle determines a trivialization.*

Write $\mathrm{Bun}_G(B)$ for the set of isomorphism classes of principal G-bundles over B. Again, this leads to a contravariant functor $\mathrm{Bun}_G : \mathbf{Top} \to \mathbf{Set}$. The above discussion gives a natural isomorphism of functors:

$$\mathrm{Bun}_{GL_n(\mathbb{R})}(B) \cong \mathrm{Vect}(B).$$

The I-invariance of Vect is therefore a special case of:

Theorem 55.7. Bun_G *is I-invariant, and hence is a homotopy functor.*

One case is easy to prove: If X is contractible, then any principal G-bundle $P \downarrow X$ is trivial. We just need a section. Since the identity map on X is homotopic to a constant map (with value $* \in X$, say), the constant map $c : X \to Q$ for any $p \in P$ over $* \in X$ satisfies $pc \simeq 1 : X \to X$. But since $P \downarrow X$ is a fibration, this implies that there is then an *actual* section (Corollary 47.3).

The other direction is harder and we will prove it in the next lecture under the hypothesis that the base is a CW complex.

Exercises

Exercise 55.8. Justify the use of the word "set" in the definitions of $\text{Vect}_n(X)$ and $\text{Bun}_G(X)$.

Exercise 55.9. Let $p : P \to B$ be a principal G-bundle.

(a) Construct a natural trivialization of the pull-back $p^*P = P \times_B P \downarrow P$.

(b) Let F be a left G-space. Let $\overline{P}$ denote the left G-space with underlying space P and G-action given in terms of the right action on P by $g \cdot x = xg^{-1}$. Construct a natural function from the set of continuous equivariant maps $\overline{P} \to F$ to the set of continuous sections of $P \times_G F \downarrow B$. Then construct a natural map the other way. Show that these two functions are inverses.

56 G-CW complexes and the I-invariance of Bun$_G$

Let G be a topological group. We want to show that the functor $\text{Bun}_G :$ **Top**$^{op} \to$ **Set** is I-invariant, i.e. the projection $\text{pr}_1 : X \times I \to X$ induces an isomorphism $\text{Bun}_G(X) \to \text{Bun}_G(X \times I)$. Injectivity is easy: inclusion $i_0 : X \to X \times I$ splits the projection, so the contravariant functor Bun_G sends pr_1 to a split injection.

The rest of this lecture is devoted to proving surjectivity. There are various ways to do this. For the general case see [27, §4.9]. Steve Mitchell has a nice treatment in [47]. We will prove this when X is a CW complex, by adapting CW methods to the equivariant situation.

To see the point of this approach, notice that the word "free" is used somewhat differently in the context of group actions than elsewhere. The left adjoint of the forgetful functor from G-spaces to spaces sends a space X to the G-space $X \times G$ in which G acts, from the right, by $(x, g)h = (x, gh)$. If G and X are discrete, any free action of G on X has this form. But this is not true topologically: just think of the antipodal action of C_2 on the circle, for instance.

The condition that an action is principal is one way to demand that an action should be "locally" free in the stronger sense. The theory of G-CW complexes affords a different way.

G-CW complexes

We would like to set up a theory of CW complexes with an action of the group G. The relevant question is, "What is a G-cell?" There is a choice here. For us, and for the standard definition of a G-CW complex, the right thing to say is that it is a G-space of the form

$$D^n \times H\backslash G.$$

Here G acts trivially on the disk, H is a closed subgroup of G, and $H\backslash G$ is the orbit space of the action of H on G by left translation, viewed as a right G-space. The "boundary" of the G-cell $D^n \times H\backslash G$ is just $\partial D^n \times H\backslash G$ (with the usual convention that $\partial D^0 = \varnothing$).

Definition 56.1. A *relative G-CW complex* is a (right) G-space X with a filtration

$$A = X_{-1} \subseteq X_0 \subseteq X_1 \subseteq \cdots \subseteq X$$

by G-subspaces such that for all $n \geq 0$ there exists a pushout square of G-spaces

$$
\begin{array}{ccc}
\coprod \partial D_i^n \times H_i\backslash G & \longrightarrow & \coprod D_i^n \times H_i\backslash G \\
\downarrow & & \downarrow \\
X_{n-1} & \longrightarrow & X_n,
\end{array}
$$

and X has the direct limit topology.

Remarks 56.2. A CW complex is just a G-CW complex for the trivial group G. If G is discrete, the skeleton filtration provides X with the structure of a CW complex by neglect of the G-action. The subspace X_n is called the "n-skeleton" of X, even though if G is itself of positive dimension X_n may well have dimension larger than n.

If X is a G-CW complex, then X/G inherits a CW structure whose n-skeleton is given by $(X/G)_n = X_n/G$.

If $P \downarrow X$ is a principal G-bundle, a CW structure on X lifts to a G-CW structure on P.

The action of G on a G-CW complex is principal if and only if all the isotropy groups are trivial.

A good source for much of this is [32]; see for example Remark 2.8 there.

Theorem 56.3 (Illman [28], Verona). *If G is a compact Lie group and M a smooth manifold on which G acts by diffeomorphisms, then M admits a G-CW structure.*

It's quite challenging in general to write down a G-CW structure even in simple cases, such as when the manifold is the unit sphere in an orthogonal representation of G. But sometimes it's easy. For example, the

CW structure on S^{n-1} described in Lecture 15, with $\mathrm{Sk}_k S^{n-1} = S^k$ for $0 \le k \le n-1$, is clearly a C_2-CW structure on the C_2 space S^{n-1} in which the nontrivial element of C_2 acts by the antipodal map. The CW structure described in Exercise 15.8 is another example of a free G-CW complex.

For another example, regard S^{n-1} as the unit sphere in $\mathbb{R}^n$ and let C_2 act by reversing the sign of the last coordinate. A C_2-CW structure on this C_2-space may be given by picking any CW structure on S^{n-2} and attaching $D^{n-1} \times C_2$.

Proof of I-invariance

Recall that our goal is to prove that every principal G-bundle $p : P \to X \times I$ is pulled back from some principal G-bundle over X, at least under the hypothesis that X is a CW complex. Actually there's no choice here; since $pr_1 \circ in_0 = 1$, P must be pulled back from $in_0^* P$, that is, from the restriction of P to $X \times 0$.

For notational convenience, let us write $Y = X \times I$. Remember that we are assuming that X is a CW complex. We will filter Y by subcomplexes, as follows. Let $Y_0 = X \times 0$; in general, we define

$$Y_n = X_n \times 0 \cup X_{n-1} \times I.$$

We may construct Y_n from Y_{n-1} via a pushout:

$$
\begin{array}{ccc}
\coprod(\partial D^{n-1} \times I \cup D^{n-1} \times 0) & \longrightarrow & \coprod(D^{n-1} \times I) \\
\downarrow & & \downarrow \\
Y_{n-1} & \longrightarrow & Y_n \,.
\end{array}
$$

The restriction of P to Y_n is a principal bundle with total space

$$P_n = p^{-1}(Y_n)\,.$$

So $P_0 \downarrow Y_0$ is just $in_0^* P \downarrow X$.

We will show that P and $pr_1^* in_0^* P$ are isomorphic over Y. For this it will be enough to construct an equivariant map $P \to in_0^* P$ covering the projection map $pr_1 : Y \to X$. We'll do this by inductively constructing compatible equivariant maps $P_n \to P_0$ covering the composites $Y_n \hookrightarrow Y \to X$, starting with the identity map $P_0 \to in_0^* P$ covering the isomorphism $Y_0 \to X$.

We can build P_n from P_{n-1} by lifting the pushout construction of Y_n from Y_{n-1}. Since $D^{n-1} \times I$ is contractible, we can pick a section of the pullback of P to it. That gives us the vertical maps in the pushout diagram

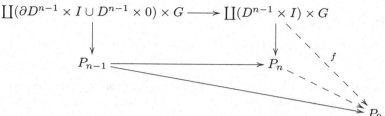

So to extend $P_{n-1} \to P_0$ to $P_n \to P_0$, we must construct an equivariant map f as shown.

Since the source of f is a disjoint union of G-spaces of the form $(D^{n-1} \times I) \times G$, it's enough to define a lifting in

$$
\begin{array}{ccc}
\partial D^{n-1} \times I \cup D^{n-1} \times 0 & \longrightarrow & P_0 \\
\downarrow & \nearrow & \downarrow \\
D^{n-1} \times I & \longrightarrow & Y_0
\end{array}
$$

and then extend by equivariance. But

$$
(D^{n-1} \times I, \partial D^{n-1} \times I \cup D^{n-1} \times 0) \cong (D^{n-1} \times I, D^{n-1} \times 0)
$$

so the dotted lifting exists, since $P_0 \to Y_0$ is a fibration. $\qquad\square$

Exercises

Exercise 56.4 (The clutching construction). Take two copies of the unreduced cone $CX = X \times I / X \times 0$ and form the unreduced suspension SX of X by identifying the two copies of $[x, t]$ for all $t > 1/2$ and $x \in X$. Let $f : X \to \mathrm{GL}_n(\mathbb{R})$ be a continuous map. Use it to identify the restrictions to $t > 1/2$ of the trivial n-plane bundles over the two cones, to produce a vector bundle over SX.

(a) Show that any vector bundle over SX is isomorphic to one produced by this "clutching construction."

(b) Describe the map $S^{n-1} \to \mathrm{GL}_n(\mathbb{R})$ corresponding to the tangent bundle of S^n.

(c) Given a topological group G, a similar construction takes a map $X \to G$ and produces a principal G-bundle over SX. Describe it.

Exercise 56.5. Let M be a compact smooth manifold (which perforce admits the structure of a CW complex) of dimension at most 4. Show that

$$
\mathrm{Bun}_{SU(2)}(M) \cong H^4(M; \mathbb{Z}).
$$

57   The classifying space of a group

Representability

Now that we know that $\text{Bun}_G(-)$ is a homotopy functor, we may ask for even more.

Theorem 57.1 (Classifying spaces). *Let G be a topological group and ξ : $E \downarrow B$ a principal G-bundle such that E is weakly contractible (just as a space, forgetting the G-action). For any CW complex X, the map*

$$[X, B] \to \text{Bun}_G(X)$$

sending a map $f : X \to B$ to the isomorphism class of $f^*\xi$ is bijective.

So Bun_G is representable, at least on the category of CW complexes! This theorem as two parts: surjectivity and injectivity. Both are proved using the following proposition.

Proposition 57.2. *Let E be a G-space that is weakly contractible as a space. Let (P, A) be a free relative G-CW complex (that is, all the G-cells used to build P from A are free). Then any equivariant map $f : A \to E$ extends to an equivariant map $P \to E$, and this extension is unique up to an equivariant homotopy rel A.*

Proof. Just do what comes naturally, after the experience of the proof of I-invariance! $\qquad\square$

Proof of Theorem 57.1. Surjectivity is immediate; take $A = \varnothing$.

To prove injectivity, let $f_0, f_1 : P \to E$ be two equivariant maps. We wish to show that they are homotopic by an equivariant homotopy, which thus descends to a homotopy between the induced maps on orbit spaces. Our data give an equivariant map $A = P \times \partial I \to E$, which we extend to an equivariant map from $P \times I$ again using Proposition 57.2. $\qquad\square$

A G-CW model for the weakly contractible principal G-space E, if it exists, will be unique up to equivariant homotopy. The orbit space will then be a CW complex with a well-defined homotopy type. Any choice of contractible free G-CW complex will be written EG, and its orbit space BG. $EG \downarrow BG$ is the universal principal G-space, and BG classifies principal G-bundles. The space BG is called the *classifying space* of the group G.

What remains is to *construct* a G-CW complex that is both free and contractible. There are many ways to do this. One can use Brown Representability, for example [12]. We will describe two other approaches: the

"Grassmann model," below, and Graeme Segal's simplicial model, in Lecture 58, which is a refinement of the Milnor construction (e.g. [72, §14.4]).

When the group is discrete, say π, this amounts to finding a $K(\pi, 1)$: the action of π on the universal cover is "properly discontinuous," which is to say principal. So we have a bunch of examples! For instance, let $\pi = \pi_1(\Sigma)$ where Σ is any closed connected surface other than S^2 and $\mathbb{R}P^2$. Then any principal π-bundle over any CW complex B is pulled back from the universal cover of Σ under a unique homotopy class of maps $B \to \Sigma$.

If G is a compact Lie group — for example a finite group — there is a very geometric way to go about this, based on the following result.

Theorem 57.3 (Peter-Weyl, [31, Corollary IV.4.22]). *Any compact Lie group admits a finite-dimensional faithful unitary representation.*

Clearly, if P is free as a G-space then it is also free as an H-space for any subgroup H of G. It's also the case that if P is a principal G-space for a compact Lie group G then it is also a principal H space for any closed subgroup of G.

Combining these facts, we see that in order to construct a contractible space with a principal G action, for any compact Lie group G, it suffices to construct such a thing for the particular Lie groups $U(n)$.

Gauss maps

Before we look for highly connected spaces on which $U(n)$ acts, let's look at the case in which the base space is a compact Hausdorff space (for example a finite complex). In this case we can be more geometrically explicit about the classifying map.

Lemma 57.4. *Over a compact Hausdorff space, any vector bundle embeds in a trivial bundle.*

Proof. Let $\mathcal{U}$ be a trivializing open cover of the base B; since B is compact, we may assume that $\mathcal{U}$ is finite, with, say, k elements $U_1, \ldots, U_k$. We agreed that our vector bundles would always be numerable, but we don't even have to mention this here since compact Hausdorff spaces are paracompact. So we can choose a partition of unity $\{\phi_i\}$ subordinate to $\mathcal{U}$. By treating path components separately if need be, we may assume that our vector bundle $\xi : E \downarrow B$ is an n-plane bundle, with projection p. The trivializations are fiberwise isomorphisms $g_i : p^{-1}(U_i) \to \mathbb{R}^n$. We can assemble these maps using the partition of unity, and define $g : E \to (\mathbb{R}^n)^k$ as the unique map such that

$$\mathrm{pr}_i g(e) = \phi_i(p(e)) g_i(e) .$$

This is a fiberwise linear embedding. The map $e \mapsto (p(e), g(e))$ is an embedding into the trivial bundle $B \times \mathbb{R}^{nk}$. $\square$

We can now use the standard inner product on $\mathbb{R}^{nk}$ (or any other metric on $B \times \mathbb{R}^{nk}$) to form the complement of E:

Corollary 57.5. *Over a compact Hausdorff space, any vector bundle has a complement (i.e. a vector bundle $\xi^\perp$ such that $\xi \oplus \xi^\perp$ is trivial).*

Suppose our vector bundle has fiber dimension n. The image of $g(E_x)$ is an n-plane in $\mathbb{R}^{nk}$; that is, an element $f(x) \in \mathrm{Gr}_n(\mathbb{R}^{nk})$. We have produced a diagram

$$\begin{array}{ccc}
E & \xrightarrow{\ g\ } & E(\xi_{nk,n}) \\
{\scriptstyle \xi}\downarrow & & \downarrow \\
B & \xrightarrow{\ f\ } & \mathrm{Gr}_n(\mathbb{R}^{nk})
\end{array}$$

that expresses ξ as the pullback of the tautologous bundle $\xi_{nk,n}$ under a map $f : B \to \mathrm{Gr}_n(\mathbb{R}^{nk})$. This map f, covered by a bundle map, is a *Gauss map* for ξ.

The Grassmannian model

The frame bundle of the tautologous vector bundle over the Grassmannian $\mathrm{Gr}_n(\mathbb{C}^{n+k})$ is the complex Stiefel manifold

$$V_n(\mathbb{C}^{n+k}) = \{\text{isometric embeddings } \mathbb{C}^n \hookrightarrow \mathbb{C}^{n+k}\} .$$

Ehresmann's Theorem 42.5 (for example) tells us that the projection map

$$V_n(\mathbb{C}^{n+k}) \downarrow \mathrm{Gr}_n(\mathbb{C}^{n+k})$$

sending an embedding to its image is a fiber bundle, so we have a principal $U(n)$-bundle.

How connected is this complex Stiefel variety? $U(q)$ acts transitively on the unit sphere in $\mathbb{C}^q$ and the isotropy group of the basis vector e_q is $U(q-1)$ embedded in $U(q)$ in the upper left corner. So we get a tower of fiber bundles with the indicated fibers:

$$S^{2k+1} \longrightarrow U(n+k)/U(k) = V_n(\mathbb{C}^{n+k})$$

$$\downarrow$$

$$S^{2k+3} \longrightarrow U(n+k)/U(1+k) = V_{n-1}(\mathbb{C}^{n+k})$$

$$\downarrow$$

$$\vdots$$

$$\downarrow$$

$$S^{2(n+k)-1} \overset{=}{\longrightarrow} U(n+k)/U((n-1)+k) = V_1(\mathbb{C}^{n+k}).$$

The long exact homotopy sequence shows that $V_n(\mathbb{C}^{n+k})$ is $(2k)$-connected. It's a "twisted product" of the the spheres $S^{2k+1}, S^{2k+3}, \dots, S^{2(n+k)-1}$.

So forming the direct limit

$$V_n(\mathbb{C}^\infty) = \lim_{k \to \infty} V_n(\mathbb{C}^{n+k})$$

gives us a contractible CW complex with a principal action of $U(n)$. The quotient map

$$V_n(\mathbb{C}^\infty) \downarrow V_n(\mathbb{C}^\infty)/U(n) = \mathrm{Gr}_n(\mathbb{C}^\infty)$$

provides us with a universal principal $U(n)$ bundle, and hence also a universal n-plane bundle ξ_n. An element of $E(\xi_n)$ is an n-dimensional subspace of the countably infinite dimensional vector space $\mathbb{C}^\infty$. This is the "infinite Grassmannian," and it deserves the symbol $BU(n)$.

Dividing by a closed subgroup $G \subseteq U(n)$ provides us with a model for BG. To be completely honest here, we should work to endow $\mathrm{Gr}_n(\mathbb{C}^\infty)$ with a G-CW structure. We do not do that here, but see [76]. We will see a more canonical construction in the next lecture. And of course sometimes we have more direct constructions; for example the same observations show that $BO(n)$ is the space of n-planes in $\mathbb{R}^\infty$. In particular, $BC_2 = BO(1) = \mathbb{RP}^\infty$ and $BU(1) = BS^1 = \mathbb{CP}^\infty$.

Exercises

Exercise 57.6. Verify Proposition 57.2.

Exercise 57.7. Let $F(M, j)$ be the space of injections of $\{1, \dots, j\}$ into a space M: the space of *ordered configurations* in M of cardinality j. Show that $F(\mathbb{R}^\infty, j)$ is contractible. It admits a free action of the symmetric group Σ_j, so the space of subsets of cardinality j in $\mathbb{R}^\infty$ is a model for $B\Sigma_j$.

58 Simplicial sets and classifying spaces

We encountered simplicial sets at the very beginning of this course, as a step on the way to constructing singular homology. We will take this story up again here, briefly, because simplicial methods provide a way to organize the combinatorial data needed for the construction of classifying spaces and maps.

Simplex category and nerve

Recall from Lecture 3 the *simplex category* $\mathbf{\Delta}$: It has as objects the finite totally ordered sets

$$[n] = \{0, 1, \ldots, n\}, n \geq 0,$$

and as morphisms the order preserving maps. In particular the "coface" map $d^i : [n] \to [n+1]$ is the injection omitting i and the "codegeneracy" map $s^i : [n] \to [n-1]$ is the surjection repeating i. Any order-preserving map can be written as the composite of these maps, and there are famous relations (Exercise 1.8) that they satisfy. They generate the category $\mathbf{\Delta}$.

The *standard (topological) simplex* is the functor $\Delta : \mathbf{\Delta} \to \mathbf{Top}$ defined by sending $[n]$ to the "standard n-simplex" Δ^n, the convex hull of the standard basis vectors $e_0, e_1, \ldots, e_n$ in $\mathbb{R}^{n+1}$. Order-preserving maps get sent to the affine extension of the map on basis vectors. So d^i includes the ith codimension one face, and s^i collapses onto a codimension one face.

Definition 58.1. Let $\mathcal{C}$ be a category. Denote by $s\mathcal{C}$ the category of *simplicial objects* in $\mathcal{C}$, i.e., the category $\mathrm{Fun}(\mathbf{\Delta}^{op}, \mathcal{C})$. We write $X_n = X([n])$ for the "object of n-simplices."

A simplicial object can be defined by giving an object $X_n \in \mathcal{C}$ for every $n \geq 0$ along with maps $d_i : X_{n+1} \to X_n$ and $s_i : X_{n-1} \to X_n$ satisfying certain quadratic identities.

Our first example of a simplicial object is the *singular simplicial set* $\mathrm{Sin}(X)$ of a space X:

$$\mathrm{Sin}(X)_n = \mathrm{Sin}_n(X) = \mathbf{Top}(\Delta^n, X).$$

There is a categorical analogue of $\Delta : \mathbf{\Delta} \to \mathbf{Top}$. After all, the ordered set $[n]$ is a particularly simple small category: $\mathbf{\Delta}$ is a full subcategory of the category of small categories. So a small category C determines a simplicial set NC, the *nerve* of C, with

$$(NC)_n = N_n C = \mathrm{Fun}([n], C).$$

Thus $N_0 C$ is the set of objects of C; $N_1 C$ is the set of morphisms; $d_0 : N_1 C \to N_0 C$ sends a morphism to its target, and $d_1 : N_1 C \to N_0 C$ sends a morphism to its source; $s_0 : N_0 C \to N_1 C$ sends an object to its identity morphism. In general $N_n C$ is the set of n-*chains* in C: composable sequences of n morphisms. For $0 < i < n$, the face map $d_i : N_n C \to N_{n-1} C$ forms the composite of two adjacent morphisms, while d_0 omits the initial morphism and d_n omits the terminal morphism. Degeneracies interpose identity maps.

For example, a group G can be regarded as a small category, one with just one object. We denote it again by G. Then $N_n G = G^n$, and for $0 < i < n$

$$d_i(g_1, \ldots, g_n) = (g_1, \ldots, g_{i-1}, g_i g_{i+1}, g_{i+2}, \ldots, g_n),$$

while

$$d_0(g_1, \ldots, g_n) = (g_2, \ldots, g_n), \quad d_n(g_1, \ldots, g_n) = (g_1, \ldots, g_{n-1}).$$

In general, the nerve construction allows us to regard small categories as a special class of simplicial sets. This attitude is the starting point for the theory of "quasi-categories" or "∞-categories," which constitute a somewhat more general class of simplicial sets.

Realization

The functor Sin transported us from spaces to simplicial sets. Milnor [41] described how to go the other way.

Let K be a simplicial set. The *geometric realization* $|K|$ of K is

$$|K| = \left(\coprod_{n \geq 0} \Delta^n \times K_n \right) / \sim$$

where $\sim$ is the equivalence relation defined by:

$$\Delta^m \times K_m \ni (v, \phi^* x) \sim (\phi_* v, x) \in \Delta^n \times K_n$$

for all maps $\phi : [m] \to [n]$.

The equivalence relation is telling us to glue together simplices as dictated by the simplicial structure on K. To see this in action, let us look at $\phi^* = d_i : K_{n+1} \to K_n$ and $\phi_* = d^i : \Delta^n \to \Delta^{n+1}$. In this case, the equivalence relation then says that $(v, d_i x) \in \Delta^n \times K_n$ is equivalent to $(d^i v, x) \in \Delta^{n+1} \times K_{n+1}$. In other words: the ith face of the $(n+1)$ simplex labeled by x is identified with the n-simplex labeled by $d_i x$.

There's a similar picture for the degeneracies s^i, where the equivalence relation dictates that every element of the form $(v, s_i x)$ is already represented by a simplex of lower dimension. A simplex in a simplicial set is "nondegenerate" if it is not in the image of a degeneracy map. Neglecting the topology, $|X|$ is the disjoint union of (topological) simplex interiors labeled by the nondegenerate simplices of K.

Example 58.2. Let $n \geq 0$, and consider the simplicial set $\Delta(-, [n])$. This is called the "simplicial n-simplex", for good reason: Its geometric realization is canonically homeomorphic to the geometric n-simplex Δ^n.

The realization $|K|$ of a simplicial set has a naturally defined CW structure with
$$\operatorname{sk}_n |K| = \left(\coprod_{k \leq n} \Delta^k \times K_k \right) / \sim .$$
The face maps give the attaching maps; for more details, see [23, Proposition I.2.3]. This is a very combinatorial way to produce CW complexes. There is one n-simplex for each *nondegenerate* n-simplex in K; that is, for each element of K_n that is not in image of any degeneracy map.

The geometric realization functor and the singular simplicial set functor form one of the most important and characteristic examples of an adjoint pair:
$$|-| : s\mathbf{Set} \rightleftarrows \mathbf{Top} : \mathrm{Sin}.$$
The adjunction morphisms are easy to describe. For $K \in s\mathbf{Set}$, the unit for the adjunction $K \to \mathrm{Sin}|K|$ sends $x \in K_n$ to the map $\Delta^n \to |K|$ defined by $v \mapsto [(v, x)]$.

To describe the counit, let X be a space. There is a continuous map $\Delta^n \times \mathrm{Sin}_n(X) \to X$ given by $(v, \sigma) \mapsto \sigma(v)$. The equivalence relation defining $|\mathrm{Sin}(X)|$ says precisely that the map factors through the realization:

$$|\mathrm{Sin}(X)| \dashrightarrow X$$

$$\coprod \Delta^n \times \mathrm{Sin}_n(X).$$

A theorem of Milnor [41] asserts that the map $|\mathrm{Sin}(X)| \to X$ is a weak equivalence. This provides a functorial (and therefore spectacularly inefficient) CW approximation for any space.

This adjoint pair enjoys properties permitting the wholesale comparison of the homotopy theory of spaces with a combinatorially defined homotopy theory of simplicial sets [54]. For more details, see [23].

Classifying spaces

Combining the two constructions we have just discussed, we can assign to any small category C a space

$$BC = |NC|,$$

known as its *classifying space*. For example, $B[n] = \Delta^n$.

When C is a group, G, this space does in fact support a principal G-bundle. Before we explain that, let's look at the example of the group C_2 of order 2. Write t for the non-identity element of C_2. There is just one non-degenerate n simplex in NC_2 for any $n \geq 0$, namely $(t, t, \ldots, t)$. So the realization BC_2 has a single n-cell for every n. Not bad, since it's supposed to be a CW structure on $\mathbb{R}P^\infty$! Think about what the low skelata are. There's just one object, so $(BC_2)_0 = *$. There is just one nondegenerate 1-simplex, $(t) \in C_2^1$, so $(BC_2)_1$ is a circle. There's just one nondegenerate 2-simplex, $(t, t) \in C_2^2$. Its faces are

$$d_0(t, t) = t, \quad d_1(t, t) = t^2 = 1, \quad d_2(t, t) = t.$$

The middle face has been identified with $*$ since it's degenerate, and we see a standard representation of $\mathbb{R}P^2$ as a "lune" with its two edges identified. A similar analysis shows that $(BC_2)_n = \mathbb{R}P^n$ for any n.

The projection maps $C \times D \to C$ and $C \times D \to D$ together induce a natural map

$$B(C \times D) \to BC \times BD.$$

The following is a nice surprise, and requires the use of the compactly generated topology on the product.

Theorem 58.3. *The natural map $B(C \times D) \to BC \times BD$ is a homeomorphism.*

Sketch of proof. This is nontrivial — not "categorical" — because it asserts that certain limits commute with certain colimits. The underlying fact is the Eilenberg-Zilber theorem, which gives a simplicial decomposition of $\Delta^m \times \Delta^n$ and verifies the result when $C = [m]$ and $D = [n]$. The general result follows since every simplicial set is a colimit of its "diagram of simplicies," and B respects colimits. $\square$

Lemma 58.4. *The classifying space construction sends natural transformations to homotopies.*

Proof. A natural transformation of functors $C \to D$ is the same thing as a functor $C \times [1] \to D$. Since $B[1] = \Delta^1$, we can form the homotopy

$$BC \times \Delta^1 = BC \times B[1] \overset{\cong}{\leftarrow} B(C \times [1]) \to BD.$$

$\square$

Corollary 58.5. *An adjoint pair induces a homotopy equivalence on classifying spaces.*

Corollary 58.6. *If C contains an initial object or a terminal object then BC is contractible.*

Proof. Saying that $o \in C$ is initial is saying that the inclusion $o : [0] \to C$ is a left adjoint. $\square$

The translation groupoid

An action of G on a set X determines a category, a groupoid in fact, the "translation groupoid," which I will denote by GX. Its object set is X, and

$$GX(x, y) = \{g \in G : gx = y\}.$$

Composition comes from the group multiplication. This is a special case of the "Grothendieck construction." (Alexander Grothendieck (1928–2014) worked mainly at the Institut des hautes études scientific outside of Paris. He was the founder of the framework for algebraic geometry as practiced today.)

When $X = *$ we recover the category G. Another case of interest is when $X = G$ with G acting from the left by translation. The category GG is "unicursal": there is exactly one morphism between any two objects; every object is both initial and terminal. This implies that $B(GG)$ is contractible.

The association

$$X \mapsto GX \mapsto N(GX) \mapsto |N(GX)| = B(GX)$$

is functorial. In particular, right multiplication by $g \in G$ on the set G is equivariant with respect to the left action of G on it. Therefore G acts from the right on GG and hence on $B(GG)$. This is a "free" action: no $g \in G$ except the identity element fixes any simplex. This implies that $B(GG)$ admits the structure of a free G-CW complex. It's not hard to verify that $B(GG)/G = BG$, so we have succeeded in constructing a functorial classifying space for any discrete group.

Exercises

Exercise 58.7. Use Lemma 58.5 to prove that if G is any group and $g \in G$, then the map $BG \to BG$ induced by the homomorphism $c_g : G \to G$ given by conjugation, $x \mapsto gxg^{-1}$, is homotopic to the identity map. Of course the map $c_g : BG \to BG$ is not homotopic to the identity through *basepoint preserving* homotopies! On $\pi_1(BG) = \pi_0(G)$ it induces conjugation by $[g] \in \pi_0(G)$.

Exercise 58.8. (a) The classifying space of a group G comes equipped with a basepoint. Let X be a pointed connected CW complex. Construct a bijection from the set of pointed homotopy classes of pointed maps $[X, BG]_*$ to $\mathrm{Bun}_G(X)_*$, the set of isomorphism classes of pairs (P, e) where $p : P \to X$ is a principal G-bundle and $e \in p^{-1}*$; that is, e is a trivialization of the fiber of P over the basepoint.

(b) Now let G be discrete. Construct a bijection $\mathrm{Hom}(\pi_1(X, *), G) \to \mathrm{Bun}_G(X)_*$.

(c) So as a functor from the homotopy category of pointed connected CW complexes to the category of groups, π_1 is a left adjoint. Why doesn't this prove the Van Kampen theorem?

59 The Čech category and classifying maps

In this lecture I'll sketch a program due to Graeme Segal [58] for classifying principal G-bundles using the simplicial description of the classifying space proposed in the last lecture. That machinery admits an extension to general topological groups.

Top-*enrichment*

The Grassmannian model provides a classifying space for any compact Lie group. This includes finite discrete groups, which are also covered by the construction we just did. But we'd like to provide a construction with the naturality of the simplicial construction that also applies to topological groups, and that works for a general topological groups.

Definition 59.1. A category *enriched in* **Top** is a category $\mathcal{C}$ together with topologies on all the morphism sets, with the property that the composition maps are continuous.

The fact that **Top** is Cartesian closed provides us with an enrichment in **Top** of the category **Top** itself. A simpler (and smaller) example is given by any topological group (or monoid), regarded as a category with one object. Then a continuous action of G on a space X is just a functor $G \to$ **Top** that is continuous on hom spaces: a "topological functor."

The "nerve" construction now produces a simplicial *space*,

$$NG \in s\textbf{Top}$$

associated to any topological group G. The formula for geometric realization still makes perfectly good sense for a simplicial space. (It won't generally be a CW complex anymore, but it does have a useful "skeleton" filtration given by assembling only simplices of dimension up to n.) Combining the two constructions, we may form the "classifying space"

$$BG = |NG|.$$

This provides a functorially defined classifying space for topological groups.

Internal categories

To justify this language, we should produce a principal G-bundle over this space with contractible total space. This construction requires one further invasion of topology into category theory (or vice versa), namely, an "internal category" in **Top**.

Definition 59.2. **Top**-*category* is a pair of spaces C_0 and C_1 (to be thought of as the space of objects and the space of morphisms), together with continuous structure maps

$$\text{source, target} : C_1 \rightrightarrows C_0, \quad \text{identity} : C_0 \to C_1$$

$$\text{composition} : C_1 \times_{C_0} C_1 \to C_1$$

satisfying the axioms of a category.

If the object space is discrete, this is just an enrichment in **Top**. But there are other important examples. The simplest one is entirely determined by a space X: write cX for it. Just take it $(cX)_0 = (cX)_1 = X$ with the "identity" map $(cX)_0 \to (cX)_1$ given by the identity map.

The nerve and classifying space constructions carry over without change to this new setting. $(NC)_0$ will no longer be discrete. The classifying space of cX is just X, for example. The observation that an adjoint pair yields a homotopy equivalence still holds.

Now suppose that G acts on a space X. The construction of GX carried out in the previous lecture provides us with a **Top**-category. Its classifying space maps to that of G, since X maps to a point.

Proposition 59.3. *If G is a Lie group (and much more generally as well) the map $B(GG) \to BG$ is a principal G-bundle, and $B(GG)$ is contractible.*

So this gives the classifying space of G, functorially in G. It's not hard to see that in fact

$$B(GX) = B(GG) \times_G X.$$

This degree of generality provides an inductive way to construct Eilenberg Mac Lane spaces explicitly. Begin with any discrete abelian group π. Apply the classifying space construction we've just described, to obtain a $K(\pi, 1)$. Now being abelian is equivalent to the multiplication map $\pi \times \pi \to \pi$ being a homomorphism. So we may leverage the functoriality of B, and the fact that it commutes with products, and form

$$B\pi \times B\pi \cong B(\pi \times \pi) \to B\pi.$$

This provides on $B\pi$ the structure of a topological abelian group. So we can apply B again: $BB\pi = K(\pi, 2)$. And so on:

$$B^n \pi = K(\pi, n).$$

Descent

Let $\pi : Y \to X$ be a map of spaces. We can use it to define a **Top**-category, the "descent category" or "Čech category" $\check{C}(\pi)$, as follows. The space of objects is X, and the space of morphisms is $Y \times_X Y$. The structure maps are given by

$$\mathrm{id} = \Delta : Y \to Y \times_X Y, \ y \mapsto (y, y)$$
$$\mathrm{source} = \mathrm{pr}_1 : Y \times_X Y \to Y, \ (y_1, y_2) \to y_1$$
$$\mathrm{target} = \mathrm{pr}_2 : Y \times_X Y \to Y, \ (y_1, y_2) \to y_2$$
$$\mathrm{comp} : (Y \times_X Y) \times_Y (Y \times_X Y) \to Y \times_X Y, \ ((y_1, y_2), (y_2, y_3)) \mapsto (y_1, y_3).$$

There is a continuous functor

$$\check{\pi} : \check{C}(\pi) \to cX$$

determined by mapping the object space by the identity.

This construction is best understood from its motivating case. Suppose that $\mathcal{U}$ is a cover of X and let

$$Y = \coprod_{U \in \mathcal{U}} U \,,$$

mapping to X by sending $x \in U$ to $x \in X$. Then

$$Y \times_X Y = \coprod_{(U,V) \in \mathcal{U} \times \mathcal{U}} U \cap V \,,$$

the disjoint union of intersections of ordered pairs of elements of $\mathcal{U}$. Source and target just embed $U \cap V$ into U and V.

In this case let's write $\check{C}(\mathcal{U})$ for the Čech category. In good cases we can recover X from $\check{C}(\mathcal{U})$:

Proposition 59.4. *If the open cover $\mathcal{U}$ of X admits a subordinate partition of unity, then $B\check{\pi} : B\check{C}(\mathcal{U}) \to X$ is a homotopy equivalence.*

Proof. A sequence $U_0, U_1, \ldots U_n$ of elements of $\mathcal{U}$ together with a point x in their intersection determines a chain $(x \in U_0) \to (x \in U_1) \to \cdots \to (x \in U_n)$ in the category $\check{C}(\mathcal{U})$. The counit of the realization-singular adjunction then gives a map

$$\varepsilon : \Delta^n \times (U_0 \cap U_1 \cap \cdots \cap U_n) \to B\check{C}(\mathcal{U}) \,.$$

Now let $\{\phi_U : U \in \mathcal{U}\}$ be a partition of unity subordinate to $\mathcal{U}$, so that, for every $x \in X$, $\phi_U(x) = 0$ for all but finitely many $U \in \mathcal{U}$, and $\sum_U \phi_U = 1$. Pick a partial order on the set $\mathcal{U}$ that is total on any subset with nonempty intersection. For any x let $U_0(x), \ldots, U_{n(x)}(x)$ be the ordered sequence of elements of $\mathcal{U}$ that contain x. Then define

$$X \to B\check{C}(\mathcal{U})$$

by sending

$$x \mapsto \varepsilon((\phi_{U_0(x)}(x), \ldots, \phi_{U_{n(x)}(x)}(x)), x) \,.$$

It's not hard to check that this gives a well-defined map that is homotopy inverse to $B\check{\pi}$. $\qquad\square$

Remark 59.5. A final comment: In [58] Segal explains how to use these methods to construct a spectral sequence from this approach, one that includes the Serre spectral sequence and more generally the topological version of the Leray spectral sequence. We won't pursue that avenue in these lectures, though, but instead will describe two other approaches.

Transition functions, cocycles, and classifying maps

Now suppose that $\xi : P \xrightarrow{p} B$ is a principal G-bundle. We will explain how to construct an explicit map $B \to BG$ classifying ξ. Pick a trivializing open cover $\mathcal{U}$, along with trivializations $\varphi_U : p^{-1}U \to U \times G$ for $U \in \mathcal{U}$. These data determine a continuous functor

$$\check{C}(\mathcal{U}) \to G$$

as follows. There's no choice about behavior on objects. On morphisms, we use the "transition functions" associated with the given trivializations. So for $U, V \in \mathcal{U}$, the intersection $U \cap V$ is a subspace of the space of morphisms in $\check{C}(\mathcal{U})$. We map it to G by

$$x \mapsto \varphi_V(x)\varphi_U(x)^{-1} \in G.$$

The "cocycle condition" on these transition functions is the statement that together these maps constitute a functor.

Therefore we get a diagram

$$B\check{C}(\mathcal{U}) \longrightarrow BG$$

$$\Big\uparrow\Big\downarrow \simeq$$

$$X$$

and one can check that the bundle $EG \downarrow BG$ pulls back to $P \downarrow X$ under the composite $X \to BG$.

Exercises

Exercise 59.6. Suppose that $\mathcal{U} = \{U, V\}$ is an open cover of X. Show that $\check{C}(\mathcal{U})$ is the double mapping cylinder $U \cup (U \cap V) \times I \cup V$. You can deduce the Mayer-Vietoris sequence from this.

Chapter 7

Spectral sequences and Serre classes

60 Why spectral sequences?

When we're solving a complicated problem, it's smart to break the problem
into smaller pieces, solve them, and then put the pieces back together.
Spectral sequences provide a powerful and flexible tool for bridging the
"local to global" divide. They contain a lot of information, and can be
queried in a variety of ways, so we will spend quite a bit of time getting to
know them.

Homology is relatively computable precisely because you can break a
space into smaller parts and then use Mayer-Vietoris to put the pieces
back together. The long exact homology sequence (along with excision) is
doing the same thing. We have seen how useful this is, in our identification
of singular homology with the cellular homology of a CW complex. This
puts a filtration on a space X, the skeleton filtration, and then makes use
of the long exact sequences of the various pairs (X_n, X_{n-1}). Things are
particularly simple here, since $H_q(X_n, X_{n-1})$ is nonzero for only one value
of q.

There are interesting filtrations that do not have that property. For
example, suppose that $p : E \to B$ is a fibration. A CW structure on B
determines a filtration of E in which

$$F_s E = p^{-1}(\mathrm{Sk}_s B) \,.$$

Now the situation is more complicated: For each s we get a long exact sequence involving $H_*(F_{s-1}E)$, $H_*(F_s E)$, and $H_*(F_s E, F_{s-1}E)$. The structure that emerges from this tangle of long exact sequences is a "spectral
sequence." It will describe the exact relationship between the homologies
of the fiber, the base, and the total space.

We can get a somewhat better idea of how this might look by thinking of the case of a product projection, $\mathrm{pr}_1 : B \times F \to B$. Then the Künneth theorem is available. Let's assume that we are in the lucky situation in which there is a Künneth isomorphism, so that

$$H_*(B) \otimes H_*(F) \xrightarrow{\cong} H_*(E).$$

You should visualize this tensor product of graded modules by putting the the summand $H_s(B) \otimes H_t(F)$ in degree $n = s + t$ of the graded tensor product in position (s, t) in the first quadrant of the plane. Then the graded tensor product in degree n sums along each "total degree" $n = s + t$. Along the x-axis we see $H_s(B) \otimes H_0(F)$; if F is path connected this is just the homology of the base space. Along the y-axis we see $H_0(B) \otimes H_t(F) = H_t(F)$; if B is path-connected this is just the homology of the fiber. Cross-products of classes of these two types fill out the first quadrant.

The Künneth theorem can't generalize directly to nontrivial fibrations, though, because of examples like the Hopf fibration $S^3 \to S^2$ with fiber S^1. The tensor product picture looks like this:

— and definitely gives the wrong answer!

What's going on here? We can represent a generating cycle for $H_2(S^2)$ using a relative homeomorphism $\sigma : (\Delta^2, \partial\Delta^2) \to (S^2, o)$. If c_o represents the constant 2-simplex at the basepoint $o \in S^2$, $\sigma - c_o$ is a cycle representing a generator of $H_2(S^2)$. We can lift each of these simplices to simplices in S^3. But a lift of σ sends $\partial\Delta^2$ to one of the fiber circles, and the lift of $\sigma - c_o$ is no longer a cycle. Rather, its boundary is a cycle in the fiber over o, and it represents a generator for $H_1(p^{-1}(o)) \cong H_1(S^1)$.

This can be represented by adding an arrow to our picture.

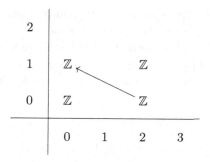

This diagram now reflects several facts: $H_1(S^1)$ maps to zero in $H_1(S^3)$ (because the representing cycle of a generator becomes a boundary!); the image of $H_2(S^3) \to H_2(S^2)$ is trivial (because no nonzero multiple of a generator of $H_2(S^2)$ lifts to a cycle in S^3); and the homology of S^3 is left with just two generators, in dimensions 0 and 3.

In terms of the filtration on the total space S^3, the lifted chain lay in filtration 2 (saying nothing, since $F_2S^3 = S^3$) but not in filtration 1. Its boundary lies two filtration degrees lower, in filtration 0. That is reflected in the differential moving two columns to the left.

The Hopf fibration $S^7 \downarrow S^4$ (which you will study in homework) shows a similar effect. The boundary of the 4-dimensional chain lifting a generating cycle lies again in filtration 0, i.e. on the fiber. This represents a drop of filtration by 4, and is represented by a differential of bidegree $(-4, 3)$.

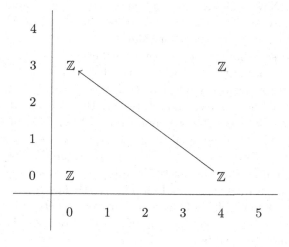

In every case, the total degree of the differential is of course -1.

The Künneth theorem provides a "first approximation" to the homology of the total space. It's generally too big, but never too small. Cancellation can occur: lifted cycles can have nontrivial boundaries, and cycles that were not boundaries in the fiber can become boundaries in the total space. More complicated cancellation can occur as well, involving the product classes.

Some history

Now I've told you almost the whole story of the Serre spectral sequence. A structure equivalent to a spectral sequence was devised by Jean Leray while he was in a prisoner of war camp during World War II. He discovered an elaborate structure determined in cohomology by a map of spaces. This was much more that just the functorial effect of the map. He worked with cohomology, and in fact invented a new cohomology theory for the purpose. He restricted himself to locally compact spaces, but on the other hand he allowed *any* continuous map — no restriction to fibrations. This is the "Leray spectral sequence." It's typically developed today in the context of sheaf theory — another local-to-global tool invented by Leray at about the same time.

Leray called his structure an *"anneau spectral"*: he was specifically interested in its multiplicative structure, and he saw an analogy between his analysis of the cohomology of the source of his map and the spectral decomposition of an operator. Before the war he had worked in analysis, especially the Navier-Stokes equation, and said that he found in algebraic topology a study that the Nazis would not be able to use in their war effort, in contrast to his expertise as a "mechanic."

It's fair to say that nobody other than Leray understood spectral sequences till well after the war was over. Henri Cartan (1904–2008) was a leading figure in post-war mathematical reconstruction. He befriended Leray and helped him explain himself better. He set his students to thinking about Leray's ideas. One was named Jean-Louis Koszul (1921–2018), and it was Koszul who formulated the algebraic object we now call a spectral sequence. Another was Jean-Pierre Serre. Serre wanted to use this method to compute things in homotopy theory proper — homotopy groups, and the cohomology of Eilenberg Mac Lane spaces. He had to recast the theory to work with singular cohomology, on much more general spaces, but in return he considered only what we now call Serre fibrations. This restriction allowed a homotopy-invariant description of the spectral sequence. Leray had used *"anneau spectral"*; Cartan used *"suite de Leray-Koszul"*; and now

Serre, in his thesis, brought the two parties together and coined the term "*suite spectral*". For more history see [37].

> La science ne s'apprend pas: elle se comprend. Elle n'est pas lettre morte et les livres n'assurent pas sa pérennité: elle est une pensée vivante. Pour s'intéresser à elle, puis la maîtriser, notre eprit doit, habilement guidé, la redécouvrir, de même que notre corps a dû revivre, dans le sien maternel, l'évolution qui créa notre espèce; non point tout ses détails, mais son schéma. Aussi n'y a-t-il qu'une façon efficace de faire acquérir par nos enfants les principes scientifiques qui sont stable, et les procédés techniques qui évoluent rapidement: c'est donner à nos enfants l'esprit de recherche. — Jean Leray [57]

61 Spectral sequence of a filtered complex

We are trying find ways to use a filtration of a space to compute the homology of that space. A simple example is given by the skeleton filtration of a CW complex. Let's recall how that goes. The singular chain complex receives a filtration by sub chain complexes by setting

$$F_s S_*(X) = S_*(\mathrm{Sk}_s X).$$

We then pass to the quotient chain complexes

$$S_*(\mathrm{Sk}_s X, \mathrm{Sk}_{s-1} X) = F_s S_*(X)/F_{s-1} S_*(X).$$

The homology of the sth chain complex in this list vanishes except in dimension s, and the group of cellular s-chains is defined by

$$C_s(X) = H_s(\mathrm{Sk}_s X, \mathrm{Sk}_{s-1} X).$$

In turn, these groups together form a chain complex with differential

$$
\begin{array}{ccc}
C_s(X) & \xrightarrow{\quad\quad\quad d \quad\quad\quad} & C_{s-1}(X) \\
\| & & \| \\
H_s(\mathrm{Sk}_s X, \mathrm{Sk}_{s-1} X) \xrightarrow{\ \partial\ } H_{s-1}(\mathrm{Sk}_{s-1} X) & \longrightarrow & H_{s-1}(\mathrm{Sk}_{s-1} X, \mathrm{Sk}_{s-2} X).
\end{array}
$$

Then $d^2 = 0$ since it factors through two consecutive maps in the long exact sequence of the pair $(\mathrm{Sk}_{s-1} X, \mathrm{Sk}_{s-2} X)$.

We want to think about filtrations

$$\cdots \subseteq F_{s-1} X \subseteq F_s X \subseteq F_{s+1} X \subseteq \cdots X$$

of a space X that don't behave so simply. But the starting point is the same: filter the singular complex accordingly:

$$F_s S_*(X) = S_*(F_s X) \subseteq S_*(X).$$

This is a *filtered (chain) complex*. To abstract a bit, suppose we are given a chain complex C_* whose homology we wish to compute by means of a filtration

$$\cdots F_{s-1}C_* \subseteq F_s C_* \subseteq F_{s+1}C_* \subseteq \cdots$$

by sub chain complexes. Note that at this point we are allowing the filtration to extend in both directions. And we need not suppose that the intersection is zero, nor that the union is all of C_*. (And C_* might be nonzero in negative degrees, as well.)

The first step is to form the quotient chain complexes,

$$\mathrm{gr}_s C_* = F_s C_* / F_{s-1}C_* \,.$$

This is a sequence of chain complexes, a *graded* object in the category of chain complexes, and is termed the "associated graded" complex.

What is the relationship between the homologies of these quotient chain complexes and the homology of C_* itself?

We'll set up grading conventions following the example of the filtration by preimages of a skeleton filtration under a fibration, as described in the previous lecture: name the coordinates in the plane (s,t), with the s-axis horizontal and the t-axis vertical. So s will be the filtration degree, and $s+t$ will be the total topological dimension. t is the "complementary degree." This suggests that we should put $\mathrm{gr}_s C_{s+t}$ in bidegree (s,t). Here then is a standard notation:

$$E^0_{s,t} = \mathrm{gr}_s C_{s+t} = F_s C_{s+t} / F_{s-1}C_{s+t} \,.$$

The differential then has bidegree $(0,-1)$. In parallel with the superscript in "E^0," this differential is written d^0.

Next we pass to homology. Let's use the notation

$$E^1_{s,t} = H_{s,t}(E^0_{*,*}, d^0)$$

for the homology of E^0. This in turn supports a differential. In the case of the skeleton filtration, this is the differential in the cellular chain complex. The definition in general is identical:

$$
\begin{array}{ccc}
E^1_{s,t} & \xrightarrow{\quad d^1 \quad} & E^1_{s-1,t} \\
\| & & \| \\
H_{s+t}(F_s/F_{s-1}) \xrightarrow{\ \partial\ } H_{s+t-1}(F_{s-1}) & \longrightarrow & H_{s+t-1}(F_{s-1}/F_{s-2}) \,.
\end{array}
$$

Thus d^1 has bidegree $(-1,0)$. Of course we will write

$$E^2_{s,t} = H_{s,t}(E^1_{*,*}, d^1) \,.$$

In the case of the skeleton filtration, $E_{s,t}^1 = 0$ unless $t = 0$, and the fact that cellular homology equals singular homology is the assertion that

$$E_{s,0}^2 = H_s(X).$$

In general the situation is more complicated because E^1 may be nonzero off the s-axis. So now the magic begins. The claim is that the bigraded group $E_{*,*}^2$ in turn supports a natural differential, written, of course, d^2, this time of bidegree $(-2, 1)$; that this pattern continues *ad infinitum*; and that in the end you get (essentially) $H_*(C_*)$. In fact the proof we gave last term that cellular homology agrees with singular homology is no more than a degenerate case of this fact.

Here's the general picture.

Theorem 61.1. *A filtered complex F_*C_* determines a natural* spectral sequence, *consisting of*

- *bigraded abelian groups $E_{s,t}^r$ for $r \geq 0$,*
- *differentials $d^r : E_{s,t}^r \to E_{s-r,t+r-1}^r$ for $r \geq 0$, and*
- *isomorphisms $E_{s,t}^{r+1} \cong H_{s,t}(E_{*,*}^r, d^r)$ for $r \geq 0$,*

such that for $r = 0, 1, 2$, $(E_{*,*}^r, d^r)$ is as described above, and that under further hypotheses "converges" to $H_*(C_*)$.

Here are further conditions that will suffice to guarantee that the spectral sequence is actually computing $H_*(C_*)$.

Definition 61.2. The filtered complex F_*C_* is *first quadrant* if

- $F_{-1}C_* = 0$,
- $H_n(\mathrm{gr}_s C_*) = 0$ for $n < s$, and
- $C_* = \bigcup F_s C_*$.

Under these conditions, E^1 is zero outside of the first quadrant, and so all the higher "pages" E^r have the same property. It's called a "first quadrant spectral sequence."

The differentials all have total degree -1, but their slopes vary. The longest possibly nonero differential emanating from (s, t) is

$$d^s : E_{s,t}^s \to E_{0,t+s-1}^s,$$

and the longest differential attacking (s, t) is

$$d^{t+1} : E_{s+t+1,0}^{t+1} \to E_{s,t}^{t+1}.$$

What this says is that for any value of (s,t), the groups $E_{s,t}^r$ stabilize for large r. That stable value is written

$$E_{s,t}^\infty.$$

Here's the rest of Theorem 61.1. It uses the natural filtration on $H_*(C_*)$ given by

$$F_s H_n(C_*) = \mathrm{im}(H_n(F_s C_*) \to H_n(C_*)).$$

Theorem 61.3. *The spectral sequence of a first quadrant filtered complex converges to* $H_*(C_*)$, *in the sense that*

$$F_{-1}H_*(C_*) = 0, \quad \bigcup_s F_s H_*(C_*) = H_*(C_*),$$

and for each s, t *there is a natural isomorphism*

$$E_{s,t}^\infty \cong \mathrm{gr}_s H_{s+t}(C_*).$$

In symbols, we may write (for any $r \geq 0$)

$$E_{*,*}^r \Longrightarrow H_*(C_*),$$

or, if you want to be explicit about the degrees and which degree is the filtration degree,

$$E_{s,t}^r \underset{s}{\Longrightarrow} H_{s+t}(C_*).$$

Notice right off that this contains the fact that cellular homology computes singular homology: In the spectral sequence associated to the skeleton filtration,

$$E_{s,t}^0 = S_{s+t}(\mathrm{Sk}_s X, \mathrm{Sk}_{s-1} X)$$

$$E_{s,t}^1 = H_{s+t}(\mathrm{Sk}_s X, \mathrm{Sk}_{s-1} X) = \begin{cases} C_s(X) & \text{if } t = 0 \\ 0 & \text{otherwise} \end{cases}$$

$$E_{s,t}^2 = \begin{cases} H_s^{\mathrm{cell}}(X) & \text{if } t = 0 \\ 0 & \text{otherwise.} \end{cases}$$

In a given total degree n there is only one nonzero group left by E^2, namely $E_{n,0}^2 = H_n^{\mathrm{cell}}(X)$. Thus no further differentials are possible:

$$E_{*,*}^2 = E_{*,*}^\infty.$$

The convergence theorem then implies that

$$\mathrm{gr}_s H_n(X) = \begin{cases} E_{n,0}^\infty = H_n^{\mathrm{cell}}(X) & \text{if } s = n \\ 0 & \text{otherwise.} \end{cases}$$

So the filtration of $H_n(X)$ changes only once:

$$0 = \cdots = F_{n-1}H_n(X) \subseteq F_n H_n(X) = \cdots = H_n(X),$$

and

$$F_n H_n(X)/F_{n-1}H_n(X) = E_{n,0}^\infty = H_n^{\mathrm{cell}}(X).$$

So

$$H_n(X) = H_n^{\mathrm{cell}}(X).$$

Before we explain how to construct the spectral sequence, let me point out one corollary at the present level of generality.

Corollary 61.4. *Let $f : C \to D$ be a map of first quadrant filtered chain complexes. If $E_{*,*}^r(f)$ is an isomorphism for some r, then $f_* : H_*(C) \to H_*(D)$ is an isomorphism.*

Proof. The map $E^r(f)$ is an isomorphism which is also also a chain map, i.e. it is compatible with the differential d^r. It follows that $E^{r+1}(f)$ is an isomorphism. By induction, we conclude that $E_{s,t}^\infty(f)$ is an isomorphism for all s, t. By Theorem 61.3, the map $\mathrm{gr}_s(f_*) : \mathrm{gr}_s H_*(C) \to \mathrm{gr}_s H(D)$ is an isomorphism. Now the conditions in Definition 61.2) let us use induction and the five lemma to conclude the proof. $\square$

Direct construction

In a later lecture I will describe a structure known as an "exact couple" that provides a construction of a spectral sequence that is both clean and flexible. But the direct construction from a filtered complex has its virtues as well. Here it is. The detailed computations are annoying but straightforward.

Define the following subspaces of $E_{s,t}^0 = F_s C_{s+t}/F_{s-1}C_{s+t}$, for $r \geq 1$.

$$Z_{s,t}^r = \{c : \exists x \in c \text{ such that } dx \in F_{s-r}C_{s+t-1}\},$$

$$B_{s,t}^r = \{c : \exists y \in F_{s+r-1}C_{s+t+1} \text{ such that } dy \in c\}.$$

So an "r-cycle" is a class that admits a representative whose boundary is r filtrations smaller; the larger r is the closer the class is to containing an actual cycle. An "r-boundary" is a class admitting a representative that is a boundary of an element allowed to lie in filtration degree $r - 1$ stages larger.

When $r = 1$, these are exactly the cycles and boundaries with respect to the differential d^0 on $E^0_{*,*}$.

We have inclusions

$$B^1_{*,*} \subseteq B^2_{*,*} \subseteq \cdots \subseteq Z^2_{*,*} \subseteq Z^1_{*,*}$$

and define

$$E^r_{s,t} = Z^r_{s,t}/B^r_{s,t}.$$

These pages are successively smaller groups of cycles modulo successively larger subgroups of boundaries. The differential d^r is of course induced from the differential d in C_*, and $H_{s,t}(E^r_{*,*}, d^r) \cong E^{r+1}_{s,t}$. In the first quadrant situation, the r-boundaries and the r-cycles stabilize to

$$Z^\infty_{s,t} = \{c : \exists\, x \in c \text{ such that } dx = 0\},$$
$$B^\infty_{s,t} = \{c : \exists\, y \in C_{s+t+1} \text{ such that } dy \in c\}.$$

The quotient, $E^\infty_{s,t}$, is exactly $F_s H_{s+t}(C_{*,*})/F_{s-1}H_{s+t}(C_{*,*})$.

Exercises

Exercise 61.5. (a) Suppose F_*C_* is a first-quadrant filtered complex such that $E^1_{*,*} = 0$. Show that $H_*(C_*) = 0$ in the following way. Let $c \in H_n(C_*)$. Show that it is represented by a cycle $z \in F_sC_n$ for some s. Then make an appropriate sequence of choices to come up with a class $y \in C_{n+1}$ with $dy = z$.

(b) Show by example that if we omit the either the first or the third condition in the definition of "first quadrant" (Definition 61.2) then the conclusion of (a) may fail.

(c) Give a counterexample to Corollary 61.4 if you keep the second and third conditions of Definition 61.2 but omit the first one.

Exercise 61.6. In Lecture 58 we studied a simplicial construction of the classifying space of topological group G. It comes equipped with a natural "skeleton" filtration, in which

$$F_s BG = \mathrm{im}\left(\coprod_{q \leq s} \Delta^q \times G^q \to BG \right).$$

Identify $F_s BG/F_{s-1}BG$. Assume that the topology of G is nice enough to give an isomorphism $H_*(F_s BG, F_{s-1}BG) \to \overline{H}_*(F_s BG/F_{s-1}BG)$, and work with coefficients in a field k. Identify the E^1 term of the spectral sequence associated with this filtration with the "bar construction" of the k-algebra $H_*(G)$. (See e.g. [33] or [77] for the bar construction.) Consequently we obtain a spectral sequence of the form

$$E_{s,t}^2 = \mathrm{Tor}_{s,t}^{H_*(G)}(k, k) \underset{s}{\Longrightarrow} H_{s+t}(BG)$$

(where s is the homological dimension and t is an internal dimension). Various cases of this spectral sequence are worth exploring! It's called variously the *bar construction*, *Rothenberg-Steenrod* [56], or *Eilenberg-Moore* spectral sequence (though be warned there is different spectral sequence [19] that is more commonly associated with Eilenberg and Moore).

62 Serre spectral sequence

Fix a fibration $p : E \to B$, with B a CW complex. We obtain a filtration on E by taking the preimage of the s-skeleton of B: $E_s = p^{-1}\mathrm{Sk}_s B$. This induces a filtration on $S_*(E)$ given by

$$F_s S_*(E) = S_*(p^{-1}\mathrm{Sk}_s(B)) \subseteq S_*(E).$$

The spectral sequence resulting from Theorem 61.1 is the *Serre spectral sequence*.

This was not Serre's construction [60], by the way; he did not employ a CW structure at all, but rather worked directly with a singular theory — but rather than simplices, he used cubes, which are well adapted to the study of bundles since a product of cubes is again a cube. We will describe a variant of Serre's construction in Lecture 66, one that is technically easier to work with and that makes manifest important multiplicative features of the spectral sequence. We will not try to dot all the i's in the construction we describe in this lecture.

In this spectral sequence,

$$E_{s,t}^1 = H_{s+t}(F_s E, F_{s-1} E).$$

Pick a cell structure

$$\coprod_{i \in I_s} S_i^{s-1} \longrightarrow \coprod_{i \in I_s} D_i^s$$

$$\downarrow \qquad\qquad \downarrow$$

$$B_{s-1} \longrightarrow B_s.$$

Let $\alpha : D_i^s \to B_s$ be one of the characteristic maps, and let F_i be the fiber over the center of the corresponding cell e_i^s in B. The pullback of $E \downarrow B$ under α is a trivial fibration since D_i^s is contractible. Now

$$\coprod_{i \in I_s} (D_i^s, S_i^{s-1}) \times F_i \to (F_s E, F_{s-1} E)$$

is a relative homeomorphism, so by excision

$$E^1_{s,t} = H_{s+t}(F_s E, F_{s-1} E) = \bigoplus_{i \in I_s} H_{s+t}((D^s_i, S^{s-1}_i) \times F_i) = \bigoplus_{i \in I_s} H_t(F_i).$$

In particular, this filtration satisfies the requirements of Definition 61.2, since $H_t(F_i) = 0$ for $t < 0$. We have a convergent spectral sequence. It remains to work out what d^1 is. I won't do this in detail but I'll tell you how it turns out.

It's important to appreciate that the fibers F_i vary from one cell to the next. If B is not path-connected, these fibers don't even have to be of the same homotopy type. If B *is* path connected, then they do, but the homotopy equivalence is determined by a homotopy class of paths from one center to the other and so is not canonical. If B is not simply connected, the functor

$$p^{-1}(-) : \Pi_1(B) \to \text{Ho}(\mathbf{Top})$$

may not be constant. But at least we see that the fibration defines functors

$$H_t(p^{-1}(-)) : \Pi_1(B) \to \mathbf{Ab} \quad \text{with} \quad b \mapsto H_t(p^{-1}(b)).$$

This is, or determines, a *local coefficient system*. We encountered these before, in our exploration of orientability. There a "local coefficient system" was a covering space with continuously varying abelian group structures on the fibers. If the space is path connected and semi-locally simply connected, there is a universal cover, and giving a covering space is equivalent to giving an action of the fundamental group on a set. CW complexes are locally contractible [24, e.g. Appendix on CW complexes, Proposition 4] and so this equivalence applies in our case.

If this local system is in fact constant (for example if B is simply connected) the differential in E^1 is none other than the cellular differential in

$$C_*(B; H_t(F))$$

(where we write F for any fiber), and so

$$E^2_{s,t} = H_s(B; H_t(F)).$$

This is the case we will mostly be concerned with. But the general case is the same, with the understanding that we mean homology of B with coefficients in the local system $H_t(p^{-1}(-))$.

Here's a base-point dependent way of thinking of how to compute homology or cohomology of a space with coefficients in a local system. We assume that our space X is path-connected and nice enough to admit a universal cover $\widetilde{X}$. Pick a basepoint $*$. Giving a local coefficient system $\mathcal{M}$ is the same as giving a $\mathbb{Z}[\pi_1(X, *)]$-module M. The fundamental group acts from the right on $\widetilde{X}$ and so on its singular chain complex. In Lecture 20 we developed the tensor product over a commutative ring. The same story produces $N \otimes_R M$ for a right R-module N and a left R-module N. You only get an abelian group (or a k-module if R is a k-algebra). Now we can say that

$$H_*(X; \mathcal{M}) = H_*(S_*(\widetilde{X}) \otimes_{\mathbb{Z}[\pi_1(X, *)]} M)$$
$$H^*(X; \mathcal{M}) = H^*(\mathrm{Hom}_{\mathbb{Z}[\pi_1(X, *)]}(S_*(\widetilde{X}), M)).$$

Here's the general result.

Theorem 62.1. *Let $p : E \to B$ be a Serre fibration, R a commutative ring, and M an R-module. There is a first quadrant spectral sequence of R-modules with*

$$E^2_{s,t} = H_s(B; H_t(p^{-1}(-); M))$$

that converges to $H_*(E; M)$. It is natural from E^2 on for maps of fibrations.

This theorem expresses one important perspective on spectral sequences: They can serve to implement a "local-to-global" strategy. A fiber bundle is locally a product. The spectral sequence explains how the "local" (in the base) homology of E gets integrated to produce the "global" homology of E itself.

Loops on spheres

Here's a first application of the Serre spectral sequence: a computation of the homology of the space of pointed loops on a sphere, ΩS^n. It is the fiber of the fibration $PS^n \to S^n$, where PS^n is the space of pointed maps $(S^n)^I_*$. The space PS^n is contractible, by the spaghetti move.

It is often said that the Serre spectral sequence is designed to compute the homology of the total space starting with the homologies of the fiber and of the base. This is not true! Rather, it establishes a relationship between these three homologies, one that can be used in many different ways. Here we know the homology of the total space (since PS^n is contractible) and of the base, and we want to know the homology of the fiber.

The case $n = 1$ is special: S^1 is a Eilenberg Mac Lane space $K(\mathbb{Z}, 1)$, so ΩS^1 is weakly equivalent to the discrete space $\mathbb{Z}$.

So suppose $n \geq 2$. Then the base is simply connected and torsion-free, so in the Serre spectral sequence

$$E^2_{s,t} = H_s(S^n; H_t(\Omega S^n)) = H_s(S^n) \otimes H_t(\Omega S^n).$$

Here's a picture, for $n = 4$.

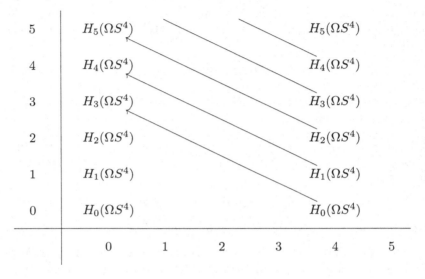

As you can see, the only possible nonzero differentials are of the form

$$d^n : E^n_{n,t} \to E^n_{0,t+n-1}.$$

So $E^2_{*,*} = E^{n-1}_{*,*}$ and $E^{n+1}_{*,*} = E^\infty_{*,*}$.

The spectral sequence converges to $H_*(PS^n)$, which is $\mathbb{Z}$ in dimension 0 and 0 elsewhere. This immediately implies that

$$H_t(\Omega S^n) = 0 \quad \text{for} \quad 0 < t < n - 1$$

since nothing could kill these groups on the fiber.

The fiber is path connected, $H_0(\Omega S^n) = \mathbb{Z}$, so we know the bottom row in E^2. $E^2_{n,0}$ must die. It can't be killed by being hit by a differential, since everything below the s-axis is trivial (and also because everything to its right is trivial). So it must die by virtue of d^n being injective on it. In fact that differential must be an isomorphism, since if it fails to surject onto $E^n_{0,n-1}$ there would be something left in $E^{n+1}_{0,n-1} = E^\infty_{0,n-1}$, and it would contribute nontrivially to $H_{n-1}(PS^n) = 0$.

This language of mortal combat gives extra meaning to the "spectral" in "spectral sequence."

So $H_{n-1}(\Omega S^n) = \mathbb{Z}$. This feeds back into the spectral sequence: $E^2_{n,n-1} = \mathbb{Z}$. Now that class has to kill or be killed. It can't be killed because everything to its right is zero, so d^n must be injective on it. And it must surject onto $E^n_{0,2(n-1)}$, for the same reason as before.

This establishes the inductive step. We have shown that all the d^n's are isomorphisms (except the ones involving $E^n_{0,0}$), and established:

Proposition 62.2. *Let* $n \geq 2$. *Then*

$$H_t(\Omega S^n) = \begin{cases} \mathbb{Z} & \text{if} \quad (n-1)|t \geq 0 \\ 0 & \text{otherwise}. \end{cases}$$

Evenness

If $E^r = E^{r+1} = \dots$ in a spectral sequence, we say that it *collapses* at E^r.

Sometimes it's easy to see that a spectral sequence collapses. For example, suppose that

$$E^r_{s,t} = 0 \text{ unless both } s \text{ and } t \text{ are even}.$$

Then all differentials in E^r and beyond must vanish, because they all have total degree -1. Actually all that is needed for this argument is that $E^r_{s,t} = 0$ unless $s + t$ is even. There may still be extension problems, though.

Exercises

Exercise 62.3. Let's get familiar with homology and cohomology with local coefficients.

(a) Let $n \geq 2$ and write C_2 for $\pi_1(\mathbb{RP}^n)$. Compute $H_*(\mathbb{RP}^n; \mathbb{Z}[C_2])$. This is the $E^2_{*,0}$ term in the Serre spectral sequence associated to the fibration $S^n \downarrow \mathbb{RP}^n$. Is everything OK?

(b) Write $\mathbb{Z}(-1)$ for the $\mathbb{Z}[C_2]$-module on which the generator acts by -1. Compute the graded groups $H_*(\mathbb{RP}^n; \mathbb{Z}(-1))$ and $H^*(\mathbb{R}^n; \mathbb{Z}(-1))$.

Remark 62.4. $\mathbb{RP}^n$ is a closed manifold. When n is odd, it is orientable and Poincaré duality relates $H_*(\mathbb{RP}^n; \mathbb{Z})$ and $H^*(\mathbb{RP}^n; \mathbb{Z})$. But when n is even, it's not; the orientation local system is given by the nontrivial $\mathbb{Z}[C_2]$-module $\mathbb{Z}(-1)$. Your solution to (b) may suggest a variant of Poincaré duality that is valid in the non-orientable case. In fact for any closed n-manifold M, there is a canonical isomorphism $H_n(M; o_M) \to H^0(M)$ and so a canonical "twisted fundamental class" $[M] \in H_n(M; o_M)$ mapping to

$1 \in H^0(M)$. An orientation of M identifies $H_n(M; o_M)$ with $H_n(M)$, and we obtain the fundamental class we worked with before. But now we can form a twisted cap product with this twisted fundamental class and obtain, for any local coefficient system $\mathcal{L}$, an isomorphism

$$H^p(M; \mathcal{L}) \to H_q(M; o_M \otimes \mathcal{L}), \quad p + q = n.$$

If $\mathcal{L}$ corresponds to a $\mathbb{Z}[\pi_1(M)]$-module L, this tensor product local coefficient system corresponds to the $\mathbb{Z}[\pi_1(M)]$ module in which $\sigma \in \pi_1(M)$ acts by $w_1(\sigma)\sigma$, where $w_1 : \pi_1(M) \to \{\pm 1\}$ gives the action corresponding to o_M. Interesting cases result from taking $\mathcal{L}$ to be trivial, and to be equal to o_M. (Note that $o_M \otimes o_M$ is trivial.)

Exercise 62.5. Suppose that $p : E \downarrow S^n$ is a fibration over the n-sphere, with fiber F. There is a natural long exact sequence involving $H_*(E)$ and $H_*(F)$, analogous to the Gysin sequence. Derive it carefully from the Serre spectral sequence. (The case $n = 1$ requires special attention.) This is the "Wang sequence."

Exercise 62.6. Let $f : S^2 \to S^2$ be a map of degree 2. Compute the homology of its homotopy fiber.

Exercise 62.7. Let $f : \mathbb{C}P^\infty \to \mathbb{C}P^\infty$ be a map inducing multiplication by 2 in H_2, and let $\mathbb{C}P^\infty \xrightarrow{\simeq} E \xrightarrow{p} \mathbb{C}P^\infty$ be a factorization of f into a homotopy equivalence followed by a fibration. Determine the behavior of the homology Serre spectral sequence for p.

63 Exact couples

Today I would like to show you a very simple piece of linear algebra called an *exact couple*. A filtered complex gives rise to an exact couple, and an exact couple gives rise to a spectral sequence. Exact couples were discovered by Bill Massey (1920–2017, Professor at Yale) independently of the French development of spectral sequences.

Definition 63.1. An *exact couple* is a diagram of abelian groups

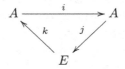

that is exact at each node.

As $(jk)(jk) = j(kj)k = 0$, the map $jk : E \to E$ is a differential, denoted d.

An exact couple determines a "derived couple"

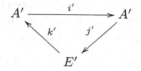

where

$$A' = \operatorname{im}(i) \quad \text{and} \quad E' = H(E, d).$$

Iterating this procedure, we get a sequence of exact couples

$$A^{(r)} \xrightarrow{\;i^{(r)}\;} A^{(r)}$$

with maps $k^{(r)}$, $j^{(r)}$ and $E^{(r)}$.

If we impose appropriate gradings, the "E" terms will form a spectral sequence.

We have to explain the maps in the derived couple.

i': this is just i restricted to $A' = \operatorname{im}(i)$. Obviously i carries $\operatorname{im}(i)$ into $\operatorname{im}(i)$.

j': Note that ja is a cycle in E: $dja = jkja = 0$. Define

$$j'(ia) = [ja].$$

To see that this is well defined, we need to see that if $ia = 0$ then ja is a boundary. By exactness there is an element $e \in E$ such that $ke = a$. Then $de = jke = ja$.

k': Let $e \in E$ be a cycle. Since $0 = de = jke$, $ke \in \operatorname{im}(i) = A'$ by exactness. Define

$$k'([e]) = ke.$$

To see that this is well defined, suppose that $e = de'$. Then $ke = kde' = kjke' = 0$.

We leave to you the exercise of checking that these maps indeed yield an exact couple.

Gradings

Now suppose we are given a filtered complex. It will define an exact couple in which A is given by the homology groups of the filtration degrees and E is given by the homology groups of the associated quotient chain complexes.

In order to accommodate this example we need to add gradings — in fact, bigradings. Here's the relevant definition.

Definition 63.2. An exact couple of bigraded abelian groups is *of type r* if the structure maps have the following bidegrees.

$$||i|| = (1, -1)$$
$$||j|| = (0, 0)$$
$$||k|| = (-r, r - 1)$$

It's clear from this that $||d|| = ||jk|| = (-r, r-1)$, the bidegree appropriate for the rth stage of a spectral sequence. We should specify the gradings on the abelian groups in the derived couple. Define $A'_{s,t}$ to sit in the factorization

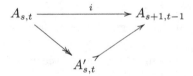

and $E'_{s,t} = H_{s,t}(E_{*,*})$. Then if $e \in E_{s,t}$, $ke \in A_{s-r,t+r-1}$, but if e is a cycle then ke lies in the subgroup $A'_{s-r-1,t-r}$, so $||k'|| = (r+1, -r)$: the derived couple is of type $(r+1)$.

Given a filtered complex

$$\cdots \subseteq F_{s-1}C_* \subseteq F_sC_* \subseteq F_{s+1}C_* \subseteq \cdots,$$

define

$$A^1_{s,t} = H_{s+t}(F_sC_*), \quad E^1_{s,t} = H_{s+t}(\mathrm{gr}_sC_*).$$

This agrees with our earlier use of the notation $E^1_{s,t}$. The structure maps are given in the obvious way: i^1 is induced by the inclusion of one filtration degree into the next (and has bidegree $(1, -1)$); j^1 is induced from the quotient map (and has bidegree $(0, 0)$); and k^1 is the boundary homomorphism in the homology long exact sequence (and has bidegree $(-1, 0)$).

Given any exact couple of type 1, (A^1, E^1), we'll write

$$A^r = (A^1)^{(r-1)}, \quad E^r = (E^1)^{(r-1)}$$

for the $(r-1)$ times derived exact couple, which is of type r.

Differentials

An exact couple can be unfolded in a series of linked exact triangles, like this (taking $r = 1$ for concreteness, and omitting the second index):

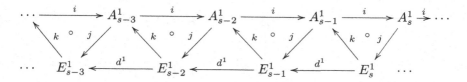

The triangles marked with ∘ are exact; the lower ones commute, and define d^1.

This image is useful in understanding the differentials in the associated spectral sequence. Start with an element $x \in E_s^1$. Suppose it's a cycle. Then its image $kx \in A_{s-1}^1$ is killed by j and hence pulls back under i, to, say, $x_1 \in A_{s-2}^1$. The image in E_{s-2}^1 of x_1 under j is a representative for $d^2[x]$. Suppose that $d^2[x] = 0$. Then we can improve the lift x_1 to one that pulls back one step further, to, say, $x_2 \in A_{s-3}^1$; and $d^3[x] = [jx_2]$. This pattern continues. The further you can pull kx back, the longer x survives in the spectral sequence. If it pulls back forever, then you appeal to a convergence condition to conclude that $kx = 0$, and x therefore lifts under j to an element $\bar{x}$ in A_s^1. The direct limit

$$L = \lim_{\to}(\cdots \to A_s^1 \to A_{s+1}^1 \to A_{s+2}^1 \to \cdots)$$

is generally what one is interested in (it's $H_*(C_*)$ in the first quadrant filtered complex situation, for example) and one may say that "x survives to" the image of $\bar{x}$ in L.

Other examples

Topology is inhabited by many spectral sequences that do not arise from a filtered complex. For example, if you have a tower of fibrations, you get an exact couple by linking together the homotopy long exact sequences of the individual fibrations. Well, almost. The problem is what happens at the bottom: groups may not be abelian, or even groups; and even if they are, you may not be able to guarantee exactness at π_0. Anyway, here's an example. Form the Whitehead tower of a space Y and map some well-pointed space X into it. We get a new tower of fibrations

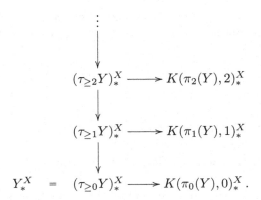

The homotopy groups of the spaces on the right form the E^1-term, and are easy to compute:

$$\pi_n(K(\pi_p(Y), p)_*^X) = [S^n \wedge X, K(\pi_p(Y), p)]_* = [X, K(\pi_p(Y), p - n)]_*$$
$$= \overline{H}^{p-n}(X; \pi_p(Y)).$$

Insofar as this is a spectral sequence at all, the E^1 term is given by

$$E_{s,t}^1 = \overline{H}^{-2s-t}(X; \pi_{-s}(Y, *)).$$

It's concentrated between the lines $t = -s$ and $t = -2s$, in the second quadrant of the plane. An element of $\pi_n(Y_*^X) = [\Sigma^n X, Y]_*$ is in filtration $s \leq 0$ if and only if it factors through $\tau_{\geq -s} Y \to Y$.

This picture is very closely related to obstruction theory, and indeed obstruction theory can be set up using it. Its failings as a spectral sequence can be repaired in various ways I won't discuss. If it can be repaired, the spectral sequence converges to $\pi_*(Y_*^X)$, or wants to.

For another example, there are many "generalized homology theories" — sequences of functors satisfying the Eilenberg-Steenrod axioms other than the dimension axiom — K-theory, bordism theories, and many others. Write $R_*(-)$ for any such theory. The skeleton filtration construction of the Serre spectral sequence can be applied to compute the R-homology of the total space of a fibration $p : E \to B$: To construct the exact couple, all you need is the long exact sequence of a pair, which is available in R-homology. You find for each t a local coefficient system $R_t(p^{-1}(-))$, and you get a spectral sequence

$$E_{s,t}^2 = H_s(B; R_t(p^{-1}(-))) \underset{s}{\Longrightarrow} R_{s+t}(E).$$

Even the case $p : E \xrightarrow{=} B$ is interesting: then the local coefficient system is guaranteed to be trivial, and we get

$$E^2_{s,t} = H_s(E; R_t(*)) \underset{s}{\Longrightarrow} R_{s+t}(E).$$

This is the "Atiyah-Hirzebruch spectral sequence," and it provides a powerful tool for computing these generalized homology theories. (Sir Michael Atiyah (1929–2019) was a prodigiously broad and creative mathematician, working at the Institute for Advanced Study, Cambridge University, and the University of Edinburgh. Fritz Hirzebruch (1927–2012) was a dominant figure in post-war German mathematics, working mainly in Bonn.)

Both of these spectral sequences require us to move out of the first quadrant setting. The Atiyah-Hirzebruch-Serre spectral sequence can fill up the right half-plane. See Lecture 70 for another example.

Exercises

Exercise 63.3. Check that the maps in the definition of the derived exact couple do indeed yield an exact couple.

Exercise 63.4. Verify that if you have a filtered complex, the the groups $E^r_{s,t}$ produced as above coincide with those described in the earlier discussion in Lecture 61.

Exercise 63.5. Let (A^1, E^1) be a bigraded exact couple, and let $(\overline{A}^1, \overline{E}^1)$ be the "truncation" at $s = n$: the exact couple mapping to (A^1, E^1) in which the maps $A^1_s \to A^1_{s+1}$ for $s \geq n$ are replaced by isomorphisms, so that $\overline{E}^1_s = 0$ for $s > n$ and $\overline{E}^1_s \to E^1_s$ is an isomorphism for $s \leq n$. Show that for all $r \geq 1$, $\overline{E}^r_s = 0$ for $s > n$, and that $\overline{E}^r_s \to E^r_s$ is surjective for $s \leq n$ and an isomorphism for $s \leq n - r + 1$. All the possibly nonzero differentials in the truncated spectral sequence thus land in the region that maps isomorphically to the original spectral sequence. The original one thus completely determines the truncated spectral sequence.

For example, this is the relationship between the Serre spectral sequence for a fibration over a CW complex and that of the restriction of the fibration to a skeleton.

Exercise 63.6. Let C be a chain complex such that C_n is free for each n (for example, the singular chains on a space or the cellular chains on a CW complex). Let p be a prime number. The *Prüfer group* is

$$\mathbb{Z}_{p^\infty} = \bigcup \mathbb{Z}/p^s \mathbb{Z} = \mathbb{Z}[1/p]/\mathbb{Z}.$$

It's filtered by

$$F_s \mathbb{Z}_{p^\infty} = \ker(p^{s+1} | \mathbb{Z}_{p^\infty}) = \mathbb{Z}/p^{s+1}\mathbb{Z}.$$

Filter $C \otimes \mathbb{Z}_{p^\infty}$ accordingly.

(a) Show that in the resulting spectral sequence

$$E^1_{s,t} = \begin{cases} H_{s+t}(C \otimes \mathbb{F}_p) & \text{if } s \geq 0 \\ 0 & \text{otherwise.} \end{cases}$$

Also describe the corresponding exact couple.

(b) This is not a first quadrant spectral sequence; which part of the (s, t)-plane does it live in? What if $C_n = 0$ for $n < 0$? Show that nevertheless it converges,

$$E^r_{s,t} \underset{s}{\Longrightarrow} H_{s+t}(C \otimes \mathbb{Z}_{p^\infty})$$

in the sense that one can define $E^\infty_{s,t}$ and these groups form the associated graded groups of a filtration on $H_{s+t}(C \otimes \mathbb{Z}_{p^\infty})$ that is "exhaustive":

$$F_{-1} = 0 \quad \text{and} \quad \lim_{s \to \infty} F_s H_*(C \otimes \mathbb{Z}_{p^\infty}) = H_*(C \otimes \mathbb{Z}_{p^\infty}).$$

The "abutment" of this spectral sequence — the group it is trying to converge to — is of interest because it determines the p-torsion in $H_*(C)$ by means of the long exact coefficient sequence

$$\cdots \to H_{n+1}(C \otimes \mathbb{Z}_{p^\infty}) \to H_n(C) \to H_n(C)[1/p] \to H_n(C \otimes \mathbb{Z}_{p^\infty}) \to \cdots.$$

(c) If $H_n(C)$ is finitely generated, show that the structure of the p-torsion in it is completely described by the spectral sequence.

(d) Show that the data contained in this spectral sequence can be captured by a simpler "singly graded spectral sequence," with

$$E^1_n = H_n(C \otimes \mathbb{Z}/p\mathbb{Z}), \quad E^{r+1}_n = H_n(E^r_*, \beta^r), \quad \beta^r : E^r_n \to E^r_{n-1}.$$

This is the "Bockstein spectral sequence." What is $\beta^1 : H_n(C \otimes \mathbb{Z}/p\mathbb{Z}) \to H_{n-1}(C \otimes \mathbb{Z}/p\mathbb{Z})$?

64 Gysin sequence, edge homomorphisms, and transgression

Now we'll discuss a general situation, a common one, that displays many of the ways in which the Serre spectral sequence relates the homology groups of fiber, total space, and base.

Suppose $p : E \to B$ is a fibration; assume the base is path-connected, and that the fiber has homology (with coefficients in a fixed PID R) isomorphic to that of S^{n-1} with $n > 1$. Let's use the Serre spectral sequence to determine how the homologies of E and of B are related. We will assume that this "spherical fibration" is orientable, and choose an orientation. This means that the local coefficient system $H_{n-1}(p^{-1}(-))$ is trivial, and provided with a trivialization: a preferred generator of $H_{n-1}(p^{-1}(b))$ that varies continuously with $b \in B$. For example, we might be looking at $S^{2k-1} \downarrow \mathbb{C}P^{k-1}$ or $S^{4k-1} \downarrow \mathbb{H}P^{k-1}$, or the complement of the zero-section in the tangent bundle of an R-oriented n-manifold.

There are just two nonzero rows in this spectral sequence. This means that there's just one possibly nonzero differential:

$$E^2_{*,*} = E^3_{*,*} = \cdots = E^n_{*,*} \,;$$

then a differential

$$d^n : E^n_{s,0} \to E^n_{s-n,n-1}$$

occurs; and then

$$E^{n+1}_{*,*} = \cdots = E^\infty_{*,*}\,.$$

Taking homology with respect to d^n gives the top row of

$$0 \longrightarrow E^\infty_{s,0} \longrightarrow E^n_{s,0} \xrightarrow{\ d^n\ } E^n_{s-n,n-1} \longrightarrow E^\infty_{s-n,n-1} \longrightarrow 0$$

$$H_s(B) \longrightarrow H_{s-n}(B).$$

To explain the rest of this diagram, path connectedness of S^{n-1} gives the isomorphism

$$E^n_{s,0} = E^2_{s,0} = H_s(B)\,,$$

and the orientation determines

$$E^n_{s-n,n-1} = E^2_{s-n,n-1} = H_{s-n}(B; H_{n-1}(S^{n-1})) = H_{s-n}(B)\,.$$

Now look at total degree s. The filtration of $H_s(E)$ changes at most twice, with associated quotients given by the E^∞ term: so there is a short exact sequence

$$0 \to E^\infty_{s-n+1,n-1} \to H_s(E) \to E^\infty_{s,0} \to 0\,.$$

These two families of exact sequences splice together to give a long exact sequence:

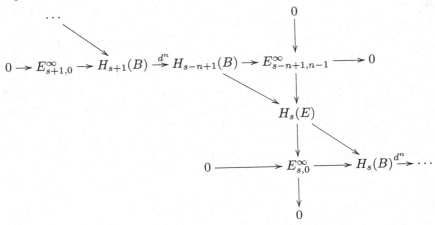

Proposition 64.1. *Let $p : E \to B$ be a Serre fibration whose fiber is an R-homology $(n-1)$-sphere, and assume it is R-oriented. There is a naturally associated long exact sequence, the* Gysin sequence

$$\cdots \to H_{s+1}(B) \to H_{s-n+1}(B) \to H_s(E) \xrightarrow{p_*} H_s(B) \to H_{s-n}(B) \to \cdots .$$

(Werner Gysin (1915–1998) described this in his thesis at ETH under Heinz Hopf.) The only part of this that we have not proven is that the middle map here is in fact the map induced by the projection p. That's the story of "edge homomorphisms," which we take up next.

First, though, an example. The Gysin sequence of the S^1-bundle $S^\infty \downarrow \mathbb{C}P^\infty$ looks like this:

This gives us another computation of $H_*(\mathbb{CP}^\infty)$: Working inductively up the tower, you compute

$$H_n(\mathbb{CP}^\infty; R) = \begin{cases} R & \text{if } 2|n \geq 0 \\ 0 & \text{otherwise}. \end{cases}$$

Edge homomorphisms

In the Serre spectral sequence for the fibration $p : E \to B$, what can we say about the evolution of the bottom edge, or of the left edge? Let's assume that the fiber is path connected and that the local coefficient system is trivial, so in

$$E_{s,t}^2 = H_s(B; H_t(F)) \underset{s}{\Longrightarrow} H_{s+t}(E)$$

the bottom edge is canonically isomorphic to $H_*(B)$.

Being at the bottom, no nontrivial differentials can ever hit it. So the successive process of taking homology will be a succession of taking kernels:

$$E_{n,0}^{r+1} = \ker(d^r : E_{n,0}^r \to E_{n-r,r-1}^r).$$

Of course when $r > n$ things quiet down. So

$$E_{n,0}^2 \supseteq E_{n,0}^3 \supseteq \cdots \supseteq E_{n,0}^{n+1} = E_{n,0}^\infty.$$

Now $H_n(E)$ enters the picture, along with its filtration. The whole of $H_n(E)$ is already hit by $H_n(p^{-1}\mathrm{Sk}_n B)$. This is confirmed by the fact that the associated graded $\mathrm{gr}_s H_n(E) = E_{s,n-s}^\infty$ vanishes for $s > n$. So $F_n H_n(E) = H_n(E)$.

Putting all this together, we get a map

$$
\begin{array}{ccc}
H_n(E) & \xrightarrow{\hspace{6cm}} & H_n(B) \\
\| & & \| \\
F_n H_n(E) \twoheadrightarrow \mathrm{gr}_n H_n(E) = E_{n,0}^\infty = E_{n,0}^{n+1} \rightarrowtail E_{n,0}^n \rightarrowtail \cdots \rightarrowtail E_{n,0}^2.
\end{array}
$$

This composite is an *edge homomorphism* for the spectral sequence. It's something you can define for any first quadrant filtered complex. In the Serre spectral sequence case, it has a direct interpretation:

Proposition 64.2. *This edge homomorphism coincides with the map* $p_* : H_n(E) \to H_n(B)$.

This explains the role of the differentials off the bottom row of the spectral sequence. They are obstructions to classes lifting to the homology of the total space. This reflects the intuition we tried to develop several lectures ago. The image of $p_* : H_n(E) \to H_n(B)$ is precisely the intersection (so to speak) of the kernels of the differentials coming off of $E^2_{n,0}$.

Before we prove this, let's notice that there is a dual picture for the vertical axis. Now all differentials leaving $E^r_{0,n}$ are trivial, so we get surjections

$$E^2_{0,n} \twoheadrightarrow E^3_{0,n} \twoheadrightarrow \cdots \twoheadrightarrow E^{n+2}_{0,n} = E^\infty_{0,n} \, .$$

On the other hand, the smallest nonzero filtration degree of $H_n(E)$ is $F_0 H_n(E)$. Thus we have another "edge homomorphism,"

$$H_n(F) = E^2_{0,n} \twoheadrightarrow E^3_{0,n} \twoheadrightarrow \cdots \twoheadrightarrow E^{n+2}_{0,n} = E^\infty_{0,n} = F_0 H_n(E) \hookrightarrow H_n(E) \, .$$

Proposition 64.3. *This edge homomorphism coincides with the map* $i_* : H_n(F) \to H_n(E)$ *induced by the inclusion of the fiber.*

So the kernel of i_* is union of the images (so to speak) of the differentials coming into $E^2_{0,n}$. The sources of these differentials represent chains in E which serve as null-homologies of cycles in F.

Proof of Propositions 64.2 and 64.3. The map of fibrations

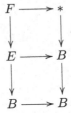

induces a commutative diagram in which the top and bottom arrows are edge homomorphisms:

$$\begin{array}{ccc} H_n(E) & \longrightarrow & H_n(B) \\ \downarrow{\scriptstyle p_*} & & \downarrow{\scriptstyle (1_B)_*} \\ H_n(B) & \longrightarrow & H_n(B) \, . \end{array}$$

So we just need to check that the bottom edge homomorphism associated to the identity fibration $1_B : B \to B$ is the identity map $H_n(B) \to H_n(B)$. This I leave to you.

The proof of Proposition 64.3 is similar. □

Very often you begin with some homomorphism, and you are interested in whether it is an isomorphism, or how it can be repaired to become an isomorphism. If you can write it as an edge homomorphism in a spectral sequence, then you can regard the spectral sequence as measuring how far from being an isomorphism your map is; it provides the reasons why the map fails to be either injective or surjective.

Transgression

There is a third aspect of the Serre spectral sequence that deserves attention, namely, the differential going clear across the spectral sequence, all the way from base to fiber. We'll study it in case the fiber and the base are both path connected and the local coefficient systems $H_t(p^{-1}(-))$ are trivial. Write F for the fiber.

The differentials

$$d^n : E^n_{n,0} \to E^n_{0,n-1}$$

are known as *transgressions*, and an element of $E^2_{n,0} = H_n(B)$ that survives to $E^n_{n,0}$ is said to be *transgressive*. The first one is a homomorphism

$$d^2 : H_2(B) \to H_1(F),$$

but after that d^n is merely an *additive relation* between $H_n(B)$ and $H_{n-1}(F)$: It has a *domain of definition*

$$E^s_{n,0} \subseteq E^2_{n,0} = H_n(B)$$

and *indeterminacy*

$$\ker(H_{n-1}(F) = E^2_{0,n-1} \twoheadrightarrow E^n_{0,n-1}).$$

Let me expand on what I mean by an additive relation. A good reference is [33, II §6].

Definition 64.4. An *additive relation* $R : A \rightharpoonup B$ is a subgroup R of $A \times B$.

For example the graph of a homomorphism $A \to B$ is an additive relation. Additive relations compose in the evident way: the composite of $R : A \rightharpoonup B$ with $S : B \rightharpoonup C$ is

$$\{(a,c) : \exists\, b \in B \text{ such that } (a,b) \in R \text{ and } (b,c) \in S\} \subseteq A \times C.$$

Every additive relation has a "converse,"

$$R^{-1} = \{(b,a) : (a,b) \in R\} : B \rightharpoonup A.$$

An additive relation has a *domain*

$$D = \{a \in A : \exists b \in B \text{ such that } (a, b) \in R\} \subseteq A$$

and an *indeterminancy*

$$I = \{b \in B : (0, b) \in R\},$$

and determines a homomorphism

$$f : D \to B/I$$

by

$$f(a) = b + I \text{ for } b \in B \text{ such that } (a, b) \in R.$$

Conversely, such a triple (D, I, f) determines an additive relation,

$$R = \{(a, b) : a \in D \text{ and } b \in f(a)\}.$$

An additive relation is defined as a subspace of $A \times B$, but any "span"

determines one by taking the image of the resulting map $C \to A \times B$.

End of digression. We have the transgression $d^n : H_n(B) \dashrightarrow H_{n-1}(F)$. Another such additive relation is determined by the span

$$
\begin{array}{ccc}
 & H_n(E, F) & \\
\swarrow{\scriptstyle p_*} & & \searrow{\scriptstyle \partial} \\
H_n(B) = H_n(B, *) & & H_{n-1}(F).
\end{array}
$$

Proposition 64.5. *These two linear relations coincide.*

Proof sketch. This phenomenon is actually how we began our discussion of spectral sequences. Let $x \in H_n(B)$. Since $n > 0$ we can just as well regard it as a class in $H_n(B, *)$. Represent it by a cycle $c \in Z_n(B, *)$. (In the Hopf fibration case this simplifies the representative by making the constant cycle optional.) Lift it to a chain in the total space E. In general, this chain will not be a cycle (as we saw in the Hopf fibration). The differentials record this boundary; let us recall the explicit construction of the differential at the end of Lecture 61. Saying that the class x survives to E^n is the same as saying that we can find a lift to a chain c in E, with $dc \in S_{n-1}(F)$, that is, to a relative cycle in $S_{n-1}(E, F)$. Then $d^n(x)$ is represented by the class $[dc] \in H_{n-1}(F)$. This is precisely the transgression. $\quad\square$

Exercise 64.6. (a) Show that if $p : E \to B$ is a fibration and each fiber has the homology of a point then p induces an isomorphism in homology.

(b) Show that any weak equivalence $f : X \to Y$ induces a homology isomorphism. Hint: Consider the homotopy fiber at a point in Y, and use **(a)**.

65 Serre exact sequence and the Hurewicz theorem

Serre exact sequence

Suppose $\pi : E \to B$ is a fibration over a path-connected base. Pick a point $* \in E$, use its image $* \in B$ as a basepoint in B, write $F = \pi^{-1}(*) \subseteq E$ for the fiber over $*$, and equip it with the point $* \in E$ as a basepoint. Suppose also that F is path connected.

Pick a coefficient ring R. Everything we've done works perfectly with coefficients in R — all abelian groups in sight come equipped with R-module structures. Let's continue to suppress the coefficient ring from the notation. Suppose that the low-dimensional homology of both fiber and base vanishes:

$$H_s(B) = 0 \quad \text{for} \quad 0 < s < p$$
$$H_t(F) = 0 \quad \text{for} \quad 0 < t < q.$$

Assume that $\pi_1(B, *)$ act trivially on $H_*(F)$, so the Serre spectral sequence (now with coefficients in R!) takes the form

$$E^2_{s,t} = H_s(B; H_t(F)) \underset{s}{\Longrightarrow} H_{s+t}(E).$$

Our assumptions imply that $E^2_{0,0} = R$ is all alone; otherwise everything with $s < p$ vanishes and everything with $t < q$ vanishes.

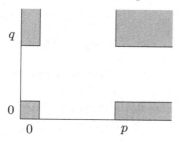

For a while, the only possibly nonzero differentials are the transgressions

$$d^s : E^s_{s,0} \to E^s_{0,s-1}.$$

The result, in this range, is an exact sequence

$$0 \to E_{s,0}^{\infty} \to H_s(B) \xrightarrow{d^s} H_{s-1}(F) \to E_{0,s-1}^{\infty} \to 0 .$$

Again, in this range, these end terms are the only two possibly nonzero associated quotients in $H_n(E)$ — there is a short exact sequence

$$0 \to E_{0,n}^{\infty} \to H_n(E) \to E_{n,0}^{\infty} \to 0 .$$

— and splicing things together we arrive at a long exact sequence

$$H_{p+q-1}(F) \xrightarrow{i_*} H_{p+q-1}(E) \xrightarrow{p_*} H_{p+q-1}(B)$$

$$H_{p+q-2}(F) \xleftarrow{i_*} H_{p+q-2}(E) \xrightarrow{p_*} H_{p+q-2}(B)$$

$$H_{p+q-3}(F) \xrightarrow{i_*} \cdots .$$

This is the *Serre exact sequence*: in this range of dimensions homology and homotopy behave the same! We can't extend it further to the left because the kernel of the edge homomorphism $H_{p+q-1}(F) \to H_{p+q-1}(E)$ has two sources: the image of $d^p : E_{p,q}^p \to E_{0,p+q-1}^p$, and the image of $d^{p+q} : E_{p+q,0}^{p+q} \to E_{0,p+q-1}^{p+q}$.

Comparison with homotopy

The Serre exact sequence mimics the homotopy long exact sequence of the fibration.

Proposition 65.1. *The Hurewicz map participates in a commutative ladder*

$$
\begin{array}{ccccccccc}
\pi_{p+q-1}(F) & \xrightarrow{i_*} & \pi_{p+q-1}(E) & \xrightarrow{\pi_*} & \pi_{p+q-1}(B) & \longrightarrow & \pi_{p+q-2}(F) & \longrightarrow & \cdots \\
\downarrow{h} & & \downarrow{h} & & \downarrow{h} & & \downarrow{h} & & \\
H_{p+q-1}(F) & \xrightarrow{i_*} & H_{p+q-1}(E) & \xrightarrow{\pi_*} & H_{p+q-1}(B) & \longrightarrow & H_{p+q-2}(F) & \longrightarrow & \cdots .
\end{array}
$$

Proof. The left two squares commutes by naturality of the Hurewicz map. The right square commutes because, according to our geometric interpretation of the transgression, both boundary maps arise in the same way:

$$
\begin{array}{ccccc}
\pi_n(B) & \xleftarrow{\cong} & \pi_n(E,F) & \xrightarrow{\partial} & \pi_{n-1}(F) \\
\downarrow{h} & & \downarrow{h} & & \downarrow{h} \\
H_n(B) & \xleftarrow{\cong} & H_n(E,F) & \xrightarrow{\partial} & H_{n-1}(F) .
\end{array}
$$

The isomorphism $\pi_n(E, F) \to \pi_n(B)$ is Lemma 47.7. $\qquad\qquad\square$

Let us now specialize to the case of the path-loop fibration

$$\Omega X \to PX \to X$$

where X is a simply-connected pointed space. The coefficient system is trivial. Suppose that in fact $\overline{H}_i(X) = 0$ for $i < n$. Since the spectral sequence converges to the homology of a point, we find that $\overline{H}_i(\Omega X) = 0$ for $i < n - 1$. The Serre exact sequence, or direct use of the spectral sequence as in the computation of $H_*(\Omega S^n)$, shows this:

Lemma 65.2. *Let X be an $(n - 1)$-connected pointed space. The transgression relation provides an isomorphism*

$$\overline{H}_i(X) \to \overline{H}_{i-1}(\Omega X)$$

for $i \leq 2n - 2$.

For example, if X is simply connected, we get a commutative diagram

$$
\begin{array}{ccc}
\pi_2(X) & \xrightarrow{\ \cong\ } & \pi_1(\Omega X) \\
\downarrow & & \downarrow \\
H_2(X) & \xrightarrow{\ \cong\ } & H_1(\Omega X).
\end{array}
$$

Since ΩX is an H-space its fundamental group is abelian, so Poincaré's theorem 31.6 shows that the Hurewicz homomorphism on the right is an isomorphism. Therefore the map on the left is. This is a case of the Hurewicz theorem! In fact, continuing by induction we discover a proof — Serre's proof — of the general case of the Hurewicz theorem.

Theorem 65.3 (Hurewicz). *Let $n \geq 1$. Suppose X is a pointed space that is $(n - 1)$-connected: $\pi_i(X) = 0$ for $i < n$. Then $\overline{H}_i(X) = 0$ for $i < n$ and the Hurewicz map $\pi_n(X)^{ab} \to H_n(X)$ is an isomorphism.*

Going relative

Any topological concept seems to get more useful if you can extend it to a relative form. So let (B, A) be a pair of spaces. To make the construction for the Serre spectral sequence that we proposed earlier work, we should assume that this is a relative CW complex. Suppose that $E \downarrow B$ is a

fibration. The pullback or restriction

provides us with a "fibration pair" (E, E_A). Suppose that B is path-connected and A nonempty, pick a basepoint $* \in A$, write F for the fiber of $E \downarrow B$ over $*$ (which is of course also the fiber of $E|_A \downarrow A$ over $*$), and suppose that $\pi_1(B, *)$ acts trivially on $H_*(F)$. With these assumptions, pulling back skelata of B rel A yields the *relative Serre spectral sequence*

$$E_{s,t}^2 = H_s(B, A; H_t(F)) \underset{s}{\Longrightarrow} H_{s+t}(E, E_A).$$

Let's apply this right away to prove a relative version of the Hurewicz theorem. We will develop conditions under which

$$h : \pi_i(X, A) \to H_i(X, A)$$

is an isomorphism for all $i \leq n$. We will of course assume that X is path connected and that A is nonempty, which together imply that $H_0(X, A) = 0$. Since $\pi_1(X, A)$ is in general only a pointed set let's begin by assuming that it vanishes. This implies that A is also path connected and that $\pi_1(A) \to \pi_1(X)$ is surjective. The induced map on abelianizations is then also surjective, so by Poincaré's theorem $H_1(A) \to H_1(X)$ is surjective and so $H_1(X, A) = 0$.

Moving up to the next dimension, we may hope that $h : \pi_2(X, A) \to H_2(X, A)$ is then an isomorphism, but $\pi_2(X, A)$ is not necessarily abelian so this can't be right in general. This can be fixed — in fact if we kill the action of $\pi_1(A)$ on $\pi_2(X, A)$ it becomes abelian and the resulting homomorphism to $H_2(X, A)$ is an isomorphism (see [63, Ch. 5, Sec. 7]). But we'll be assuming that $\pi_1(X) = 0$ in a minute anyway, so let's just go ahead now and assume that $\pi_1(A) = 0$. The long exact homotopy sequence then shows that $\pi_2(X, A)$ is a quotient of $\pi_2(X)$ and so is abelian. We'll show that $h : \pi_2(X, A) \to H_2(X, A)$ is then an isomorphism.

We will use the fact (from Exercise 47.9) that the projection map induces a isomorphism

$$\pi_n(E, E_A) \overset{\cong}{\to} \pi_n(B, A)$$

for any $n \geq 1$. In particular, let F be the homotopy fiber of the inclusion map $A \hookrightarrow X$: that is, the pullback in

$$\begin{array}{ccc} F & \longrightarrow & PX \\ \downarrow & & \downarrow \\ A & \longrightarrow & X \, . \end{array}$$

The path space PX is contractible, so from the long exact homotopy sequence for the pair (PX, F) we find that the maps on the top row of the following commutative diagram are isomorphisms.

$$
\begin{array}{ccccc}
\pi_{n-1}(F) & \xleftarrow{\cong} & \pi_n(PX, F) & \xrightarrow{\cong} & \pi_n(X, A) \\
\downarrow{h} & & \downarrow{h} & & \downarrow{h} \\
\overline{H}_{n-1}(F) & \xleftarrow{\cong} & H_n(PX, F) & \xrightarrow{p_*} & H_n(X, A) .
\end{array}
$$

Returning to our $n = 2$ case, the left arrow is an isomorphism by Poincaré's theorem, since F is path connected and by our assumptions its fundamental group is abelian. What remains in this case then is to show that homology behaves like homotopy, in the sense that $H_2(PX, F) \to H_2(X, A)$ is an isomorphism.

In general, if we assume that, for some $n \geq 3$, $\pi_i(X, A) = 0$ for $i < n$, then the absolute case of the Hurewicz theorem implies that the left Hurewicz homomorphism is an isomorphism, and we are left wanting to show that $p_* : H_n(PX, F) \to H_n(X, A)$ is an isomorphism.

For this we can appeal to the relative Serre spectral sequence for the fibration pair $(PX, F) \downarrow (X, A)$. It takes the form

$$
E^2_{s,t} = H_s(X, A; H_t(\Omega X)) \underset{s}{\Longrightarrow} H_{s+t}(PX, F) = \overline{H}_{s+t-1}(F)
$$

provided the coefficient system is trivial. Since $H_0(\Omega X) = \mathbb{Z}[\pi_1(X)]$, we are pretty much forced to assume that X is simply connected if we want simple coefficients.

The universal coefficient theorem gives us a handle on the E^2 term:

$$
H_s(X, A; H_t(\Omega X)) \cong H_s(X, A) \otimes H_t(\Omega X) \oplus \operatorname{Tor}(H_{s-1}(X, A), H_t(\Omega X)) .
$$

Now is the time to think about using induction on n: This will allow us to use the assumption that $\pi_i(X, A) = 0$ for $i < n - 1$ to conclude that $H_i(X, A) = 0$ for $i < n - 1$ and that $\pi_{n-1}(X, A) \xrightarrow{\cong} H_{n-1}(X, A)$; but we have the additional assumption that $\pi_{n-1}(X, A) = 0$ as well, so $H_{n-1}(X, A) = 0$ too. The induction begins with the case $n = 2$.

So when $s < n$ both end terms vanish, and the entire spectral sequence is concentrated along and to the right of $s = n$.

We glean two facts from this vanishing result: First, $H_i(PX, F) = 0$ for $i < n$, so $\overline{H}_i(F) = 0$ for $i < n - 1$. We knew this already from the absolute Hurewicz theorem.

The second fact is that $E^2_{n,0}$ survives intact to $E^\infty_{n,0}$: Nothing can hit it, and it can hit nothing. This is also the only nonzero group along the total degree line n, so (using what we know about the bottom edge homomorphism) the projection map induces an isomorphism $H_n(PX, F) \to H_n(X, A)$. This is a spectral sequence "corner argument."

Putting this together:

Theorem 65.4 (Relative Hurewicz theorem). *Let X be a space and A a subspace. Assume both of them are simply connected, and let $n \geq 2$. Assume that $\pi_i(X, A) = 0$ for $2 \leq i < n$. Then $H_i(X, A) = 0$ for $i < n$, and the relative Hurewicz map*

$$\pi_n(X, A) \to H_n(X, A)$$

is an isomorphism.

With more care (see [63, Ch. 7, Sec. 5]) you can avoid the simple connectivity assumption. However, with it in place, you get a converse statement: Suppose that both X and A are simply connected, let $n \geq 2$, and assume that $H_q(X, A) = 0$ for $q < n$. Simple connectivity of X implies that $\pi_1(X, A)$ is trivial, so we have the hypotheses of the relative Hurewicz theorem with $n = 2$, and conclude from $H_2(X, A) = 0$ that $\pi_2(X, A) = 0$. Continuing in this manner, we have:

Corollary 65.5. *Let X be a space and A a subspace. Assume both of them are simply connected, and let $n \geq 2$. Assume that $H_i(X, A) = 0$ for $2 \leq i < n$. Then $\pi_i(X, A) = 0$ for $i < n$, and the relative Hurewicz map*

$$\pi_n(X, A) \to H_n(X, A)$$

is an isomorphism.

By replacing a general map by a relative CW complex, up to weak homotopy, we find the following important corollary (which we state without the simple connectivity assumptions needed to apply our work so far).

Corollary 65.6 (Whitehead theorem). *Let $f : X \to Y$ be a map of path connected spaces and let $n \geq 1$. If $f_* : \pi_q(X) \to \pi_q(Y)$ is an isomorphism for $q < n$ and an epimorphism for $q = n$ then $f_* : H_q(X) \to H_q(Y)$ is an isomorphism for $q < n$ and an epimorphism for $q = n$. The converse holds if both X and Y are simply connected.*

Taking $n = \infty$ gives the further corollary:

Corollary 65.7. *Any weak equivalence induces an isomorphism in homology. Conversely, if X and Y are simply connected then any homology isomorphism $f : X \to Y$ is a weak equivalence.*

Combining this with "Whitehead's little theorem," we conclude that if a map between simply connected CW complexes induces an isomorphism in homology then it is a homotopy equivalence.

Exercises

Exercise 65.8. (a) Show that $H_{n+1}(K(A,n); \mathbb{Z}) = 0$ for any abelian group and any $n \geq 2$. Give a counter-example for $n = 1$.

(b) Conclude that if X is $(n-1)$ connected, with $n \geq 2$, then the Hurewicz map $\pi_{n+1}(X) \to H_{n+1}(X)$ is surjective. Give a counter-example for $n = 1$.

66 Double complexes and the Dress spectral sequence

A certain very rigid way of constructing a filtered complex occurs quite frequently — and, indeed, the Serre or even the Leray spectral sequence can be constructed in this way. It leads to an easy treatment of the multiplicative properties of the Serre spectral sequence (as well as, in due course, an account of the behavior of Steenrod operations in it).

Double complexes

A *double complex* is a bigraded abelian group $A = A_{*,*}$ together with differentials $d_h : A_{s,t} \to A_{s-1,t}$ and $d_v : A_{s,t} \to A_{s,t-1}$ that commute:

$$d_v d_h = d_h d_v.$$

For the moment we might as well assume that $A_{s,t}$ is "first quadrant":

$$A_{s,t} = 0 \quad \text{unless} \quad s \geq 0 \quad \text{and} \quad t \geq 0.$$

An example is provided by the tensor product of two chain complexes C_* and D_*: define

$$A_{s,t} = C_s \otimes D_t, \quad d_h(a \otimes b) = da \otimes b, \quad d_v(a \otimes b) = a \otimes db.$$

The graded tensor product is then the "total complex," which in general is the chain complex tA given by

$$(tA)_n = \bigoplus_{s+t=n} A_{s,t}$$

with differential determined by sending $a \in A_{s,t}$ to

$$da = d_h a + (-1)^s d_v a.$$

Then

$$d^2 a = d(d_h a + (-1)^s d_v a)$$
$$= (d_h^2 a + (-1)^s d_h d_v a) + (-1)^{s-1}(d_v d_h a + (-1)^s d_v^2 a) = 0 .$$

Define a filtration on the chain complex tA as follows:

$$F_p(tA)_n = \bigoplus_{s+t=n,\, s \leq p} A_{s,t} \subseteq (tA)_n .$$

Let's compute the low pages of the resulting spectral sequence. For a start,

$$E_{s,t}^0 = \mathrm{gr}_s(tA)_{s+t} = (F_s/F_{s-1})_{s+t} = A_{s,t} .$$

The differential in this associated graded object is determined by the vertical differential in A:

$$d^0 a = \pm d_v a .$$

Then

$$E_{s,t}^1 = H_{s,t}(E^0, d^0) = H_{s,t}(A; d_v) ,$$

which we might write as $H_{s,t}^v(A)$.

Now d^1 is the part of the differential d that decreases s by 1: for a d_v cycle in $A^{s,t}$,

$$d^1[a] = [d_h a] .$$

So

$$E_{s,t}^2 = H_{s,t}^h(H^v(A)) \underset{s}{\Longrightarrow} H_{s+t}(tA) .$$

But we can do something else as well. A double complex A can be "transposed" to produce a new double complex A^T with

$$A_{t,s}^\mathsf{T} = A_{s,t}$$

and for $a \in A_{t,s}^\mathsf{T}$

$$d_h^\mathsf{T}(a) = (-1)^s d_v a, \quad d_v^\mathsf{T}(a) = (-1)^t d_h a .$$

When we set the signs up like that,

$$tA^\mathsf{T} \cong tA$$

as complexes. The double complex A^T has its own filtration and its own spectral sequence,

$$^\mathsf{T}E_{t,s}^2 = H_{t,s}^v(H^h(A)) \underset{t}{\Longrightarrow} H_{s+t}(tA) ,$$

converging to the same thing.

If $A_{*,*}$ has a compatible multiplication — and we'll let you decide what that means — then the associated spectral sequences are multiplicative, as can easily be seen from the direct construction given in §61.

Dress spectral sequence

Andreas Dress ([16]; see also [53]) developed the following variation of the approach to the Serre spectral sequence originally employed by Serre himself. He proposed to model a general fibration — indeed, a general map — by the product projections

$$\mathrm{pr}_1 : \Delta^s \times \Delta^t \to \Delta^s \,.$$

He used these models to form a "singular" construction associated to any map $\pi : E \to B$.

$$\mathrm{Sin}_{s,t}(\pi) = \left\{ (f, \sigma) : \begin{array}{ccc} \Delta^s \times \Delta^t & \xrightarrow{f} & E \\ {\scriptstyle \mathrm{pr}_1} \downarrow & & \downarrow {\scriptstyle \pi} \\ \Delta^s & \xrightarrow{\sigma} & B \end{array} \quad \text{commutes} \right\} \,.$$

Since $\Delta^s \times \Delta^t \downarrow \Delta^s$ is surjective, σ is determined by f. Commutativity says that the map σ is "fiberwise."

This construction sends any map $\pi : E \to B$ to a functor

$$\mathrm{Sin}_{*,*}(\pi) : \mathbf{\Delta}^{op} \times \mathbf{\Delta}^{op} \to \mathrm{Set} \,,$$

a "bisimplicial set."

Continuing to imitate the construction of singular homology, we will next apply the free R-module functor to this, to get a bisimplicial R-module $R\mathrm{Sin}_{*,*}(\pi)$. The final step is to define boundary maps by taking alternating sums of the face maps. This provides us with a double complex, that I will write $S_{*,*}(\pi)$.

There are two associated spectral sequences. One of them is a singular homology version of the Leray spectral sequence, and specializes to the Serre spectral sequence in case π is a fibration. The other serves to identify what the first one converges to. I will sketch the arguments.

Let's compute the spectral sequence attached to the transposed double complex first. For this, observe that an element of $\mathrm{Sin}_{s,t}(\pi)$ may be regarded as a pair of dotted arrows in the commutative diagram

$$\begin{array}{ccc} \Delta^s & \xdashrightarrow{\hat{f}} & E^{\Delta^t} \\ {\scriptstyle \sigma} \downarrow & & \downarrow {\scriptstyle \pi} \\ B & \xrightarrow{c} & B^{\Delta^t} \end{array}$$

where c denotes the inclusion of the constant maps. If we form the pullback E_t' in

$$
\begin{array}{ccc}
E_t' & \longrightarrow & E^{\Delta^t} \\
\downarrow & & \downarrow{\scriptstyle \pi} \\
B & \xrightarrow{\ c\ } & B^{\Delta^t}
\end{array}
$$

this is saying that $\mathrm{Sin}_{s,t}(\pi) = \mathrm{Sin}_s(E_t')$, so

$$S_{s,t}(\pi) = S_s(E_t').$$

But the map $E_t' \to E^{\Delta^t}$ is a weak equivalence (because $c : B \to B^{\Delta^t}$ is), so

$$S_*(E_t') \to S_*(E)$$

is a quasi-isomorphism. This shows that

$$^{\mathsf{T}}E_{s,t}^1 = H_s(E)$$

for every $t \geq 0$.

Now we should think about what the differential in the t direction does. Each face map will induce the identity, so the alternating sums will induce alternately 0 and the identity. The result is that

$$
^{\mathsf{T}}E_{s,t}^2 = \begin{cases} H_s(E) & \text{if } t = 0 \\ 0 & \text{otherwise}. \end{cases}
$$

The spectral sequence collapses at this point, and we learn that there is a canonical isomorphism

$$H_*(tS_{*,*}(\pi)) = H_*(E).$$

This is then what the un-transposed spectral sequence will converge to. So how does it begin?

Fix a singular simplex $\sigma : \Delta^s \to B$, and pull $E \downarrow B$ back along it. Any $f : \Delta^s \times \Delta^t \to E$ compatible with σ then factors uniquely as

$$
\begin{array}{c}
\overbrace{\phantom{\Delta^s \times \Delta^t \dashrightarrow \sigma^{-1}E \longrightarrow E}}^{f} \\
\Delta^s \times \Delta^t \dashrightarrow \sigma^{-1}E \longrightarrow E \\
\searrow{\scriptstyle \mathrm{pr}_1} \quad \downarrow{\scriptstyle \pi_\sigma} \qquad \downarrow{\scriptstyle \pi} \\
\Delta^s \xrightarrow{\ \sigma\ } B.
\end{array}
$$

Adjointing this, we find that the set of such f's forms the set of singular t-simplices in a space of sections:

$$\mathrm{Sin}_t \Gamma(\Delta^s, \sigma^{-1}E).$$

Forming the free R-module and then taking the corresponding chain complex gives a chain complex for each $\sigma \in \mathrm{Sin}_s(B)$, namely

$$S_*(\Gamma(\Delta^s, \sigma^{-1}E)).$$

So

$$E^1_{s,t} = \bigoplus_{\sigma: \Delta^s \to B} H_t(\Gamma(\Delta^s, \sigma^{-1}E)).$$

A map $\phi : [s'] \to [s]$ in the simplex category determines a map

$$\phi^* : \Gamma(\Delta^s, \sigma^{-1}E) \to \Gamma(\Delta^{s'}, (\sigma \circ \phi)^{-1}E)$$

and thereby a map $\phi^* : E^1_{s,t} \to E^1_{s',t}$: we have a simplicial R-module. The differential d^1 is the alternating sum of the face maps in this simplicial structure, and E^2 is the homology of the resulting chain complex. This much you can say for a general map π; this is a singular homology form of the Leray spectral sequence.

If π is a fibration, the map $\sigma^{-1}E \downarrow \Delta^s$ is a fibration, and hence trivial because Δ^s is contractible. So the space of sections is then just the space of maps from the base to the fiber. Write F_σ for the fiber over the barycenter of Δ^s, so that

$$\Gamma(\Delta^s, \sigma^{-1}E) \simeq F_\sigma^{\Delta^s} \simeq F_\sigma$$

and

$$E^1_{s,t} \simeq \bigoplus_{\sigma \in \mathrm{Sin}_s(B)} H_t(F_\sigma).$$

The resulting E^2-term is the homology of B with coefficients in a corresponding local coefficient system:

$$E^2_{s,t} = H_s(B; H_t(p^{-1}(-))).$$

There are many advantages to this construction. It is transparently natural in the fibration and it exists for *any* map. It presents the spectral sequence as one associated to a double complex, and when we turn to cohomology, in the next lecture, the multiplicative structure of the associated spectral sequence will be easy to establish.

Exercises

Exercise 66.1. Let R be any ring and C_* a chain complex of projective (or even just flat) R-modules, and let M be an R-module. Construct a "universal coefficient spectral sequence"

$$E^2_{s,t} = \mathrm{Tor}^R_s(H_t(C_*), M) \Longrightarrow H_{s+t}(C_* \otimes_R M)$$

in the following manner. Let $M \leftarrow P_*$ be a projective resolution of M as an R-module. Form the double complex $C_* \otimes_R P_*$, and study the associated pair of spectral sequences.

Observe that this returns a short exact sequence as in Theorem 24.1 if R is a PID.

67 Cohomological spectral sequences

Upper indexing

We have set everything up for homology, but of course there are cohomology versions of everything as well. Given a filtered space

$$\cdots \subseteq F_{-1}X \subseteq F_0X \subseteq F_1X \subseteq \cdots$$

we filtered the singular chains $S_*(X)$ by

$$F_sS_*(X) = S_*(F_sX).$$

Now we will filter the cochains with values in M by

$$F_{-s}S^*(X; M) = \ker(S^*(X; M) \to S^*(F_{s-1}X; M)).$$

Note the $-s$; this is necessary to produce an *increasing* filtration of $S^*(X; M)$. Note also the $s - 1$. This will make the indexing of the multiplicative structure better. For example, most of our filtered spaces will have $F_{-1} = \varnothing$, in which case $F_0S^*(X; M) = S^*(X; M)$ and all the other filtration degrees are subcomplexes of this. Because we all have a bias towards positive numbers, it's standard and convenient to change notation to "upper indexing" as follows:

$$F^s = F_{-s}.$$

Then F^* is a *decreasing* filtration: $F^s \supseteq F^{s+1}$. If $F_{-1}X = \varnothing$, then $F^0S^*(X; M) = S^*(X; M)$ and $F^sS^*(X; M)$ consists of the cochains that vanish on $F_{s-1}X$.

The singular cochain complex as normally written is the outcome of a similar sign reversal; so the differential is of degree $+1$. The combination of these two reversals produces a spectral sequence with the following "cohomological" indexing:

$$d_r : E_r^{s,t} \to E_r^{s+r,t-r+1}.$$

To set this up slightly more generally, suppose that C^* is a cochain complex equipped with a decreasing filtration F^*C^*. Write

$$\text{gr}^s C^n = F^s C^n / F^{s+1} C^n.$$

Call it *first quadrant* if

- $F^0 C^* = C^*$,
- $H^n(\text{gr}^s C^*) = 0$ for $n < s$,
- $\bigcap F^s C^* = 0$.

Filter the cohomology of C^* by

$$F^s H^n(C^*) = \ker(H^n(C^*) \to H^n(F^{s-1} C^*)).$$

Theorem 67.1. *Let C^* be a cochain complex with a first quadrant decreasing filtration. There is a naturally associated convergent cohomological spectral sequence*

$$E_r^{s,t} \underset{s}{\Longrightarrow} H^{s+t}(C)$$

with

$$E_1^{s,t} = H^{s+t}(\text{gr}^s C^*)$$

and

$$E_\infty^{s,t} = \text{gr}^s H^{s+t}(C^*).$$

In particular we have the *cohomology Serre spectral sequence* of a fibration $p : E \to B$:

$$E_2^{s,t} = H^s(B; H^t(p^{-1}(-))) \underset{s}{\Longrightarrow} H^{s+t}(E).$$

Product structure

One of the reasons for passing to cohomology is to take advantage of the cup-product. It turns out that the cup product behaves itself in the cohomology Serre spectral sequence of a fibration $p : E \to B$. With a commutative coefficient ring R understood, the local coefficient system

$H^*(p^{-1}(-))$ is now a contravariant functor from $\Pi_1(B)$ to graded commutative R-algebras. Such coefficients produce bigraded R-algebra

$$E_2^{s,t} = H^s(B; H^t(p^{-1}(-)))$$

that is graded commutative in the sense that

$$yx = (-1)^{|x||y|}xy,$$

where $|x|$ and $|y|$ denote total degrees of elements. The entire spectral sequence is then "multiplicative" in the following sense.

- Each $E_r^{*,*}$ is a commutative bigraded R-algebra.
- d_r is a derivation: $d_r(xy) = (d_r x)y + (-1)^{|x|}x(d_r y)$.
- The isomorphism $E_{r+1}^{*,*} \cong H^{*,*}(E_r^{*,*})$ is one of bigraded algebras.
- $E_2^{*,*} = H^*(B; H^*(p^{-1}(-)))$ as bigraded R-algebras.
- The filtration on $H^*(E)$ satisfies

$$F^s H^n(E) \cdot F^{s'} H^{n'}(E) \subseteq F^{s+s'} H^{n+n'}(E).$$

- The isomorphisms

$$E_\infty^{s,t} \cong \mathrm{gr}^s H^{s+t}(E)$$

together form an isomorphism of bigraded R-algebras.

Theorem 67.2. *Let $p : E \to B$ be a Serre fibration, and assume given a commutative coefficient ring R. There is a naturally associated multiplicative cohomological first quadrant spectral sequence of R-modules*

$$E_2^{s,t} = H^s(B; H^t(p^{-1}(-))) \underset{s}{\Longrightarrow} H^{s+t}(E).$$

One of the virtues of the construction of the Serre (or more generally Leray) spectral sequence by the method described in Lecture 66 is that the multiplicative structure arises in a natural and explicit way. The bisimplicial set $S_{*,*}(\pi)$ gives rise to a bicosimplicial R-algebra $\mathrm{Map}(S_{*,*}(\pi), R)$, where the R-algebra structure is obtained by simply multiplying in R. Then applying the Alexander-Whitney map in both directions produces a (noncommutative but associative) algebra structure on a double complex, and the resulting filtered complex has the structure of a filtered differential graded algebra. The multiplicative structure of the spectral sequence is then easy to produce, and extends to a description of the effect of Steenrod operations in it as well [62]. The construction from a CW filtration of the base requires us to choose a skeletal approximation of the diagonal. Anyway, I will not make a further attempt to justify the multiplicative behavior of the Serre spectral sequence.

Instead, let's look at an example: The cohomology Gysin sequence for a fibration $p : E \to B$ whose fibers are R-homology $(n-1)$-spheres with compatible R-orientations takes the form

$$\cdots \to H^{s-n}(B) \xrightarrow{\pm e(\xi)\cdot} H^s(B) \xrightarrow{p^*} H^s(E) \xrightarrow{p_*} H^{s-n+1}(B) \to \cdots.$$

The identity of the middle map with p^* follows from the edge-homomorphism arguments of Lecture 64 but reformulated in cohomology. How about the other two maps?

Euler class

To understand them let's look at the cohomological Serre spectral sequence giving rise to the Gysin exact sequence. It has two nonzero rows, $E_r^{*,0}$ and $E_r^{*,n-1}$. The multiplicative structure provides $E_r^{*,n-1}$ with the structure of a module over $E_r^{*,0}$. The assumed orientation of the spherical fibration determines a distinguished class σ in the R-module $E_2^{0,n-1} = H^0(B; H^{n-1}(F))$ (one that evaluates to 1 on each orientation class — remember, the base may not be connected!), and $E_2^{*,n-1}$ is free as $E_2^{*,0} = H^*(B)$-module on this generator.

The transgression of this element,

$$e = d_n \sigma \in E_n^{n,0} = H^n(B) ,$$

is a canonically defined class, called the *Euler class* of the R-oriented spherical fibration.

This class determines the entire transgression $H^*(B) \to H^*(B)$ in the Gysin sequence:

$$x \mapsto d_n(x \cdot \sigma) = (-1)^{|x|} x e = \pm e x$$

by the Leibniz formula, since $d_n x = 0$.

The Euler class is a "characteristic class," in the sense that if we use $f : B' \to B$ to pull the spherical fibration $\xi : E \downarrow B$ back to $f^* \xi : E' \downarrow B'$ (along with the chosen orientation), then

$$f^*(e(\xi)) = e(f^* \xi) .$$

In particular E might be the complement of the zero section of an R-oriented real n-plane bundle. The universal case is then $\xi_n : ESO(n) \downarrow BSO(n)$, and we receive a canonical cohomology class

$$e_n = e(\xi_n) \in H^n(BSO(n); R) .$$

If we use coefficients in $\mathbb{F}_2$, every n-plane bundle is canonically oriented and we receive a class $e_n \in H^n(BO(n); \mathbb{F}_2)$.

In a sense the Euler class is the fundamental characteristic class: it rules all others. To illustrate its importance, notice that if the spherical fibration $p : E \to B$ has a section $s : B \to E$ then the map $p^* : H^*(B) \to H^*(E)$ is a split injection. The Gysin sequence becomes a short exact sequence; $p_* = 0$. Said differently, the edge homomorphism story shows that in that case all differentials hitting the base are trivial; in particular $e(\xi) = 0$. So if $e(\xi) \neq 0$ then the bundle doesn't admit a section. If the bundle was the complement of the zero section in an R-oriented vector bundle, $e(\xi)$ is an obstruction to the existence of a nowhere zero section.

The Euler class gets its name from the following theorem.

Theorem 67.3 (e.g. [46, Corollary 11.12]). *Let M be an R-oriented closed manifold. Then evaluating the Euler class of the tangent bundle τ on the fundamental class of M produces the image in R of the Euler characteristic of M:*

$$< e(\tau), [M] >= \chi(M) \in R.$$

Remark 67.4. If ξ is an oriented n-plane bundle over a finite CW complex B of dimension at most n, then it turns out that the Euler class is the only obstruction to compressing a classifying map $B \to BSO(n)$ through a map to $BSO(n-1)$: it is a complete obstruction to a section. Thus for example the Euler characteristic of a closed oriented n-manifold vanishes if and only if the manifold admits a nowhere vanishing vector field. Since $2e(\xi) = 0$ if n is odd, it follows that any oriented odd dimensional manifold has vanishing Euler characteristic (as follows also from Poincaré duality), and so admits a nowhere vanishing vector field.

Integration along the fiber

How about the last map, $H^s(E) \to H^{s-n+1}(B)$? This is a "wrong-way" or "umkher" map — it moves in the opposite direction from $p^* : H^s(B) \to H^s(E)$ — and also decreases dimension by the dimension of the fiber. In fact let $p : E \to B$ be any fibration such that $H^t(p^{-1}(-)) = 0$ for all $t \geq n$, and suppose we are given a map of local systems

$$H^{n-1}(p^{-1}(-)) \to R$$

to the trivial local system of R-modules. For example the fibers might be closed $(n - 1)$-manifolds, equipped with compatible R-orientations.

Now we have a new edge, an upper edge, and our map is given by a new edge homomorphism:

$$H^s(E) \xrightarrow{\quad\quad\quad\quad p_* \quad\quad\quad\quad} H^{s-n+1}(B)$$

$$F^0H^s(E) \rightarrowtail F^{s-n+1}H^s(E) \longrightarrow E_\infty^{s-n+1,n-1} \rightarrowtail E_2^{s-n+1,n-1}.$$

This edge homomorphism can sometimes be given geometric meaning as well. With real coefficients, for example, we can use deRham cohomology, and regard the map p_* as "integration along the fiber." We'll see another interpretation of the umkher map in terms of the Pontryagin-Thom construction in Lecture 76.

The multiplicative structure of the spectral sequence implies that the umkher map p_* is a module homomorphism for the graded algebra $H^*(B)$:

$$p_*((p^*x) \cdot y) = x \cdot p_* y.$$

This important formula has various names: "Frobenius reciprocity," or the "projection formula."

Loop space of S^n again

Let's try to compute the cup product structure in the cohomology of ΩS^n, again using the Serre spectral sequence for $PS^n \downarrow S^n$. One way to analyze this would be to set up the cohomology version of the Wang sequence, subject of a homework problem. But let's just use the spectral sequence directly. Take $n > 1$.

To begin,

$$E_2^{s,t} = H^s(S^n; H^t(\Omega S^n)) = H^s(S^n) \otimes H^t(\Omega S^n).$$

There are two nonzero columns. Write $\iota_n \in H^n(S^n)$ for the dual of the orientation class. The cohomology transgression $d_n : E_2^{0,n-1} \to E_2^{n,0}$ must be an isomorphism. Write $x \in H^{n-1}(\Omega S^n)$ for the unique class mapping to ι_n.

As in the homology calculation (or because of it) we know that $H^{k(n-1)}(\Omega S^n)$ is an infinite cyclic group. A first question then is: Is the the cup k-th power x^k a generator?

First assume that n is odd, so that $|x| = n - 1$ is even. Then by the Leibniz rule

$$d_n x^2 = 2(d_n x)x = 2\iota_n x.$$

This is twice the generator of $E_2^{n,n-1}$. In order to kill the generator itself, we must be able to divide x^2 by 2 in $H^{2(n-1)}(\Omega S^n)$. So there is a unique element, call it γ_2, such that $2\gamma_2 = x^2$, and it serves as a generator for the infinite cyclic group $H^{2(n-1)}(\Omega S^n)$.

With this in the bag, let's observe that the transgression of x^k is

$$d_n x^k = k(d_n x)x^{k-1} = k\iota_n x^{k-1}.$$

For example

$$d_n x^3 = 3\iota_n x^2 = 3 \cdot 2\iota_n \gamma_2.$$

Since $\iota_n \gamma_2$ is a generator of $E_2^{n,2(n-1)}$, the element x^3 must be divisible by $3 \cdot 2 = 3!$: there is a unique element of $H^{3(n-1)}(\Omega S^n)$, call it γ_3, such that $x^3 = 3!\gamma_3$.

This evidently continues: $H^{k(n-1)}(\Omega S^n)$ is generated by a class γ_k such that $x^k = k!\gamma_k$. This implies that these generators satisfy the product formula

$$\gamma_j \gamma_k = (j,k)\gamma_{j+k}, \quad (j,k) = \frac{(j+k)!}{j!k!}.$$

This is a *divided power algebra*, denoted by $\Gamma[x]$:

$$H^*(\Omega S^n) = \Gamma[x] \quad \text{for } n \text{ odd}, \quad |x| = n - 1.$$

The answer is the same for any coefficients. With rational coefficients, these divided classes are already present, so

$$H^*(\Omega S^n; \mathbb{Q}) = \mathbb{Q}[x].$$

Then $H^*(\Omega S^n; \mathbb{Z})$, being torsion-free, sits inside this as the sub-algebra generated additively by the classes $x^k/k!$.

Now let's turn to the case in which n is even. Then $|x|$ is odd, so by commutativity $2x^2 = 0$. But $H^{2(n-1)}(\Omega S^n)$ is torsion-free, so $x^2 = 0$.

So we need a new indecomposable element in $H^{2(n-1)}(\Omega S^n)$: Call it y. Choose the sign so that

$$d_n y = \iota_n x \in E_n^{n,n-1}.$$

Now $|y| = 2(n-1)$ is even, so

$$d_n y^k = k\iota_n y^{k-1}x$$

and

$$d_n(xy^k) = \iota_n y^k - x \cdot ky^{k-1}\iota_n x = \iota_n y^k$$

(since $x^2 = 0$). Reasoning as before, we find that

$$H^*(\Omega S^n) = E[x] \otimes \Gamma[y] \quad \text{for } n \text{ even}, \quad |x| = n - 1, \quad |y| = 2(n - 1).$$

Exercises

Exercise 67.5. Compute the homology of the unit sphere bundle of a closed oriented surface.

Exercise 67.6. Analyze the cohomology Serre spectral sequence for the same fibration that you studied in Exercise 62.7: Differentials? Extensions?

Exercise 67.7. Let $f : S^2 \to S^2$ be a map of degree 2, as in Exercise 62.6, and let F be its homotopy fiber. Compute the homology and cohomology of F, with coefficients in $\mathbb{Z}$ and in $\mathbb{F}_2$. What is the ring structure in cohomology? The action map $\Omega S^2 \times F \to F$ makes $H_*(F)$ into a module for the algebra $H_*(\Omega S^2)$. What is this module structure?

68 Serre classes

Let X be a simply connected space. Suppose that $\overline{H}_q(X)$ is a torsion group for all q: every element $x \in H_q(X)$ is killed by some positive integer. This is the same as saying that X has the same rational homology as a point. Is every homotopy group then also a torsion group, or can rational homotopy make an appearance? What if the reduced homology was all p-torsion (i.e. every element is killed by some power of p) — must $\pi_*(X)$ also be entirely p-torsion? What if the homology is assumed to be of finite type (finitely generated in every dimension) — must the same be true of homotopy? Serre explained how things like this can be checked, without explicit computation (which is often not an option!) by describing what is required of a class $\mathcal{C}$ of abelian groups that allow it to be considered "negligible."

Definition 68.1. A class $\mathcal{C}$ of abelian groups is a *Serre class* if $0 \in \mathcal{C}$, and, for any short exact sequence $0 \to A \to B \to C \to 0$, A and C lie in $\mathcal{C}$ if and only if B does.

Here are some immediate consequences of this definition.

- A Serre class is closed under isomorphisms.
- A Serre class is closed under formation of subgroups and quotient groups.

- Let $A \overset{i}{\to} B \overset{p}{\to} C$ be exact at B. If $A, C \in \mathcal{C}$, then $B \in \mathcal{C}$: In

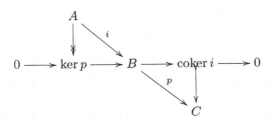

the row is exact and the indicated factorizations exist since $pi = 0$; the surjectivity and injectivity express exactness.

Here are the main examples.

Example 68.2. The class of trivial abelian groups; the class $\mathcal{C}_{\text{fin}}$ of all finite abelian groups; the class $\mathcal{C}_{\text{fg}}$ of all finitely generated abelian groups; the class of all abelian groups.

Example 68.3. $\mathcal{C}_{\text{tors}}$, the class of all torsion abelian groups. To see that this is a Serre class, start with a short exact sequence

$$0 \to A \overset{i}{\to} B \overset{p}{\to} C \to 0 .$$

It's clear that if B is torsion then so are A and C. Conversely, suppose that A and C are torsion groups. Let $b \in B$. Then $p(nb) = np(b) = 0$ for some $n > 0$, since C is torsion; so there is $a \in A$ such that $i(a) = nb$. But A is torsion too, so $ma = 0$ for some $m > 0$, and hence $mnb = 0$.

Example 68.4. Fix a prime p. The class of p-torsion groups forms a Serre class. More generally, let $\mathcal{P}$ be a set of primes. Define $\mathcal{C}_{\mathcal{P}}$ to be the class of torsion abelian groups A such that if p divides the order of $a \in A$ for some $p \in \mathcal{P}$ then $a = 0$. If $\mathcal{P} = \varnothing$ this is just $\mathcal{C}_{\text{tors}}$. Write $\mathcal{C}_p$ for $\mathcal{C}_{\{p\}}$. This is the class of torsion abelian groups without p-torsion. Since $\mathbb{Z}_{(p)}$ is a direct limit of copies of $\mathbb{Z}$ with bonding maps running through the natural numbers prime to p, $A \in \mathcal{C}_p$ if and only if $A \otimes \mathbb{Z}_{(p)} = 0$. These are the kinds of groups you're willing to ignore if you are only interested in "p-primary" information.

Example 68.5. The intersection of a collection of Serre classes is again a Serre class. For example, $\mathcal{C}_{\text{fin}} \cap \mathcal{C}_p$ is the class of finite abelian groups of order prime to p.

The definition of a Serre class is set up so that it makes sense to work "modulo C." So we'll say that A is "zero mod C" if $A \in C$. A homomorphism is a "mod C monomorphism" if its kernel lies in C; a "mod C epimorphism" if its cokernel lies in C; and a "mod C isomorphism" if both kernel and cokernel lie in C. So for example $f : A \to B$ is a mod C_{tors} isomorphism exactly when $f \otimes 1 : A \otimes \mathbb{Q} \to B \otimes \mathbb{Q}$ is an isomorphism of rational vector spaces.

Lemma 68.6. *Let C be a Serre class. The classes of mod C monomorphisms, epimorphisms, and isomorphisms contain all isomorphisms and are closed under composition. The class of mod C isomorphisms satisfies 2-out-of-3.*

Proof. Form

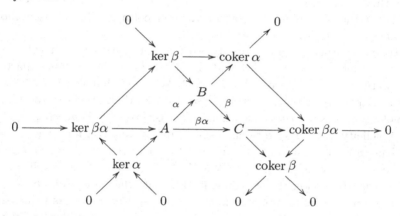

and check that the outside path is exact.   □

Here are some straightforward consequences of the definition. Let C be a Serre class.

- Let C_* be a chain complex. If $C_n \in C$ then $H_n(C_*) \in C$.
- Suppose $F_* A$ is a filtration on an abelian group. If $A \in C$, then $\operatorname{gr}_s A \in C$ for all s. If the filtration is finite (i.e. $F_m = 0$ and $F_n = A$ for some m, n) and $\operatorname{gr}_s A \in C$ for all s, then $A \in C$.
- Suppose we have a spectral sequence $\{E^r_{s,t}\}$. If $E^2_{s,t} \in C$, then $E^r_{s,t} \in C$ for $r \geq 2$. If $\{E^r\}$ is a first quadrant spectral sequence (so that $E^\infty_{s,t}$ is defined and achieved at a finite stage) it follows that $E^\infty_{s,t} \in C$. Thus if the spectral sequence comes from a first quadrant filtered complex C and $E^2_{s,t} \in C$ for all $s + t = n$, then $H_n(C) \in C$.

The first implication in homology is this: Suppose that $A \subseteq X$ is a pair of path-connected spaces. If two of $\overline{H}_n(A), \overline{H}_n(X), H_n(X, A)$ are zero mod $\mathcal{C}$ for all n, then so is the third. More generally, if you have a ladder of abelian groups (a map of long exact sequences) and two out of every three consecutive rungs are mod $\mathcal{C}$ isomorphisms then so is the third: a mod $\mathcal{C}$ five-lemma.

Serre rings and Serre ideals

To apply this theory to the Serre spectral sequence we need to know that our class is compatible with tensor product. Let's say that a Serre class $\mathcal{C}$ is a *Serre ring* if whenever both A and B are in $\mathcal{C}$, $A \otimes B$ and $\mathrm{Tor}(A, B)$ are too. It's a *Serre ideal* if we only require one of A and B to lie in $\mathcal{C}$ to have this conclusion.

All of the examples given above are Serre rings. The ones without finiteness assumptions are Serre ideals.

Here's another closure property we might investigate, and will need. Suppose that $\mathcal{C}$ is a Serre ring and $A \in \mathcal{C}$. Form the classifying space or Eilenberg Mac Lane space $BA = K(A, 1)$. We know that $H_1(K(A, 1)) = A$ (for example by Poincaré's theorem) so it lies in $\mathcal{C}$. How about the higher homology groups? If they are again in $\mathcal{C}$, the Serre ring is *acyclic*.

Acyclicity is a computational issue. Suppose $\mathcal{C} = \mathcal{C}_{\mathrm{fin}}$ for example. By the Künneth theorem (and the fact that $\mathcal{C}_{\mathrm{fin}}$ is a Serre ring), it's enough to consider finite cyclic groups. What is $H_*(BC_n)$, where C_n is a cyclic group of order n? To answer this we can embed C_n into the circle group S^1 as nth roots of unity. The group of complex numbers of norm 1 acts principally on the unit vectors in $\mathbb{C}^\infty$, and that space, S^∞, is contractible. So $\mathbb{CP}^\infty = BS^1$. The subgroup $C_n \subset S^1$ acts principally on this contractible space as well, so

$$BC_n = C_n \backslash S^\infty = (C_n \backslash S^1) \times_{S^1} S^\infty$$

fibers over $\mathbb{CP}^\infty$ with fiber $C_n \backslash S^1 \cong S^1$. Let's study the resulting Serre spectral sequence, first in homology.

In it, $E^2_{s,t} = H_s(\mathbb{CP}^\infty) \otimes H_t(S^1)$. The only possible differential is d^2. The one thing we know about $K(C_n, 1)$ is that is fundamental group is C_n — abelian, so $H_1(K(C_n, 1)) = C_n$. The only way to accomplish this in the spectral sequence is by $d^2 a = n\sigma$, where $\sigma \in H_1(S^1)$ is one of the generators.

This implies that in the cohomology spectral sequence $d_2 e = nx$, where e generates $H^1(S^1)$ and x generates $H^2(\mathbb{CP}^\infty)$. Then the multiplicative structure takes over: $d_2(x^i e) = nx^{i+1}$.

The effect is that $E_3^{s,t} = 0$ for $t > 0$. The edge homomorphism $H^*(\mathbb{CP}^\infty) \to H^*(BC_n)$ is thus surjective, and we find

$$H^*(BC_n) = \mathbb{Z}[x]/(nx), \quad |x| = 2.$$

Passing back to homology, we find that $\overline{H}_i(BC_n)$ is cyclic of order n if i is a positive odd integer and zero otherwise. In particular, it is finite, so $\mathcal{C}_{\text{fin}}$ is acyclic.

Since any torsion abelian group A is the direct limit of the directed system of its finite subgroups, we find that $\overline{H}_q(K(A,1))$ is then torsion as well: so $\mathcal{C}_{\text{tors}}$ is also acyclic.

The calculation also shows that the class of finite p-groups and the class $\mathcal{C}_p$ are acyclic.

To deal with $\mathcal{C}_{\text{fg}}$, we just have to add the infinite cyclic group, whose homology is certainly finitely generated in each degree. So all our examples of Serre rings are in fact acyclic.

Serre classes in the Serre spectral sequence

Let $\mathcal{C}$ be a Serre ideal. If $H_n(X)$ and $H_{n-1}(X)$ are zero mod $\mathcal{C}$ then $H_n(X; M)$ is zero mod $\mathcal{C}$ for any abelian group M, by the universal coefficient theorem. If $\mathcal{C}$ is only a Serre ring, we still reach this conclusion provided $M \in \mathcal{C}$.

The convergence theorem for the Serre spectral sequence shows this:

Proposition 68.7 (Mod $\mathcal{C}$ Vietoris-Begle Theorem). *Let $\pi : E \to B$ be a fibration such that B and the fiber F are path connected, and suppose $\pi_1(B)$ acts trivially on $H_*(F)$. Let $\mathcal{C}$ be a Serre ideal and suppose that $H_t(F) \in \mathcal{C}$ for all $t > 0$. Then $\pi_* : H_n(E) \to H_n(B)$ is a mod $\mathcal{C}$ isomorphism for all n.*

Proof. The universal coefficient theorem guarantees that $E_{s,t}^2 = H_s(B; H_t(F)) \in \mathcal{C}$ as long as $t > 0$. The same is thus true of $E_{s,t}^r$ and hence of $E_{s,t}^\infty$, so the edge homomorphism $\pi_* : H_n(E) \to H_n(B)$ is a mod $\mathcal{C}$ isomorphism. $\qquad\square$

This theorem admits a refinement that will be useful in proving the mod $\mathcal{C}$ Hurewicz theorem. For one thing, we would like a result that works for a Serre ring, not merely an Serre ideal, in order to cover cases like $\mathcal{C}_{\text{fg}}$.

Proposition 68.8. *Let $\pi : E \to B$ be a fibration such that B is simply connected and the fiber F is path connected. Let C be a Serre ring and suppose that*

- $H_s(B) \in C$ *for all s with $0 < s < n$, and*
- $H_t(F) \in C$ *for all $0 < t < n - 1$.*

Then $\pi_* : H_i(E, F) \to H_i(B, *)$ is an isomorphism mod C for all $i \leq n$.

Proof. We appeal to the relative Serre spectral sequence

$$E_{s,t}^2 = \overline{H}_s(B; H_t(F)) \underset{s}{\Longrightarrow} H_{s+t}(E, F).$$

At $E_{s,t}^2$, both the $s = 0$ column and the $s = 1$ column vanish. Also, $E_{s,t}^2 \in C$ for (s, t) in the rectangle

$$2 \leq s \leq n - 1, \quad 1 \leq t \leq n - 2.$$

In total degree i, $i \leq n$, the only group not vanishing mod C is $E_{i,0}^2$. So the edge homomorphism $\pi_* : H_i(E, F) \to \overline{H}_i(B)$ is a mod C isomorphism. $\quad\square$

Theorem 68.9 (Mod C Hurewicz theorem). *Assume that C is an acyclic Serre ring. Let X be a simply connected space and let $n \geq 2$. Then $\pi_q(X) \in C$ for all $q < n$ if and only if $\overline{H}_q(X) \in C$ for all $q < n$, and in that case the Hurewicz map $\pi_n(X) \to H_n(X)$ is a mod C isomorphism.*

We'll present the proof in the next lecture. For now, a small selection of corollaries:

Corollary 68.10. *Let X be a simply connected space and $n \geq 2$ or $n = \infty$.*
(1) *$H_q(X)$ is finitely generated for all $q < n$ if and only if $\pi_q(X)$ is finitely generated for all $q < n$.*
(2) *Let p be a prime number. $H_q(X)$ is p-torsion for all $q < n$ if and only if $\pi_q(X)$ is p-torsion for all $q < n$.*
(3) *If $\overline{H}_q(X; \mathbb{Q}) = 0$ for $q < n$, then $\pi_q(X) \otimes \mathbb{Q} = 0$ for $q < n$, and $h : \pi_n(X) \otimes \mathbb{Q} \to H_n(X; \mathbb{Q})$ is an isomorphism.*

Exercise

Exercise 68.11. Suppose that X is simply connected space. Show that $H_*(X)$ is of finite type (finitely generated as abelian group in each dimension) if and only if $H_*(\Omega X)$ is. Similarly, show that $\overline{H}_*(X)$ is entirely p-torsion if and only if $\overline{H}_*(\Omega X)$ is entirely p-torsion.

Exercise 68.12. Show that if A is a finitely generated abelian group then $H_*(K(A,n))$ is of finite type for any $n > 1$. (We did the case $n = 1$ in class.) Show that if A is p-torsion then the same is true for $\overline{H}_i(K(A,n))$ for any n and i.

69   Mod $\mathcal{C}$ Hurewicz and Whitehead theorems

Proof of Theorem 68.9. This follows the proof of the Hurewicz theorem, but some extra care is needed. Again we use induction and the path-loop fibration. Again, it will suffice to show that if $\pi_q(X) \in \mathcal{C}$ for $q < n$ then $\pi_n(X) \to H_n(X)$ is an isomorphism — now mod $\mathcal{C}$. To start the induction, with $n = 2$, we can appeal to the Hurewicz isomorphism: the map $\pi_2(X) \to H_2(X)$ is an actual isomorphism.

The inductive step uses the commutative diagram

$$
\begin{array}{ccccc}
\pi_q(X) & \xleftarrow{\;\cong\;} & \pi_q(PX, \Omega X) & \xrightarrow{\;\cong\;} & \pi_{q-1}(\Omega X) \\
\Big\downarrow{\scriptstyle h} & & \Big\downarrow{\scriptstyle h} & & \Big\downarrow{\scriptstyle h} \\
\overline{H}_q(X) & \longleftarrow & H_q(PX, \Omega X) & \xrightarrow{\;\cong\;} & H_{q-1}(\Omega X).
\end{array}
$$

Two thing need checking: (1) the map $H_n(PX, \Omega X) \to \overline{H}_n(X)$ is an isomorphism mod $\mathcal{C}$, and (2) the map $h : \pi_{n-1}(\Omega X) \to H_{n-1}(\Omega X)$ is an isomorphism mod $\mathcal{C}$.

Neither of these facts follow from an inductive hypothesis if $\pi_2(X) \neq 0$ (unless $\mathcal{C}$ is the trivial class), but we begin by showing that they do follow from the inductive hypothesis if $\pi_2(X) = 0$.

Suppose $\pi_2(X) = 0$, so that ΩX is simply connected. Since $\pi_i(\Omega X) = \pi_{i+1}(X)$ we know it lies in $\mathcal{C}$ for $i < n-1$. The inductive hypothesis applies to ΩX and shows that $\overline{H}_i(\Omega X) \in \mathcal{C}$ for $i < n-1$ and that $h : \pi_{n-1}(\Omega X) \to H_{n-1}(\Omega X)$ is a mod $\mathcal{C}$ isomorphism. The inductive hypothesis also applies to X of course, and shows that $\overline{H}_i(X) \in \mathcal{C}$ for $i < n$. So we are in position to apply Proposition 68.8 from last lecture to see fact (1).

But if $\pi_2(X) \neq 0$, ΩX is not simply connected. To deal with that, let's take the 2-connected cover in the Whitehead tower: This is a fibration $Y \downarrow X$ with fiber $K = K(\pi_2(X), 1)$. This is where the acyclic condition comes in: since $\pi_2(X) \in \mathcal{C}$, $H_i(K) \in \mathcal{C}$ for $i > 0$. The long exact sequence for the pair (Y, K) shows that

$$
\overline{H}_i(Y) \to H_i(Y, K)
$$

is a mod C isomorphism. We will apply Proposition 68.8 to $(Y,K) \downarrow (X,*)$, using the fact that X is simply connected and $H_i(X) \in C$ for $0 < i < n$. We find that

$$H_i(Y,K) \to H_i(X,*)$$

is a mod C isomorphism for $i \leq n$. Therefore the projection map $\overline{H}_i(Y) \to \overline{H}_i(X)$ is a mod C isomorphism for $i \leq n$.

The map $\pi_i(Y) \to \pi_i(X)$ is an isomorphism for $i \geq 2$, so our hypothesis applies to Y, and we can perform the inductive step on it instead of on X. $\square$

Corollary 69.1. *Let X be a simply connected space, p a prime, and $n \geq 2$. Then $\pi_i(X) \otimes \mathbb{Z}_{(p)} = 0$ for all $i < n$ if and only if $\overline{H}_i(X; \mathbb{Z}_{(p)}) = 0$ for all $i < n$, and in that case*

$$h : \pi_n(X) \otimes \mathbb{Z}_{(p)} \to H_n(X; \mathbb{Z}_{(p)})$$

is an isomorphism.

Proof. The acyclic Serre ring C_p consists of abelian groups such that $A \otimes \mathbb{Z}_{(p)} = 0$. $\square$

Now for the relative version!

Theorem 69.2 (Relative mod C Hurewicz theorem). *Let C be an acyclic Serre ideal, and (X,A) a pair of spaces, both simply connected. Fix $n \geq 1$. Then $\pi_i(X,A) \in C$ for all i with $2 \leq i < n$ if and only if $H_i(X,A) \in C$ for all i with $2 \leq i < n$, and in that case $h : \pi_n(X,A) \to H_n(X,A)$ is a mod C isomorphism.*

The proof follows the same line as in the absolute case. But note the requirement here, in the relative case, that C is a Serre *ideal*. Let me just point out where that assumption is required. We use the same diagram, in which F is the homotopy fiber of the inclusion $A \hookrightarrow X$:

$$
\begin{array}{ccccc}
\pi_{n-1}(F) & \xleftarrow{\cong} & \pi_n(PX,F) & \xrightarrow{\cong} & \pi_n(X,A) \\
\downarrow h & & \downarrow h & & \downarrow h \\
\overline{H}_{n-1}(F) & \xleftarrow{\cong} & H_n(PX,F) & \xrightarrow{p_*} & H_n(X,A).
\end{array}
$$

In the proof that p_* is an isomorphism, we'll again use the relative Serre spectral sequence, but now the E^2 term is $E^2_{s,t} = H_s(X,A; H_t(X))$, and we have no control over $H_t(X)$: all our assumptions related to the relative homology.

And this leads on to a mod C Whitehead theorem:

Theorem 69.3 (Mod $\mathcal{C}$ Whitehead theorem). *Let $\mathcal{C}$ be an acyclic Serre ideal, and $f : X \to Y$ a map of simply connected spaces. Fix $n \geq 2$. The following are equivalent.*

(1) $f_* : \pi_i(X) \to \pi_i(Y)$ *is a mod $\mathcal{C}$ isomorphism for $i \leq n - 1$ and a mod $\mathcal{C}$ epimorphism for $i = n$, and*

(2) $f_* : H_i(X) \to H_i(Y)$ *is a mod $\mathcal{C}$ isomorphism for $i \leq n - 1$ and a mod $\mathcal{C}$ epimorphism for $i = n$.*

The theory of Serre classes is quite beautiful, but it does not relate easily to the standard way of working with homology with coefficients. The following lemma forms the link between mod p homology and the mod $\mathcal{C}_p$ Whitehead theorem.

Lemma 69.4. *Let X and Y be spaces whose p-local homology is of finite type, and suppose $f : X \to Y$ induces an isomorphism in mod p homology. Then it induces a mod $\mathcal{C}_p$ isomorphism in integral homology.*

Proof. Since $\mathbb{Z}_{(p)}$ is flat, a homomorphism $f : A \to B$ is a mod $\mathcal{C}_p$ isomorphism if and only if $f \otimes 1 : A \otimes \mathbb{Z}_{(p)} \to B \otimes \mathbb{Z}_{(p)}$ is an isomorphism.

A finitely generated module over $\mathbb{Z}_{(p)}$ is trivial if it's trivial mod p. So we want to show that the kernel and cokernel of $f_* : H_*(X) \to H_*(Y)$ are trivial after tensoring with $\mathbb{F}_p$.

Form the mapping cone Z of the map f. By assumption it has trivial mod p reduced homology. Since $\mathbb{Z}_{(p)}$ is Noetherian, $H_*(Z; \mathbb{Z}_{(p)})$ is of finite type. The universal coefficient theorem shows that $\overline{H}_*(Z; \mathbb{Z}_{(p)}) \otimes \mathbb{F}_p$ embeds in $\overline{H}_*(Z; \mathbb{F}_p)$, which is trivial, so we conclude that $\overline{H}_*(Z) \otimes \mathbb{Z}_{(p)} = \overline{H}_*(Z; \mathbb{Z}_{(p)}) = 0$, and hence that $f_* \otimes 1 : H_*(X) \otimes \mathbb{Z}_{(p)} \to H_*(Y) \otimes \mathbb{Z}_{(p)}$ is an isomorphism. $\qquad\square$

Corollary 69.5. *Let X and Y be simply connected spaces whose p-local homology is of finite type, and suppose $f : X \to Y$ induces an isomorphism in mod p homology. Then $f_* : \pi_*(X) \otimes \mathbb{Z}_{(p)} \to \pi_*(Y) \otimes \mathbb{Z}_{(p)}$ is an isomorphism.*

This is every topologist's favorite theorem! Absent the fundamental group, you can treat primes one by one.

Some calculations

Let's first compute the homology — well, at least the rational homology — of the Eilenberg Mac Lane space $K(A, n)$, for A finitely generated.

By the Künneth isomorphism it suffices to do this for A cyclic. When A is any torsion group, the mod C_{tors} Hurewicz theorem shows that $\overline{H}_*(K(A,n);\mathbb{Q}) = 0$. So we will focus on $K(\mathbb{Z},n)$.

The case $n = 1$ is the circle, whose cohomology is an exterior algebra on one generator of dimension 1: $H^*(K(\mathbb{Z},1);\mathbb{Q}) = E[\iota_1]$, $|\iota_1| = 1$.

We know what $H^*(K(\mathbb{Z},2);\mathbb{Q})$ is, too, but let's compute it in a way that starts an induction. It also follows the path laid down by Serre in his computation of the mod 2 cohomology of $K(A,n)$, using the fiber sequence

$$K(A,n-1) \to PK(A,n) \to K(A,n).$$

When $n = 2$ there are only two rows — this is a spherical fibration. The class ι_1 must transgress to a generator, call it $\iota_2 \in H^2(K(\mathbb{Z},2);\mathbb{Q})$. Proceeding inductively, using $d_2(\iota_2^k \iota_1) = \iota_2^{k+1}$, you find that

$$H^*(K(\mathbb{Z},2);\mathbb{Q}) = \mathbb{Q}[\iota_2].$$

When $n = 3$, there is a polynomial algebra in the fiber. Again the fundamental class must transgress to a generator, $\iota_3 = d_3\iota_2 \in H^3(K(\mathbb{Z},3);\mathbb{Q})$. The Leibniz formula gives $d^3(\iota_2^k) = k\iota_3\iota_2^{k-1}$. This differential is an isomorphism: this is where working over $\mathbb{Q}$ separates from working anywhere else. So we discover that

$$H^*(K(\mathbb{Z},3);\mathbb{Q}) = E[\iota_3].$$

This starts the induction, and leads to

$$H^*(K(\mathbb{Z},n);\mathbb{Q}) = \begin{cases} E[\iota_n] & \text{if } n \text{ is odd} \\ \mathbb{Q}[\iota_n] & \text{if } n \text{ is even}. \end{cases}$$

In both cases, the cohomology is free as a graded commutative $\mathbb{Q}$-algebra.

Proposition 69.6. *The homotopy group $\pi_i(S^n)$ is finite for all i except for $i = n$ and if n is even for $i = 2n - 1$, when it is finitely generated of rank 1.*

Proof. The case $n = 1$ is special and simple, so suppose $n \geq 2$. Let

$$S^n \to K(\mathbb{Z},n)$$

represent a generator of $H^n(S^n)$. It induces an isomorphism in π_n and in H_n.

When n is odd, it induces an isomorphism in rational homology, and therefore in rational homotopy.

When n is even, we should compute the cohomology of the fiber F. The class ι_n on the base survives to a generator of $H^n(S^n;\mathbb{Q})$, but ι_n^2 must die. The only way to kill it is by a transgression from a class $\iota_{2n-1} \in H^{2n-1}(F)$: $d_{2n}\iota_{2n-1} = \iota_n^2$. Then the Leibniz formula gives $d_{2n}(\iota_n^k \iota_{2n-1}) = \iota_n^{k+2}$, leaving precisely the cohomology of S^n. So the fiber has the same rational cohomology as $K(\mathbb{Z}, 2n-1)$. The generator ι_{2n-1} gives a map $F \to K(\mathbb{Z}, 2n-1)$ that induces an isomorphism in rational homology, and hence in rational homotopy.

You might ask: Why couldn't this cancellation happen some other way? You can complete this argument, but perhaps you'll prefer a different approach. Loop the Barratt-Puppe sequence back one notch, to a fiber sequence $K(\mathbb{Z}, n-1) \to F \to S^n$, and work directly in homology. Now $(n-1)$ is odd, so the entire E^2 term has just four generators. The generator $x \in H_n(S^n)$ must transgress to the fiber (else F would have the wrong homology in dimension $n-1$, or using the relationship between the transgression and the boundary map in homotopy), and what's left at E^{n+1} is just a $\mathbb{Q}$ for $E_{0,0}^2$ and a $\mathbb{Q}$ for $E_{n,n-1}^2$. □

We can identify an element of infinite order in $\pi_{4k-1}(S^{2k})$ in several ways. Here's one. The space $S^m \times S^n$ admits a CW structure with $(m+n-1)$-skeleton given by the wedge $S^m \vee S^n$. There is thus a map

$$\omega : S^{m+n-1} \to S^m \vee S^n$$

that serves as the attaching map for the top cell. Given homotopy classes $\alpha \in \pi_m(X)$ and $\beta \in \pi_n(X)$, we an form the composite

$$S^{m+n-1} \xrightarrow{\omega} S^m \vee S^n \xrightarrow{\alpha \vee \beta} X \vee X \xrightarrow{\nabla} X.$$

This defines the *Whitehead product*

$$[-,-] : \pi_m(X) \times \pi_n(X) \to \pi_{m+n-1}(X).$$

When $m = 1$, this is determined by the action of $\pi_1(X)$ on $\pi_n(X)$:

$$[\alpha, \beta] = \alpha \cdot \beta - \beta$$

(where $|\alpha| = 1$, and if $|\beta| = 1$ this is to be interpreted as $\alpha\beta\alpha^{-1}\beta^{-1}$). In general, it provides homotopy groups with the structure of a graded Lie algebra: for $\alpha, \beta, \gamma \in \pi_*(X)$ of degrees $p, q, r > 1$, then [80, §X.7]

$$[\alpha, \beta] = (-1)^{pq}[\beta, \alpha]$$

and

$$(-1)^{rp}[[\alpha, \beta], \gamma] + (-1)^{pq}[[\beta, \gamma], \alpha] + (-1)^{qr}[[\gamma, \alpha], \beta] = 0.$$

Now we can define the *Whitehead square*

$$w_n = [\iota_n, \iota_n] \in \pi_{2n-1}(S^n).$$

When $n = 2k$, it generates an infinite cyclic subgroup.

The same calculation works for a while locally at a prime. Let's look at S^3 for definiteness. Follow the Barratt-Puppe sequence back one stage, to get a fibration sequence

$$K(\mathbb{Z}, 2) \to \tau_{\geq 4} S^3 \to S^3.$$

In the spectral sequence, with integral coefficients,

$$E_2^{*,*} = E[\sigma] \otimes \mathbb{Z}[\iota_2].$$

The class ι_2 must transgress to σ (at least up to sign), and then

$$d_2(\iota_2^k) = k\sigma\iota_2^{k-1}.$$

This map is always injective, leaving

$$E_3^{3,2k-2} = \mathbb{Z}/k\mathbb{Z}$$

and nothing else except for $E_3^{0,0} = \mathbb{Z}$. The result is that

$$H_{2k}(\tau_{\geq 4} S^3) = \mathbb{Z}/k\mathbb{Z}, \quad k \geq 1.$$

The first time p-torsion appears is in dimension $2p$: $H_{2p}(\tau_{\geq 4} S^3) = \mathbb{Z}/p\mathbb{Z}$. This is the mod $\mathcal{C}_p$ Hurewicz dimension, so $\pi_i(S^3)$ has no p-torsion in dimension less than $2p$, and

$$\pi_{2p}(S^3) \otimes \mathbb{Z}_{(p)} = \mathbb{Z}/p\mathbb{Z}.$$

When $p = 2$, this group is generated by the suspension of the Hopf map.

70   Freudenthal, James, and Bousfield

Suspension

The transgression takes on a particularly simple form if the total space is contractible.

Remember the adjoint pair

$$\Sigma : \mathbf{Top}_* \rightleftarrows \mathbf{Top}_* : \Omega.$$

The adjunction morphisms

$$\sigma : X \to \Omega\Sigma X, \quad \mathrm{ev} : \Sigma\Omega X \to X$$

are given by

$$\sigma(x)(t) = [x, t], \quad \mathrm{ev}(\omega, t) = \omega(t).$$

Proposition 70.1. *Let X be a path connected space. The transgression relation*

$$\overline{H}_n(X) \rightharpoonup \overline{H}_{n-1}(\Omega X)$$

associated to the path loop fibration $p : PX \to X$ is the converse of the relation defined by the map

$$\overline{H}_{n-1}(\Omega X) = \overline{H}_n(\Sigma\Omega X) \xrightarrow{\text{ev}_*} \overline{H}_n(X).$$

Proof. Recall that the transgression relation is given (in this case) by the span

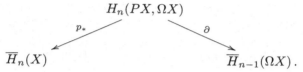

It consists of the subgroup

$$\{(x,y) \in \overline{H}_n(X) \times \overline{H}_{n-1}(\Omega X) : \exists\, z \in H_n(PX, \Omega X)$$
$$\text{such that } p_* z = x \text{ and } \partial z = y\}.$$

We are claiming that this is the same as the subgroup

$$\{(x,y) \in \overline{H}_n(X) \times \overline{H}_{n-1}(\Omega X) : \exists\, w \in \overline{H}_n(\Sigma\Omega X)$$
$$\text{such that } \text{ev}_* w = x \text{ and } iw = y\}$$

determined by the span

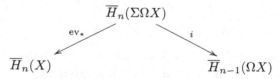

where $i : \overline{H}_{n-1}(\Omega X) \xrightarrow{\cong} \overline{H}_n(\Sigma\Omega X)$ is the canonical isomorphism.

To see this, we just have to remember how the boundary map and the isomorphism i are related. This is a general point. So suppose we have a space X and a subspace A, so we are interested in $i : \overline{H}_n(\Sigma X) \to \overline{H}_{n-1}(X)$ and the boundary map $\partial : H_n(X, A) \to \overline{H}_{n-1}(A)$. The latter may be described geometrically in the following way. Form the mapping cylinder M of the inclusion map $A \to X$. Then $A \hookrightarrow M$ is a cofibration with cofiber ΣA, and we have the span

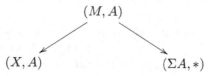

in which the left arrow is a homology isomorphism. The boundary map is induced by this span, together with the isomorphism i.

Specializing to the pair $(PX, \Omega X)$ gives commutativity of part of the diagram

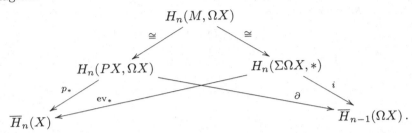

The other part follows from homotopy commutativity of

$$(M, \Omega X) \longrightarrow (\Sigma \Omega X, *) \qquad \sigma; (\omega, t) \longmapsto *; [\omega, t]$$
$$\downarrow \qquad \qquad \downarrow ev \qquad \qquad \downarrow \qquad \qquad \downarrow$$
$$(PX, \Omega X) \xrightarrow{\ p\ } (X, *) \qquad \sigma(0); \omega \longmapsto \sigma(0); * \qquad *; \omega(t).$$

Notation: the first entry is the map on $PX = \{\sigma : I \to X \text{ such that } \sigma(0) = *\}$; the second entry is the map on $\Omega X \times I$. A homotopy between the two branches is given at time s by

$$\sigma; \omega \mapsto \sigma(s); (t \mapsto \omega(st)).$$

This diagram shows that the two relations are identical. □

The evaluation map $\Sigma \Omega X \to X$ also admits an interesting interpretation in cohomology, with coefficients in an abelian group π:

$$\overline{H}^n(X; \pi) \longrightarrow \overline{H}^{n-1}(\Omega X; \pi)$$
$$\downarrow \cong \qquad \qquad \qquad \downarrow \cong$$
$$\qquad \qquad \qquad [\Omega X, K(\pi, n-1)]_*$$
$$\qquad \qquad \qquad \downarrow \cong$$
$$[X, K(\pi, n)]_* \xrightarrow{\ \Omega\ } [\Omega X, \Omega K(\pi, n)]_*$$

commutes.

Our identification of the evaluation map as the converse of a transgression allows us to invoke the Serre exact sequence. After all, if the total space is contractible, every third term in the Serre exact sequence vanishes, and the remaining map, the transgression, is an isomorphism. In fact, in that case we get just a little extra, the last clause in the following proposition, which we state in the generality of working modulo a Serre ring.

Proposition 70.2. *Let $\mathcal{C}$ be a Serre ring. Let $n \geq 1$ and suppose X is simply connected and that $\overline{H}_i(X) \in \mathcal{C}$ for all $i < n$. Then the evaluation map $ev_* : \overline{H}_{i-1}(\Omega X) \to \overline{H}_i(X)$ is an isomorphism mod $\mathcal{C}$ for $i < 2n - 1$ and an epimorphism mod $\mathcal{C}$ for $i = 2n - 1$.*

This result leads the way to the "suspension theorem" of Hans Freudenthal (1905–1990; German, working in Amsterdam, escaped from a labor camp during World War II). The relevant adjunction morphism is now the "suspension"

$$\sigma_X : X \to \Omega\Sigma X.$$

The formalism of adjunction guarantees commutativity of

$$
\begin{array}{ccc}
\Sigma X & \xrightarrow{\Sigma\sigma_X} & \Sigma\Omega\Sigma X \\
 & \searrow{\scriptstyle 1} & \big\downarrow{\scriptstyle ev_{\Sigma X}} \\
 & & \Sigma X,
\end{array}
$$

which shows for a start that σ_X induces a split monomorphism in reduced homology. But we also know from 70.2 that if X is $(n-1)$ connected, the evaluation map in

$$
\begin{array}{ccc}
\overline{H}_i(X) & \xrightarrow{(\sigma_X)_*} & \overline{H}_i(\Omega\Sigma X) \\
 & \searrow{\scriptstyle =} & \big\downarrow{\scriptstyle (ev_{\Sigma X})_*} \\
 & & \overline{H}_{i+1}(\Sigma X)
\end{array}
$$

is an isomorphism mod $\mathcal{C}$ for $i < 2n$: so the same is true for $(\sigma_X)_*$. Now we can apply the mod $\mathcal{C}$ Whitehead theorem to conclude:

Theorem 70.3 (Mod $\mathcal{C}$ Freudenthal suspension theorem). *Let $\mathcal{C}$ be an acyclic Serre ideal and $n \geq 1$. Let X be a simply connected space such that $\overline{H}_i(X)$ is zero mod $\mathcal{C}$ for $i < n$. Then the suspension map*

$$\pi_i(X) \to \pi_i(\Omega\Sigma X) = \pi_{i+1}(\Sigma X)$$

is a mod $\mathcal{C}$ isomorphism for $i < 2n - 1$ and a mod $\mathcal{C}$ epimorphism for $i = 2n - 1$.

Corollary 70.4. *Let $n \geq 2$. The suspension map*

$$\pi_i(S^n) \to \pi_{i+1}(S^{n+1})$$

is an isomorphism for $i < 2n - 1$ and an epimorphism for $i = 2n - 1$.

For example, $\pi_2(S^2) \to \pi_3(S^3)$ is an isomorphism (the degree is a stable invariant), while $\pi_3(S^2) \to \pi_4(S^3)$ is only an epimorphism: the Hopf map $S^3 \to S^2$ suspends to a generator of $\pi_4(S^3)$, which as we saw has order 2.

In any case, the Freudenthal suspension theorem show that the sequence

$$\pi_k(X) \to \cdots \to \pi_{n+k}(\Sigma^n X) \to \pi_{n+1+k}(\Sigma^{n+1} X) \to \cdots$$

stabilizes. The direct limit is the *reduced kth stable homotopy group* of the pointed space X, $\pi_k^s(X)$. These functors turn out to form a generalized homology theory. The coefficients form a graded commutative ring, the *stable homotopy ring*

$$\pi_*^s = \pi_*^s(S^0) = \lim_{n \to \infty} \pi_{*+n}(S^n).$$

The group $\pi_{k+n}(S^n)$ contributes to π_k^s, and the index k is often referred to as the "stem" or "stem degree."

EHP sequence

The homotopy groups of spheres are related to each other via the suspension maps, but it turns out that there is more, based on the following consequence of a splitting theorem due to Ioan James.

Proposition 70.5 (e.g. [59, §7.9]). *Let $n \geq 2$. There is a map $h : \Omega S^n \to \Omega S^{2n-1}$ that induces an isomorphism in $H_{2n-2}(-)$.*

Granting this, we can compute the entire effect in cohomology. When n is even, say $n = 2k$, the generator $y \in H^{4k-2}(\Omega S^{4k-1})$ hits the divided power generator in $H^{4k-2}(\Omega S^{2k})$, and hence embeds $H^*(\Omega S^{4k-1})$ into $H^*(\Omega S^{2k})$ isomorphically in dimensions divisible by $(4k - 2)$. The induced map in homology thus has the same behavior. It follows from the spectral sequence that the homotopy fiber has the homology of S^{2k-1}. But the suspension map $S^{2k-1} \to \Omega S^{2k}$ certainly composes into ΩS^{4k-1} to a null map, and hence lifts to a map to the homotopy fiber inducing a homology isomorphism. By Whitehead's theorem, it is a weak equivalence.

The Whitehead square $w_{2k} = [\iota_{2k}, \iota_{2k}] : S^{4k-1} \to S^{2k}$ has the property that the composite

$$\Omega S^{4k-1} \xrightarrow{\Omega w_{2k}} \Omega S^{2k} \xrightarrow{h} \Omega S^{4k-1}$$

is an isomorphism in homology away from 2. So, using the multiplication in ΩS^{2k}, there is a map,

$$S^{2k-1} \times \Omega S^{4k-1} \to \Omega S^{2k}$$

that induces an isomorphism in homology away from 2, and hence by the mod $\mathcal{C}$ Whitehead theorem in homotopy away from 2. For this reason, even spheres are not very interesting homotopy theoretically away from 2.

When n is odd, $y \in H^{2n-2}(\Omega S^{2n-1})$ maps to the divided square of $x \in H^{n-1}(\Omega S^n)$. This implies that

$$\gamma_k(y) = \frac{y^k}{k!} \mapsto \frac{(x/2)^k}{k!} = \frac{(2k)!}{2^k k!}\gamma_{2k}(x)\,.$$

A little thought shows that the numerator and denominator here contain the same power of 2, so the map is still an isomorphism in $\mathbb{Z}_{(2)}$ homology in dimensions divisible by $2n-2$, and hence the fiber has the 2-local homology of S^{n-1}. The mod $\mathcal{C}_2$ Whitehead theorem shows that it also has the 2-local homotopy of S^{n-1}. We conclude:

Theorem 70.6. *For any positive even integer n there is a fiber sequence*

$$S^{n-1} \xrightarrow{e} \Omega S^n \xrightarrow{h} \Omega S^{2n-1}\,.$$

Localized at 2, this sequence exists for n odd as well.

The map e is suspension ("Einhangung" in German) and h is the James-Hopf map. Moving one step back in the Barratt-Puppe sequence, we have a map $p : \Omega^2 S^{2n-1} \to S^{n-1}$. It's denoted by p since it is related to the Whitehead product; for example, composed with the double suspension map $e^2 : S^{2n-3} \to \Omega^2 S^{2n-1}$ gives us the Whitehead square $w_{n-1} \in \pi_{2n-3}(S^{n-1})$. The long exact homotopy sequence gives us the *EHP sequence*

of homotopy groups (localized at 2 if n is odd).

These sequences link together to form an exact couple! You can see this clearly from the diagram of fiber sequences obtained by looping down the sequences of Theorem 70.6. Locally at 2, we have a diagram in which each L is a fiber sequence.

$$\cdots \longrightarrow \Omega^{s-1}S^{s-1} \longrightarrow \Omega^s S^s \longrightarrow \Omega^{s+1}S^{s+1} \longrightarrow \cdots \longrightarrow \Omega^\infty S^\infty$$

$$\Omega^{s-1}S^{2s-3} \qquad \Omega^s S^{2s-1} \qquad \Omega^{s+1}S^{2s+1}.$$

The limiting space $\Omega^\infty S^\infty$ has homotopy equal to π_*^s.

The resulting spectral sequence, the *EHP spectral sequence*, has the form

$$E_{s,t}^1 = \pi_{2s+1+t}(S^{2s+1}) \underset{s}{\Longrightarrow} \pi_{s+t}^s .$$

Here's a picture, taken from [39]. In it, 2^3 represents the elementary abelian group of order 8 and "∞" means an infinite cyclic group. A superscript 2 means the index of the image is 2. Differentials entering or exiting this range are not shown.

(t)	1	2	3	4	5	6	7	8	9	10	11
9	2^2	2^3	2^4	2^4	2^3	2^3	2^3	2^3	2^3	2^3	2^3
8	2	2	2^3	2^3	2^2	2^2	2^2	2^2	2^2	2^2	2^2
7		2	8	16	16	16	16	16	16	16	16
6		2	2	2	2	2	2	2	2	2	2
5	2	2									
4	2	2									
3	4	8	8	8	8	8	8	8	8	8	8
2	2	2	2	2	2	2	2	2	2	2	2
1	2	2	2	2	2	2	2	2	2	2	2
0	∞	∞	∞	∞	∞	∞	∞	∞	∞	∞	∞

Stem of Input: 0 (row), with stems 1–11 across the bottom.

	1	2	3	4	5	6	7	8	9	10	11
Sphere of Origin:	S^2	S^3	S^4	S^5	S^6	S^7	S^8	S^9	S^{10}	S^{11}	S^{12}
Input Sphere:	S^3	S^5	S^7	S^9	S^{11}	S^{13}	S^{15}	S^{17}	S^{19}	S^{21}	S^{23}

This spectral sequence contains an immense amount of data. For example, the differential $d^1 : E_{2,0}^1 \to E_{1,0}^1$ tells us that a class (called η_2) is born in the 1-stem on S^2 with Hopf invariant 1 in $\pi_3(S^3) = \mathbb{Z}$, and that it suspends to a class (called η_3) in $\pi_4(S^3)$ whose order is 2 because $P(\iota_5) = 2\eta_2$ and $EP = 0$. The class η_3 persist to a class $\eta \in \pi_1^s$. Similarly, the differential $d^2 : E_{5,0}^2 \to E_{3,1}^2$ tells us that a class x_4 is born in $\pi_8(S^4)$ with Hopf invariant $\eta_7 \in \pi_8(S^7)$; it suspends nontrivially to a class $x_5 \in \pi_9(S^5)$, but to 0 in $\pi_{10}(S^6)$ because $P(\iota_{11}) = x_5$.

According to Exercise 63.5, the EHP spectral sequence may be truncated to obtain a spectral sequence converging to $\pi_*(S^n)$ rather than π_*^s.

Bousfield localization

I can't leave the subject of Serre classes without mentioning a more recent and more geometric approach to localization in algebraic topology, due to Bousfield, following diverse early ideas of Dennis Sullivan, Mike Artin and Barry Mazur, and Frank Adams. (A. K. ("Pete") Bousfield (1941–2020), a student of Dan Kan's at MIT, worked at UIC.)

Theorem 70.7 (Bousfield, [9]). *Let E_* be any generalized homology theory and X any CW complex. There is a space $L_E X$ and a map $X \to L_E X$ that is terminal in the homotopy category among E_*-equivalences from X.*

So $L_E X$ is as far away (to the right) from X as possible while still receiving an E_*-equivalence from it. The localization strips away all features not detected by E-homology.

The class of maps given by E_*-equivalences determines a class of objects: A space W is E_*-*local* if for every E_*-equivalence $X \to Y$ between CW complexes the induced map $[X, W] \leftarrow [Y, W]$ is bijective. You can't tell two E_*-equivalent spaces apart by mapping them into an E_*-local space.

Theorem 70.8 (Addendum to Theorem 70.7). *For any CW complex X, $L_E X$ is E_*-local, and the localization map $X \to L_E X$ is initial among maps to E_*-local spaces.*

The functor L_E is "Bousfield localization" at the homology theory E_*. The subcategory of E_*-local spaces affords the ultimate extension of the Whitehead theorem:

Lemma 70.9. *Any E_*-equivalence $f : X \to Y$ between E_*-local CW complexes is a homotopy equivalence.*

Proof. Take $W = X$ in the definition of "E_*-local": then the identity map $X \to X$ lifts in the homotopy category uniquely through a map $g : Y \to X$. By construction $gf = 1_X$. But then both fg and 1_Y lift $f : X \to Y$ across f, and hence must be equal by uniqueness. $\qquad\square$

Another example is given by rational homology $H\mathbb{Q}_*$.

Proposition 70.10. *A simply connected CW complex is $H\mathbb{Q}_*$-local if and only if its homology in each positive dimension is a rational vector space.*

In this case we can also compute the homotopy: For a simply connected CW complex X, $\pi_*(X) \to \pi_*(L_{H\mathbb{Q}}X)$ simply tensors the homotopy with $\mathbb{Q}$. This is the beginning of an extensive development of "rational homotopy theory," pioneered independently by Daniel Quillen and Dennis Sullivan. The entire homotopy theory of simply connected rational spaces of finite type over $\mathbb{Q}$ is equivalent to the opposite of the homotopy theory of commutative differential graded $\mathbb{Q}$-algebras that are simply connected and of finite type. The quest for analogous completely algebraic descriptions of other sectors of homotopy theory has been a major research objective over the past half century.

Bousfield localization at $H\mathbb{F}_p$ is trickier, because the map from S^n to the Moore space M with homology given by the p-adic integers $\mathbb{Z}_p$ in dimension n is an isomorphism in mod p homology. In fact $L_{H\mathbb{F}_p}S^n = M$: so in this case Bousfield localization behaves like a completion.

When the fundamental group is nontrivial, localization at $H\mathbb{Z}$ can lead to unexpected results. For example, let Σ_∞ be the group of permutations of a countably infinite set that move only finitely many elements. Then

$$L_{H\mathbb{Z}}B\Sigma_\infty \simeq \Omega_0^\infty S^\infty,$$

a single component of the union of $\Omega^s S^s$'s. This is the "Barratt-Priddy-Quillen theorem."

For another example, let R be a ring and $GL_\infty(R)$ the increasing union of the groups $GL_n(R)$. The homotopy groups of the space $L_{H\mathbb{Z}}BGL_\infty(R)$ formed Quillen's first definition of the higher algebraic K-theory of R.

Exercises

Exercise 70.11. We found that $\pi_4(S^3) = \mathbb{Z}/2\mathbb{Z}$. Explain why this shows that $\pi_4(S^2) = \mathbb{Z}/2\mathbb{Z}$ as well, and describe a non-null map $S^4 \to S^2$.

Chapter 8

Characteristic classes, Steenrod operations, and cobordism

71 Chern classes, Stiefel-Whitney classes, and the Leray-Hirsch theorem

A good supply of interesting geometric objects is provided by the theory of principal G-bundles, for a topological group G. For example giving a principal $GL_n(\mathbb{C})$-bundle over X is the same thing as giving a complex n-plane bundle over X.

Principal bundles reflect a great deal of geometric information in their topology. This is a great asset, but it can make them correspondingly hard to visualize. It's reasonable to hope to construct invariants of principal G-bundles of some more understandable sort. A good candidate is a cohomology class.

So let's fix an integer n and an abelian group A, and try to associate, in some way, a class $c(\xi) \in H^n(Y; A)$ to any principal G-bundle ξ over Y. To make this useful, this association should be *natural*: given $f : X \to Y$ and a principal G-bundle ξ over Y, we can pull ξ back under f to a principal G-bundle $f^*\xi$ over X, and find ourselves with two classes in $H^n(X; A)$: $f^*c(\xi)$ and $c(f^*(\xi))$. Naturality insists that these two classes coincide. This means, incidentally, that $c(\xi)$ depends only on the isomorphism class of ξ. Let $\mathrm{Bun}_G(X)$ denote the set of isomorphism classes of principal G-bundles over X; it is a contravariant functor of X. We have come to a definition:

Definition 71.1. Let G be a topological group, A an abelian group, and $n \geq 0$. A *characteristic class* for principal G-bundles with values in $H^n(-; A)$ is a natural transformation of functors $\mathbf{Top}^{op} \to \mathbf{Set}$:

$$c : \mathrm{Bun}_G(-) \to H^n(-; A).$$

Cohomology classes are more formal or algebraic, and are correspondingly relatively easy to work with. $\mathrm{Bun}_G(X)$ is often hard (or impossible) to compute, partly because it has no algebraic structure and partly exactly because its elements are interesting geometrically, while $H^n(X; A)$ is relatively easy to compute but its elements are not very geometric. A characteristic class provides a bridge between these two, and information flows across this bridge in both directions. It gives computable information about certain interesting geometric objects, and provides a geometric interpretation of certain formal or algebraic things.

Example 71.2. The Euler class is the first and most fundamental characteristic class. Let R be a commutative ring. The Euler class takes an R-oriented real n-plane bundle ξ and produces an n-dimensional cohomology class $e(\xi)$, given by the transgression of the class in $H^0(B; H^{n-1}(\mathbb{S}(\xi)))$ that evaluates to 1 on every orientation class. Naturality of the Gysin sequence shows that this assignment is natural. There are really only two cases: $R = \mathbb{Z}$ and $R = \mathbb{F}_2$. A $\mathbb{Z}$-orientation of a vector bundle is the same thing as an orientation in the usual sense, and the Euler class is a natural transformation

$$e : \mathrm{Vect}_n^{or}(X) = \mathrm{Bun}_{SO(n)}(X) \to H^n(X; \mathbb{Z}).$$

Any vector bundle is canonically $\mathbb{F}_2$-oriented, so the mod 2 Euler class is a natural transformation

$$e : \mathrm{Vect}_n(X) = \mathrm{Bun}_{O(n)}(X) \to H^n(X; \mathbb{F}_2).$$

On CW complexes, $\mathrm{Bun}_G(-)$ is representable: there is a "universal" principal G-bundle $\xi_G : EG \downarrow BG$ such that

$$[X, BG] \to \mathrm{Bun}_G(X), \quad f \mapsto f^* \xi_G$$

is a bijection. A characteristic class $\mathrm{Bun}_G(-) \to H^n(-; A)$ is the same thing as a class in $H^n(BG; A)$, or, since cohomology is also representable, as a homotopy class of maps $BG \to K(A, n)$.

Thus for example the set of all integral characteristic classes of complex line bundles is given by $H^*(BU(1)) = \mathbb{Z}[e]$. Is there an analogous classification of characteristic classes for higher dimensional complex bundles? How about real bundles?

Chern classes

We'll begin with complex vector bundles. Any complex vector bundle (numerable of course) admits a Hermitian metric, well defined up to homotopy. This implies that $\mathrm{Bun}_{U(n)}(X) \to \mathrm{Bun}_{GL_n(\mathbb{C})}(X)$ is bijective;

$BU(n) \to BGL_n(\mathbb{C})$ is a homotopy equivalence. I will tend to favor $U(n)$ and $BU(n)$.

A finite dimensional complex vector space V determines an orientation of the underlying real vector space: Pick an ordered basis $(e_1, \ldots, e_n)$ for V over $\mathbb{C}$, and provide V with the ordered basis over $\mathbb{R}$ given by $(e_1, ie_1, \ldots, e_n, ie_n)$. The group $\mathrm{Aut}_{\mathbb{C}}(V)$ acts transitively on the space of complex bases. But choosing a basis for V identifies $\mathrm{Aut}(V)$ with $\mathrm{GL}_n(\mathbb{C})$, which is path connected. So the set of ordered real bases obtained in this way are all in the same path component of the set of all oriented real bases, and hence defines an orientation of V.

This construction yields a natural transformation $\mathrm{Vect}_{\mathbb{C}}(-) \to \mathrm{Vect}_{\mathbb{R}}^{or}(-)$. In particular, the real 2-plane bundle underlying a complex line bundle has a preferred orientation — the one determined in each fiber ξ_x by the ordered basis (v, iv) where $v \neq 0$ in ξ_x. A complex line bundle ξ over B thus has a well-defined Euler class $e(\xi) \in H^2(B; \mathbb{Z})$.

Theorem 71.3 (Chern classes). *There is a unique family of characteristic classes for complex vector bundles that assigns to a complex n-plane bundle ξ over X its kth Chern class $c_k^{(n)}(\xi) \in H^{2k}(X; \mathbb{Z})$, $k \in \mathbb{N}$, such that:*

- $c_0^{(n)}(\xi) = 1$.
- $c_1^{(1)}(\xi) = -e(\xi)$.
- *The* Whitney sum formula *holds: if ξ is a p-plane bundle and η is a q-plane bundle, then*

$$c_k^{(p+q)}(\xi \oplus \eta) = \sum_{i+j=k} c_i^{(p)}(\xi) \cup c_j^{(q)}(\eta) \in H^{2k}(X; \mathbb{Z}).$$

Moreover, if ξ_n is the universal n-plane bundle, then

$$H^*(BU(n); \mathbb{Z}) \cong \mathbb{Z}[c_1^{(n)}, \ldots, c_n^{(n)}]$$

where $c_k^{(n)} = c_k^{(n)}(\xi_n)$.

This result says that all characteristic classes for complex vector bundles are given by polynomials in the Chern classes, and that there are no universal algebraic relations among the Chern classes. (Shiing-Shen Chern (1911–2004) was a father of twentieth century differential geometry, and a huge force in the development of mathematics in China.)

Remark 71.4. Since $BU(n)$ supports the universal n-plane bundle ξ_n, the Chern classes $c_k^{(n)} = c_k^{(n)}(\xi_n)$ are themselves universal, pulling back to the Chern classes of any other n-plane bundle.

The $(p + q)$-plane bundle $\xi_p \times \xi_q = \mathrm{pr}_1^* \xi_p \oplus \mathrm{pr}_2^* \xi_q$ over $BU(p) \times BU(q)$ is classified by a map $\mu : BU(p) \times BU(q) \to BU(p + q)$. The Whitney sum formula computes the effect of μ on cohomology:

$$\mu^*(c_k^{(n)}) = \sum_{i+j=k} c_i^{(p)} \times c_j^{(q)} \in H^{2k}(BU(p) \times BU(q)),$$

where, you'll recall, $x \times y = \mathrm{pr}_1^* x \cup \mathrm{pr}_2^* y$.

The Chern classes are "stable" in the following sense. Let ϵ be the trivial one-dimensional complex vector bundle over X and let ξ be an n-dimensional vector bundle over X. What is $c_k^{(n+q)}(\xi \oplus q\epsilon)$? The trivial bundle is obtained by pulling back under $X \to *$:

$$
\begin{array}{ccc}
X \times \mathcal{C}^q = E(q\epsilon) & \longrightarrow & \mathcal{C}^q \\
\downarrow & & \downarrow \\
X & \longrightarrow & *.
\end{array}
$$

By naturality, we find that $c_j^{(n)}(n\epsilon) = 0$ for $j > 0$. The Whitney sum formula therefore implies that

$$c_k^{(n+q)}(\xi \oplus q\epsilon) = c_k^{(n)}(\xi).$$

Thus the Chern class only depends on the "stable equivalence class" of the vector bundle. Also, the map $BU(n - 1) \to BU(n)$ classifying $\xi_{n-1} \oplus \epsilon$ sends $c_k^{(n)}$ to $c_k^{(n-1)}$ for $k < n$ and $c_n^{(n)}$ to 0.

For this reason, we will drop the superscript on $c_k^{(n)}(\xi)$, and simply write $c_k(\xi)$.

Grothendieck's construction

Let $\xi : E \xrightarrow{p} X$ be a complex n-plane bundle. Associated to it is a fiber bundle whose fiber over $x \in X$ is $\mathbb{P}(p^{-1}(x))$, the projective space of the vector space given by the fiber of ξ over x. This "projectivization" can also be described using the $GL_n(\mathbb{C})$ action on $\mathbb{CP}^{n-1} = \mathbb{P}(\mathbb{C}^n)$ induced from its action on $\mathbb{C}^n$, and forming the balanced product

$$\mathbb{P}(\xi) = P \times_{GL_n(\mathbb{C})} \mathbb{CP}^{n-1},$$

where $P \downarrow X$ is the principalization of ξ.

Let us attempt to compute the cohomology of $\mathbb{P}(\xi)$ using the Serre spectral sequence:

$$E_2^{s,t} = H^s(X; H^t(\mathbb{CP}^{n-1})) \Rightarrow H^{s+t}(\mathbb{P}(\xi)).$$

We claim that this spectral sequence almost completely determines the cohomology of $\mathbb{P}(\xi)$ as a ring. Here is a general theorem that tells us what to look for, and what we get.

Theorem 71.5 (Leray-Hirsch). *Let $\pi : E \to B$ be a fibration and R a commutative ring. Assume that B is path connected, so that the fiber is well defined up to homotopy. Call it F, and suppose that for each t the R-module $H^t(F)$ is free of finite rank. Finally, assume that the restriction $H^*(E) \to H^*(F)$ is surjective. (One says that the fibration is "totally non-homologous to zero.") Because $H^t(F)$ is a free R-module for each t, the surjection $H^*(E) \to H^*(F)$ admits a splitting; pick one, say $s : H^*(F) \to H^*(E)$. The projection map renders $H^*(E)$ a module over $H^*(B)$. The $H^*(B)$-linear extension of s,*

$$\overline{s} : H^*(B) \otimes_R H^*(F) \to H^*(E)$$

is then an isomorphism of $H^*(B)$-modules.

Proof. First we claim that the group $\pi_1(B)$ acts trivially on the cohomology of $F = \pi^{-1}(*)$. The map of fibrations

$$
\begin{array}{ccc}
E & \xrightarrow{\ 1\ } & E \\
{\scriptstyle\pi}\downarrow & & \downarrow \\
B & \longrightarrow & *
\end{array}
$$

shows that the map $H_*(F) \to H_*(E)$ is equivariant with respect to the group homomorphisms $\pi_1(B) \to \pi_1(*)$. In cohomology, this says that the restriction $H^*(E) \to H^*(F)$ has image in the $\pi_1(B)$-invariant subgroup (which, by the way, is $H^0(B; H^*(\pi^{-1}(-))))$. So the assumption that this map is surjective guarantees that the action of $\pi_1(B)$ on $H_*(F)$ is trivial.

Now the edge homomorphism in the Serre spectral sequence

$$E_2^{s,t} = H^s(B; H^t(F)) \underset{s}{\Longrightarrow} H^{s+t}(E)$$

is that restriction map. Our assumption that $H^t(F)$ is free of finite rank implies (27.8) that

$$E_2^{s,t} = H^s(B) \otimes_R H^t(F)$$

as R-algebras. All the generators lie on either $t = 0$ or $s = 0$. The ones on the base survive because the differentials hit zero groups. The generators on the fiber survive by assumption. So inductively you find that $E_r = E_{r+1}$, and hence that the entire spectral sequence collapses at E_2.

We now define a new filtration on $H^*(E)$ with the advantage that it is a filtration by $H^*(B)$-modules. I call it the "Quillen filtration," after [55, p. 544], though it is probably older. It's the *increasing* filtration given in terms of the decreasing filtration associated to the spectral sequence by

$$F_t H^n(E) = F^{n-t} H^n(E).$$

For instance, $F_0 H^n(E) = F^n H^n(E) = \operatorname{im}(H^n(B) \to H^n(E))$; or

$$F_0 H^*(E) = \operatorname{im}(H^*(B) \to H^*(E)).$$

On the level of associated graded modules,

$$\operatorname{gr}_t H^n(E) = F^{n-t} H^n(E)/F^{n-t+1} H^n(E) = E_\infty^{n-t,t}$$

— that is, the tth row: so

$$\operatorname{gr}_t H^*(E) = E_\infty^{*,t}$$

which, in case the spectral sequence collapses at E_2, is $H^*(B; H^t(F))$. So for us it is $H^*(B) \otimes_R H^t(F)$.

Now we can think about the map $\bar{s} : H^*(B) \otimes H^*(F) \to H^*(E)$. Filter $H^*(B) \otimes H^*(F)$ by degree in $H^*(F)$:

$$F_t(H^*(B) \otimes H^*(F)) = H^*(B) \otimes \bigoplus_{i \leq t} H^i(F).$$

The map $\bar{s}$ respects filtrations and is an isomorphism on associated graded modules: so it is an isomorphism.   □

Returning now to the example of the projectivization of a vector bundle, $\mathbb{P}(\xi) \downarrow X$, the hypotheses of the Leray-Hirsch Theorem are satisfied except perhaps surjectivity of the restriction to the fiber.

Here's where the representation of a cohomology class as a characteristic class comes in useful. The cohomology of the fiber over $x \in X$ is generated as an R-module by powers of the Euler class of the canonical line bundle λ_x over $\mathbb{P}(\xi_x)$. Since $i^* : H^*(E) \to H^*(\mathbb{CP}^{n-1})$ is an R-algebra map, it will suffice to see that $e(\lambda_x)$ is in the image of i^*. Since the Euler class is natural, the natural thing to do is to construct a line bundle over the whole of $\mathbb{P}(\xi)$ that restricts to λ_x on ξ_x. And indeed these line bundles over fibers assemble themselves into a tautologous line bundle, call it λ, over $\mathbb{P}(\xi)$.

So we have an expression for $H^*(\mathbb{P}(\xi))$ as a module over $H^*(X)$:

$$H^*(\mathbb{P}(\xi)) = H^*(X)\langle 1, e, e^2, \ldots, e^{n-1} \rangle,$$

where $e = e(\lambda) \in H^2(\mathbb{P}(\xi))$. This gives us some information about the algebra structure in $H^*(\mathbb{P}(\xi))$, but not complete information. What is lacking is an expression for e^n in terms of the basis given by lower powers of e. The Euler class e satisfies a unique monic polynomial equation $c_\xi(e) = 0$, where $c_\xi(t)$ is the "Chern polynomial"

$$c_\xi(t) = t^n + c_1 t^{n-1} + \cdots + c_{n-1} t + c_n$$

with $c_k \in H^{2k}(X)$.

The naturality of this construction guarantees that the c_k's are natural in the n-plane bundle ξ; they are characteristic classes. We will see that they satisfy the axioms for Chern classes set out above.

Note that the Whitney sum formula has a nice expression in terms of the Chern polynomials:

$$c_\xi(t)c_\eta(t) = c_{\xi \oplus \eta}(t).$$

Stiefel-Whitney classes

Exactly parallel theorems hold for real n-plane bundles, with mod 2 coefficients:

Theorem 71.6 (Stiefel-Whitney classes). *There is a unique family of characteristic classes for real vector bundles that assigns to a real n-plane bundle ξ over X its "kth Stiefel-Whitney class" $w_k(\xi) \in H^k(X; \mathbb{F}_2)$, $k \in \mathbb{N}$, such that:*

- $w_0(\xi) = 1$.
- *If $n = 1$ then $w_1(\xi) = e(\xi)$.*
- *The* Whitney sum formula *holds: if ξ is a p-plane bundle and η is a q-plane bundle, then*

$$w_k(\xi \oplus \eta) = \sum_{i+j=k} w_i(\xi) \cup w_j(\eta) \in H^{2k}(X; \mathbb{F}_2).$$

Moreover, if ξ_n is the universal n-plane bundle, then

$$H^*(BO(n); \mathbb{F}_2) \cong \mathbb{F}_2[w_1, \ldots, w_n]$$

where $w_k = w_k(\xi_n)$.

And the same construction produces them:

$$H^*(\mathbb{P}(\xi); \mathbb{F}_2) = H^*(B; \mathbb{F}_2)[e]/(e^n + w_1 e^{n-1} + \cdots + w_{n-1}e + w_n)$$

for unique elements $w_i \in H^i(B; \mathbb{F}_2)$.

Remark 71.7. The Euler class depends only on the sphere bundle of the vector bundle ξ, but these constructions appear to depend heavily on the existence of an underlying vector bundle. This is a genuine dependence in the case of Chern classes, but it turns out that the Stiefel-Whitney classes depend only on the sphere bundle. We'll explain this later, in Proposition 77.9.

Remark 71.8. In the complex case, the triviality of the local coefficient system can be verified in other ways as well. After all, the action of $\pi_1(X)$ on the fiber $H^*(\mathbb{CP}^{n-1})$ is compatible with the action of $\pi_1(BU(n))$ on the homology of the fiber of the projectivized universal example. But since $U(n)$ is connected, its classifying space is simply connected.

You can't make this argument in the real case, but then you don't have to since we are looking at an action of $\pi_1(B)$ on a one-dimensional vector space over $\mathbb{F}_2$.

Example 71.9. Complex projective space $\mathbb{CP}^n$ is a complex manifold, and its tangent bundle is thereby endowed with a complex structure. A standard argument (Example 54.7) shows that

$$\tau_{\mathbb{CP}^n} = \mathrm{Hom}(\lambda, \lambda^\perp).$$

Adding $\epsilon = \mathrm{Hom}(\lambda, \lambda)$, we find

$$\tau_{\mathbb{CP}^n} \oplus \epsilon = (n+1)\lambda^{-1},$$

where λ^{-1} is the dual of the line bundle λ: $\lambda^{-1} = \mathrm{Hom}(\lambda, \epsilon)$. Its Euler class is $-e$, where e is the Euler class of λ. Thus by the Whitney sum formula

$$c_\tau(t) = c_{\tau \oplus \epsilon}(t) = c_{\lambda^{-1}}(t)^{n+1} = (1 - et)^{n+1}$$

and so

$$c_k(\tau_{\mathbb{CP}^n}) = (-1)^k \binom{n+1}{k} e^k.$$

Exercises

Exercise 71.10. Put an upper bound on the number of everywhere independent vector fields you can put on $\mathbb{RP}^{n-1}$. In particular, for many n you can rule out the possibility that $\mathbb{RP}^{n-1}$ is parallelizable. See Exercise 54.14. Note also that this result gives an upper bound for the number of everywhere independent vector fields on S^{n-1}. The definitive result is due to Adams [2].

Exercise 71.11. An *immersion* of manifold M into manifold N is a smooth map $f : M \to N$ that induces an injection on tangent bundles. Draw some immersions of the circle in $\mathbb{R}^2$. The "Whitney-Graustein theorem" classifies them. Find out about the Boys surface, an immersion of $\mathbb{RP}^2$ into $\mathbb{R}^3$. Whitney proved that any closed n-manifold immerses in codimension $n - 1$.

An immersion $f : M \to N$ determines an embedding $\tau_M \hookrightarrow f^* \tau_N$ of vector bundles over M. In particular, an immersion $M \to \mathbb{R}^n$ determines an embedding $\tau_M \hookrightarrow n\epsilon$. Not every m-plane bundle admits an $(n - m)$-dimensional complement; this provides obstructions to possible immersions. In fact Smale-Hirsch theory shows that this is the only obstruction; any map $f : M \to N$ that is covered by a fiberwise linear embedding of the tangent bundles (not necessarily given by the derivative of the map!) f is homotopic to an immersion.

Use this idea, together with Stiefel-Whitney classes, to put a lower bound on the codimension of an immersion of $\mathbb{RP}^n$ into Euclidean space. (Determination the minimal codimension of an immersion of $\mathbb{RP}^n$ in general is a very difficult computational problem, still not completely resolved.)

Using these examples, find, for each n, a closed n-manifold that does not immerse into $\mathbb{R}^{2n-\alpha(n)-1}$, where $\alpha(n)$ is the sum of the digits in the binary expansion of n.

This is a best possible result, according to a theorem of Ralph Cohen (following work of Ed Brown and Frank Peterson): every closed n-manifold immerses in codimension $n - \alpha(n)$.

72 $H^*(BU(n))$ and the splitting principle

Here's another characterization of the Chern classes.

Theorem 72.1. *Let $n \geq 1$. There is a unique family of characteristic classes $c_i(\xi) \in H^{2i}(B(\xi))$, $1 \leq i \leq n$, for complex n-plane bundles ξ, such that if ξ is isomorphic to $\zeta \oplus (n - i)\epsilon$ then*

$$c_i(\xi) = (-1)^i e(\zeta)$$

where $e(\zeta)$ is the Euler class of the oriented real $2i$-bundle underlying ζ. These classes generate all characteristic classes for n-plane bundles and there are no universal algebraic relations among them.

We will prove this by computing the cohomology of $BU(n)$, by induction on n. Here's how $BU(n)$ and $BU(n - 1)$ are related. Embed $U(n - 1) \hookrightarrow U(n)$ by

$$A \mapsto \begin{bmatrix} A & 0 \\ 0 & 1 \end{bmatrix}.$$

This subgroup is exactly the set of matrices fixing the last basis vector e_n in $\mathbb{C}^n$. The orbit of e_n under the defining action of $U(n)$ on $\mathbb{C}^n$ is the subspace

S^{2n-1} of unit vectors in $\mathbb{C}^n$, which is thus identified with the homogeneous space $U(n)/U(n-1)$.

Make a choice of $EU(n)$ — a contractible space on which $U(n)$ acts principally — the Stiefel model $V_n(\mathbb{C}^\infty)$ for example. The orbit space is then the Grassmann model for $BU(n)$. The subgroup $U(n-1)$ also acts principally on $EU(n)$, so we get a model for $BU(n-1)$:

$$
BU(n-1) = EU(n)/U(n-1) = (EU(n) \times_{U(n)} U(n))/U(n-1)
$$
$$
= EU(n) \times_{U(n)} (U(n)/U(n-1)) = EU(n) \times_{U(n)} S^{2n-1}.
$$

This establishes $p : BU(n-1) \to BU(n)$ as the unit sphere bundle in the universal complex n-plane bundle ξ_n. The map $BU(n-1) \to BU(n)$ classifies the n-plane bundle $\xi_{n-1} \oplus \epsilon$.

Here's a restatement of Theorem 72.1 in terms of universal examples.

Theorem 72.2. *There exist unique classes $c_i \in H^{2i}(BU(n))$ for $1 \le i \le n$ such that:*

(1) *the map $p_* : H^*(BU(n)) \to H^*(BU(n-1))$ sends*

$$
c_i \mapsto \begin{cases} c_i & \text{for} \quad i < n \\ 0 & \text{for} \quad i = n. \end{cases}
$$

(2) *the Euler class e of the oriented real $2n$-plane bundle underlying the universal complex n-plane bundle ξ_n is related to the top class c_n by the equation*

$$
c_n = (-1)^n e \in H^{2n}(BU(n)).
$$

Moreover,

$$
H^*(BU(n)) = \mathbb{Z}[c_1, \dots, c_n].
$$

We postpone to Lecture 74 the verification that the classes we constructed in the last lecture coincide with these.

Proof. We will study the Gysin sequence of the spherical fibration

$$
S^{2n-1} \to BU(n-1) \xrightarrow{p} BU(n).
$$

For a general oriented spherical fibration

$$
S^{2n-1} \to E \xrightarrow{p} B
$$

the Gysin sequence takes the form

$$\to H^{q-1}(E) \xrightarrow{p_*} H^{q-2n}(B) \xrightarrow{e\cdot} H^q(B) \xrightarrow{p^*} H^q(E) \xrightarrow{p_*} H^{q-2n+1}(B) \to$$

where $e \in H^{2n}(B)$ is the Euler class.

Suppose we know that $H^*(E)$ vanishes in odd dimensions. Then either the source or the target of each instance of the umkher map p_* is zero, so we receive a short exact sequence

$$0 \to H^{q-2n}(B) \xrightarrow{e\cdot} H^q(B) \xrightarrow{p^*} H^q(E) \to 0 \,.$$

This shows several things:

- $e \in H^{2n}(B)$ is a non-zero-divisor;
- p^* is surjective and induces an isomorphism $H^*(B)/(e) \to H^*(E)$;
- p^* is an isomorphism in dimensions less than $2n$;
- $H^q(B) = 0$ for q odd.

The last is clear for $q < 2n$, but feeding this into the leftmost term we find by induction that $H^q(B) = 0$ for all odd q.

Now let's suppose in addition that $H^*(E)$ is a polynomial algebra. Lift the generators to elements in $H^*(B)$. (If they all happen to lie in dimension less than $2n$, these lifts are unique.) Extending to a map of algebras gives a map $H^*(E) \to H^*(B)$. Further adjoining e gives us an algebra map

$$H^*(E)[e] \to H^*(B)$$

which when composed with p^* kills e and maps $H^*(E)$ by the identity. We claim this map is an isomorphism. To see this, filter both sides by powers of e. Modulo e this map is an isomorphism from what we observed above. On both sides, multiplication by e induces an isomorphism from one associated quotient to the next, so the map induces an isomorphism on associated graded modules. The five-lemma shows that it induces an isomorphism mod e^k for any k. But the powers of e increase in dimension, so we obtain an isomorphism in each dimension.

These observations provide the inductive step. All that remains is to start the induction. We can, if we like, use what we know about $H^*(\mathbb{CP}^\infty)$ and start with $n = 2$, though starting at $n = 1$ makes sense too, and provides another perspective on the computation of $H^*(\mathbb{CP}^\infty)$.

We define $c_n \in H^{2n}(BU(n))$ to be $(-1)^n e(\xi_n)$, also a generator. The choice of sign will make it agree with our earlier definition. $\qquad\square$

Once we verify that these classes coincide with the classes constructed in the last lecture, we will have available an important interpretation of the top Chern class: up to sign it is the Euler class of the underlying oriented real vector bundle.

The splitting principle

A wonderful fact about Chern classes is that it suffices to check relations among them on sums of line bundles. This is captured by the following theorem.

Theorem 72.3 (Splitting principle). *Let $\xi : E \downarrow X$ be a complex n-plane bundle. There exists a map $f : \mathrm{Fl}(\xi) \to X$ such that:*

(1) $f^*\xi \cong \lambda_1 \oplus \cdots \oplus \lambda_n$, where the λ_i are line bundles on $\mathrm{Fl}(\xi)$, and
(2) the map $f^* : H^*(X) \to H^*(\mathrm{Fl}(\xi))$ is monic.

Proof. We have already done the hard work, in our study of the projectivization $\pi : \mathbb{P}(\xi) \to X$. We found that the Serre spectral sequence collapses at E^2. This implies that the projection map induces a monomorphism in cohomology. We used the "tautologous" line bundle λ on $\mathbb{P}(\xi)$. The key additional point about this construction is that there is a canonical embedding $\lambda \hookrightarrow \pi^*\xi$ of vector bundles over $\mathbb{P}(\xi)$. A vector in $E(\lambda)$ is $(v \in L \subseteq \xi_x)$ (where L is a line in the fiber ξ_x). A vector in the pullback $\pi^*\xi$ is $(v \in \xi_x, L \subseteq \xi_x)$; $E(\lambda)$ is the subspace of elements such that $v \in L$.

By picking a metric on ξ we see that when pulled back to $\mathbb{P}(\xi)$ a line bundle splits off. Now just induct (using our important standing assumption that vector bundles have finite dimensional fibers).   □

It's worth being more explicit about what this "flag bundle" $\mathrm{Fl}(\xi)$ is. The complement of λ in $\pi^*\xi$ over $\mathbb{P}(\xi)$ is the the space of vectors of the form $(v \in L^\perp, L \subseteq \xi_x)$. If we iterated this construction, we will get, in the end, the space of ordered orthogonal decompositions of fibers into lines. This can be built as a balanced product. Let Fl_n be the space of "orthogonal flags," that is, decompositions of $\mathbb{C}^n$ into an ordered sequence of n 1-dimensional subspaces. There is an evident action of $U(n)$ on this space, and

$$\mathrm{Fl}(\xi) = P \times_{U(n)} \mathrm{Fl}_n$$

where $P \downarrow X$ is the principal $U(n)$ bundle associated to ξ (and a choice of Hermitian metric).

The action of $U(n)$ on Fl_n is transitive, and the isotropy subgroup of $(\mathbb{C}e_1, \ldots, \mathbb{C}e_n)$ is the subgroup of diagonal unitary matrices,

$$T^n = (S^1)^n \subseteq U(n) \,,$$

so

$$\mathrm{Fl}_n = U(n)/T^n \,.$$

In the universal case, over $BU(n)$,

$$\text{Fl}(\xi_n) = EU(n) \times_{U(n)} (U(n)/T^n) = EU(n)/T^n = BT^n$$

and this is just a product of n copies of $\mathbb{CP}^\infty$. So we have discovered that

$$H^*(BU(n)) \hookrightarrow H^*(BT^n) = \mathbb{Z}[t_1, \cdots, t_n]$$

where t_i is the Euler class of the line bundle $\text{pr}_i^* \lambda$, the pull back of the universal line bundle under the projection onto the ith factor of $\mathbb{CP}^\infty$. What is the image?

Well, the symmetric group Σ_n sits inside the unitary group as matrices with a single 1 in each column. The maximal torus T^n is sent to itself by conjugation by a permutation matrix, which has the effect of reordering the diagonal entries. In cohomology, the action permutes the generators. These permutation matrices also act by conjugation on all of $U(n)$, but there they act trivially on $H^*(BU(n))$ since any matrix is connected to the identity matrix by a path in $U(n)$. The consequence is that the image of $H^*(BU(n))$ lies in the symmetric invariants:

$$H^*(BU(n)) \hookrightarrow H^*(BT^n)^{\Sigma_n}.$$

These symmetric invariants are well-studied in Algebra! Define the *elementary symmetric polynomials* σ_i as the coefficients in the product of $t - t_i$'s:

$$\prod_{i=1}^{n}(t - t_i) = \sum_{j=0}^{n} \sigma_j t^{n-j}.$$

For example,

$$\sigma_0 = 1, \quad \sigma_1 = -\sum_{j=1}^{n} t_j, \quad \sigma_n = (-1)^n \prod_{j=1}^{n} t_j.$$

The theorem from algebra is that the elementary symmetric polynonomials are algebraically independent and generate the ring of symmetric invariants —

$$R[t_1, \ldots, t_n]^{\Sigma_n} = R[\sigma_1, \ldots, \sigma_n]$$

— over any coefficient ring R.

If we give each t_i a grading of 2, the elementary symmetric polynomials are homogeneous and $|\sigma_i| = 2i$.

So $H^*(BU(n))$ embeds into a graded algebra of exactly the same size. This does not yet show that the embedding is surjective! For each q, we know that $H^q(BU(n))$ embeds into $H^q(BT^n)^{\Sigma_n}$ as a subgroup of the same rank. If L is a free abelian group of finite rank and L' is a subgroup, the little exact sequence

$$0 \to \mathrm{Tor}_1(L/L', \mathbb{F}_p) \to L' \otimes \mathbb{F}_p \to L \otimes \mathbb{F}_p$$

shows that the p-torsion in L/L' vanishes if $L' \otimes \mathbb{F}_p \to L \otimes \mathbb{F}_p$ is injective. Now our argument above actually works for any coefficient ring, so $H^*(BU(n); \mathbb{F}_p) \to H^*(BT^n; \mathbb{F}_p)$ is monic for any prime p. Because $H^*(BU(n))$ is torsion free this says that $H^*(BU(n)) \otimes \mathbb{F}_p \to H^*(BT^n) \otimes \mathbb{F}_p$ is monic for any prime. The result is that the index of $H^*(BU(n))$ in $H^*(BT^n)^{\Sigma_n}$ is prime to p for every prime number p, and so this injection must also be surjective.

We have proven most of:

Theorem 72.4. *The inclusion $T^n \hookrightarrow U(n)$ induces an isomorphism*

$$H^*(BU(n)) \xrightarrow{\cong} H^*(BT^n)^{\Sigma_n}.$$

Under this identification, the classes c_i constructed in Theorem 72.2 map to the elementary symmetric functions.

In the context of Chern classes, the elements t_i are called "Chern roots." The extension $H^*(BU(n)) \hookrightarrow H^*(BT^n)$ adjoins the roots of the Chern polynomial

$$c(t) = t^n + c_1 t^{n-1} + \cdots + c_n.$$

Remark 72.5. Everything we have done admits a version for real vector bundles, with mod 2 coefficients. One point deserves some special attention: the argument we gave for why conjugation by a permutation induces the identity on $H^*(BU(n))$ fails because the group $O(n)$ is not path-connected. However, there is a better and more general argument available; see Exercise 58.7.

Exercises

Exercise 72.6. Let ξ be an real n-plane bundle over B. Show that if n is odd then $2e(\xi) = 0$ in $H^n(B)$. Give an example with n odd in which $e(\xi)$ is nevertheless nonzero.

73 Thom class and Whitney sum formula

We now have four perspectives on Chern classes:

(1) Axiomatic
(2) Grothendieck's definition in terms of $H^*(\mathbb{P}(\xi))$
(3) In terms of Euler classes
(4) As elementary symmetric polynomials via the splitting principle

In this lecture we will explain why these are four facets of the same gem, though at the expense of introducing a new perspective on the Euler class. Developing that perspective lets us introduce another important construction in topology, the Thom space. We'll use that to verify that (3) and (4) agree. Then we'll prove the Whitney sum formula from this perspective. We'll take the identification of Chern classes with symmetric polynomials as the starting point.

Thom space and Thom class

Let $\xi : E \xrightarrow{p} B$ be a real n-plane bundle. The *Thom space* is obtained by forming the one-point compactification of each fiber, and then identifying all the newly adjoined basepoints to a single point. If B is a compact Hausdorff space, this amounts to the one-point compactification of the total space $E(\xi)$.

Example 73.1. There is a canonical homeomorphism

$$\mathrm{Th}(\lambda^* \downarrow \mathbb{RP}^{n-1}) \to \mathbb{RP}^n .$$

It is given by sending $(\varphi \in L^*, L \subseteq \mathbb{R}^n)$ to the graph of φ in $\mathbb{R}^n \times \mathbb{R}$. This map embeds $E(\lambda^*)$ into $\mathbb{RP}^n$, and misses only the line $\mathbb{R}e_{n+1}$. This establishes $\mathbb{RP}^n$ as the one-point compactification of $E(\lambda^*)$. (It also shows that λ^* is the normal bundle of the linear embedding $\mathbb{RP}^{n-1} \hookrightarrow \mathbb{RP}^n$.)

By choosing a metric we get a different expression for the same space. Let $\mathbb{D}(\xi)$ and $\mathbb{S}(\xi) = \partial \mathbb{D}(\xi)$ denote the unit disk and unit sphere bundles. The Thom space of ξ is the quotient space

$$\mathrm{Th}(\xi) = \mathbb{D}(\xi)/\mathbb{S}(\xi) .$$

Rather than this quotient space, you may prefer to think of the pair $(\mathbb{D}(\xi), \mathbb{S}(\xi))$; it is homotopy equivalent to the pair $(E(\xi), E(\xi)\backslash Z)$, where Z is the image of the zero-section.

Note that $\mathrm{Th}(0) = B/\varnothing = B_+$, the base with a disjoint basepoint adjoined. The Thom space of the n-plane bundle over a point is $D^n/\partial D^n = S^n$.

An important point about the Thom space construction is its behavior on the product of two bundles, say ξ and η. Since

$$\partial(D^p \times D^q) = (\partial D^p \times D^q) \cup (D^p \times \partial D^q),$$

we find

$$\mathrm{Th}(\xi \times \eta) = \frac{\mathbb{D}(\xi \times \eta)}{\partial \mathbb{D}(\xi \times \eta)} = \frac{\mathbb{D}(\xi) \times \mathbb{D}(\eta)}{\mathbb{S}(\xi) \times \mathbb{D}(\eta) \cup \mathbb{D}(\xi) \times \mathbb{S}(\eta)} = \mathrm{Th}(\xi) \wedge \mathrm{Th}(\eta).$$

In particular, if η is the n-plane bundle over a point, $\xi \times \eta = \xi \oplus n\epsilon$ and

$$\mathrm{Th}(\xi \oplus n\epsilon) = \mathrm{Th}(\xi) \wedge S^n = \Sigma^n \mathrm{Th}(\xi).$$

In general, the Thom space is a "twisted n-fold suspension."

The Thom space construction is natural for bundle maps: Given $f : B' \to B$, covered by a bundle map $\xi' \to \xi$ (so that $\xi' \cong f^*\xi$) we get a canonical pointed map

$$\overline{f} : \mathrm{Th}(\xi') \to \mathrm{Th}(\xi).$$

This construction can be used to define a module structure on the cohomology of the Thom space, in the following way. Notice that the bundle $0 \times \xi$ over $B \times B$ is just the pullback of ξ under $\mathrm{pr}_2 : B \times B \to B$. The diagonal map $\Delta : B \to B \times B$ satisfies $\mathrm{pr}_2 \circ \Delta = 1_B$, and is therefore covered by a bundle map $\xi \to 0 \times \xi$, which then induces a twisted diagonal map

$$\mathrm{Th}(\xi) \to \mathrm{Th}(0) \wedge \mathrm{Th}(\xi) = B_+ \wedge \mathrm{Th}(\xi).$$

This in turn induces a "relative cup product" in cohomology:

$$\cup : H^*(B) \otimes \overline{H}^*(\mathrm{Th}(\xi)) \to \overline{H}^*(\mathrm{Th}(0) \wedge \mathrm{Th}(\xi)) \to \overline{H}^*(\mathrm{Th}(\xi)).$$

Since the diagonal map is associative and unital, this map defines on $\overline{H}^*(\mathrm{Th}(\xi))$ the structure of a module over the graded ring $H^*(B)$.

Here is the essential fact about the Thom space.

Proposition 73.2 (Thom isomorphism theorem). *Let R be a commutative ring and let ξ be an R-oriented real n-plane bundle over B. There is a unique class $U \in \overline{H}^n(\mathrm{Th}(\xi); R)$ that restricts on each fiber to the dual of the orientation class, and the map*

$$- \cup U : H^*(B) \to \overline{H}^*(\mathrm{Th}(\xi))$$

is an isomorphism.

Proof. The proof is very simple, if you grant yet another relative form of the Serre spectral sequence. This time I want to have a fibration $p : E \to B$ — say a fiber bundle — together with a subbundle $p_0 : E_0 \to B$. Then there is spectral sequence

$$E_2^{s,t} = H^s(B; H^t(p^{-1}(-), p_0^{-1}(-))) \underset{s}{\Longrightarrow} H^{s+t}(E, E_0).$$

It is a module over the multiplicative spectral sequence

$$E_2^{s,t} = H^s(B; H^t(p^{-1}(-))) \underset{s}{\Longrightarrow} H^{s+t}(E).$$

We will apply this to the fiber bundle pair $(\mathbb{D}(\xi), \mathbb{S}(\xi))$. The fiber pair is then (D^n, S^{n-1}), which has cohomology in just one dimension! This spectral sequence has just one row: the nth row. It collapses at E_2, there are no extension problems, and we get a canonical isomorphism

$$H^s(B; H^n(p^{-1}(-), p_0^{-1}(-))) \to H^{s+n}(\mathbb{D}(\xi), \mathbb{S}(\xi)) = \overline{H}^{s+n}(\mathrm{Th}(\xi)).$$

The assumed orientation identifies the local coefficient system with the constant system R. The generator of $E_2^{0,n}$ survives to a class U that restricts as stated, and the multiplicative structure of the spectral sequence implies that this is an isomorphism of modules over $H^*(B)$. $\square$

Thom and Euler

We now use this construction to define a new class in $H^n(B)$ associated to the oriented n-plane bundle ξ, by means of the composite

$$\pi : B \to \mathbb{D}(\xi) \to \mathrm{Th}(\xi).$$

The first map is the zero-section, homotopy inverse to the projection map. The second one is the collapse map. The Thom class $U \in \overline{H}^n(\mathrm{Th}(\xi))$ pulls back under this map to a class in $\overline{H}^n(B)$.

This class is at least up to sign the Euler class as we defined it earlier:

Lemma 73.3. *This class coincides up to sign with the Euler class:* $\pi^*U = \pm e$.

Proof. We will verify that they generate the same submodule. If the coefficient ring is of characteristic 2, that gives the result. Otherwise, these classes are both in the image of integral classes, and that again gives the result.

Work in the universal case. As a notational choice, we will work over $\mathbb{Z}$, so we are looking at ξ_n over $BSO(n)$. We've seen that the total space of its sphere bundle is $BSO(n-1)$. The Serre spectral sequence for this fibration shows that the kernel of the projection map

$p^* : H^n(BSO(n-1)) \to H^n(BSO(n))$ is the image of the transgression $H^{n-1}(S^{n-1}) \to H^n(BSO(n-1))$. So the kernel is cyclic and generated by the Euler class. On the other hand, we have the cofibration sequence

$$BSO(n-1) \to BSO(n) \xrightarrow{\pi} MSO(n),$$

where we are using Thom's notation $MSO(n) = \mathrm{Th}(\xi_n)$. The Thom class $U \in H^n(MSO(n))$ generates this group (by the Thom isomorphism theorem) so its image in $H^n(BSO(n))$ also generates $\ker(H^n(BSO(n)) \to H^n(BSO(n-1)))$.

We will see, as a consequence of a computation of $H^*(BSO(n); \mathbb{Z}[1/2])$ in Theorem 74.5 that this kernel is infinite cyclic if n is even, so then the generator is at least well defined up to sign. For homework you will show that $2e = 0$ if n is even, so the generator is then unique.   □

But in fact, it's better just to take $\pi^* U$ as the *definition* of the Euler class. With that definition, we get a new construction of the Gysin sequence: It's the long exact cohomology sequence of the pair $(\mathrm{Th}(\xi), B)$, aided by the Thom isomorphism:

$$\longrightarrow H^{s-1}(B) \xrightarrow{p^*} H^{s-1}(E) \xrightarrow{\delta} \overline{H}^s(\mathrm{Th}(\xi)) \xrightarrow{\pi^*} H^s(B) \xrightarrow{p^*} H^s(E) \longrightarrow$$

with p_*, $\cong$, $\cdot e$ and $H^{s-n}(B)$.

This is a long exact sequence of modules over $H^*(B)$. This gives a different perspective on integration along the fiber:

$$(p_* x) \cup U = \delta x.$$

We'll just use this definition going forward. Notice that with this definition, the Euler class is multiplicative for Whitney sum. We should be careful about orientations. The direct sum of oriented vector spaces V and W has an orientation given by putting a positive ordered basis for V first and follow it by a positive ordered basis for W. This convention orients the Whitney sum of two vector bundles over a space.

Proposition 73.4. *Let ξ and η be oriented vector bundles over spaces X and Y.*

$$e(\xi \times \eta) = e(\xi) \times e(\eta).$$

Proof. First, $U_\xi \wedge U_\eta \in \overline{H}^{p+q}(\text{Th}(\xi) \wedge \text{Th}(\eta))$ is a Thom class for $\xi \times \eta$, since its restriction to each fiber is dual to the direct sum orientation. Then the collapse maps are compatible:

$$
\begin{array}{ccc}
\text{Th}(\xi \times \eta) & \xrightarrow{\;=\;} & \text{Th}(\xi) \wedge \text{Th}(\eta) \\
\Big\uparrow{\scriptstyle \pi_{X \times Y}} & & \Big\uparrow{\scriptstyle \pi_X \wedge \pi_Y} \\
(X \times Y)_+ & \xrightarrow{\;=\;} & X_+ \wedge Y_+
\end{array}
$$

commutes, and in cohomology we chase

$$
\begin{array}{ccc}
U_\xi \wedge U_\eta & \longmapsto & U_{\xi \times \eta} \\
\Big\downarrow & & \Big\downarrow \\
e(\xi) \times e(\eta) & \longmapsto & e(\xi \times \eta)
\end{array}
$$

to see the result. $\square$

If we take $X = Y$ here and pull back along the diagonal, $\xi \times \eta$ goes to the Whitney sum and $e(\xi) \times e(\eta)$ goes to the cup-product:

$$e(\xi \oplus \eta) = e(\xi) \cdot e(\eta).$$

Euler class and symmetric polynomials

One of our descriptions of the Chern classes was this: If an n-plane bundle ξ splits $\zeta \oplus (n - k)\epsilon$, then $c_k(\xi) = (-1)^k e(\zeta)$. Let's check that this holds for the classes defined by means of elementary symmetric functions. It might be clearest if we look at the universal example, where the splitting map $f : BT^n \to BU(n)$ pulls ξ_n back to the direct sum of line bundles $\lambda_1 \oplus \cdots \oplus \lambda_n$ and induces an isomorphism $f^* : H^*(BU(n)) \to H^*(BT^n)^{\Sigma_n}$. Let's do the case $k = n$ first, so I want to show that $(-1)^n e(\xi_n)$ maps to σ_n. Using multiplicativity of the Euler class,

$$f^* e(\xi_n) = e(\lambda_1 \oplus \cdots \oplus \lambda_n) = e(\lambda_1) \cdots e(\lambda_n).$$

With the notation $t_i = e(\lambda_i)$, this shows that

$$f^*((-1)^n e(\xi_n)) = (-1)^n t_1 \cdots t_n = \sigma_n.$$

For smaller k, we'll use the fact that the maximal tori $T^k \subseteq U(k)$ are compatible as k increases. This gives the commutative diagram

$$
\begin{array}{ccc}
H^{2k}(BU(n)) & \longrightarrow & H^{2k}(BT^n)^{\Sigma_n} \\
\Big\downarrow & & \Big\downarrow \\
H^{2k}(BU(k)) & \longrightarrow & H^{2k}(BT^k)^{\Sigma_k}.
\end{array}
$$

The elementary symmetric polynomial definition of c_k specifies that it maps to σ_k along the top arrow. We want to see that this class maps to $(-1)^k e(\xi_k) \in H^{2k}(BU(k))$. Well, by the $k = n$ case that we just did, we know that that class maps to σ_k along the bottom. So what remains is to check that $\sigma_k \in H^{2k}(BT^n)^{\Sigma_n}$ maps to the class of the same name in $H^{2k}(BT^k)^{\Sigma_k}$.

To keep things straight, let's write $\sigma_k^{(n)}$ for the first class and $\sigma_k^{(k)}$ for the second. The restriction $H^*(BT^n) \to H^*(BT^k)$ sends t_i to t_i if $i \leq k$ and to 0 if $i > k$. So

$$\sum_{i=0}^n \sigma_i^{(n)} t^{n-i} \quad = \quad \prod_{j=1}^n (t - t_j)$$

$$\downarrow$$

$$\left(\sum_{i=0}^k \sigma_i^{(k)} t^{k-i} \right) t^{n-k} \quad = \quad \left(\prod_{j=1}^k (t - t_j) \right) t^{n-k},$$

and comparing coefficients we see that $\sigma_k^{(n)} \mapsto \sigma_k^{(k)}$.

The Whitney sum formula

By our discussion above, the Whitney sum formula of Theorem 71.3 reduces to proving the following identity:

$$\sigma_k^{(p+q)} = \sum_{i+j=k} \sigma_i^{(p)} \cdot \sigma_j^{(q)}$$

inside $\mathbb{Z}[t_1, \ldots, t_p, t_{p+1}, \ldots, t_{p+q}]$. Here, $\sigma_i^{(p)}$ is thought of as a polynomial in $t_1, \ldots, t_p$, while $\sigma_j^{(q)}$ is thought of as a polynomial in $t_{p+1}, \ldots, t_{p+q}$. To derive this equation, simply compare coefficients in the following:

$$\sum_{k=0}^{p+q} \sigma_k^{(p+q)} t^{p+q-k} = \prod_{i=1}^{p+q} (t - t_i)$$

$$= \prod_{i=1}^p (t - t_i) \cdot \prod_{j=p+1}^{p+q} (t - t_j)$$

$$= \left(\sum_{i=0}^p \sigma_i^{(p)} t^{p-i} \right) \left(\sum_{j=0}^q \sigma_j^{(q)} t^{q-j} \right)$$

$$= \sum_{k=0}^{p+q} \left(\sum_{i+j=k} \sigma_i^{(p)} \sigma_j^{(q)} \right) t^{p+q-k}.$$

Hassler Whitney once called this his hardest theorem. Apparently he didn't have the splitting principle working for him.

Exercises

Exercise 73.5. Let ξ be a complex n-plane bundle and λ a complex line bundle, both over a space B. Find an explicit expression for the Chern classes of $\lambda \otimes \xi$ in terms of the Chern classes of ξ and the Euler class of λ.

74 Closing the Chern circle, and Pontryagin classes

Back to Grothendieck

Now we'll use the splitting principle to show that the Chern classes (defined as corresponding to the elementary symmetric polynomials) participate in a monic polynomial satisfied by the Euler class of the tautologous bundle over the projectivization of a vector bundle. This will complete the identification of the various versions of Chern classes.

So we have an n-plane bundle ξ over B, and consider the projectivization $\pi : \mathbb{P}(\xi) \to B$. We observed in the last lecture that the tautological bundle λ embeds (canonically) into the pullback $\pi^*\xi$. Let $\overline{\lambda}$ denote the complex conjugate or inverse line bundle, so that $\overline{\lambda} \otimes \lambda = \epsilon$. Tensoring $\pi^*\xi$ with $\overline{\lambda}$ thus results in a bundle with a trivial summand; that is, with a nowhere vanishing section. Its Euler class therefore vanishes. We will compute what that Euler class is, using the splitting principle.

The splitting principle allows us to assume that ξ is a sum of line bundles, say $\xi = \lambda_1 \oplus \cdots \oplus \lambda_n$. Then

$$\overline{\lambda} \otimes \pi^*\xi = \bigoplus_{i=1}^{n} \overline{\lambda} \otimes \pi^*\lambda_i \, .$$

By multiplicativity of the Euler class, we find

$$e(\overline{\lambda} \otimes \pi^*\xi) = \prod_{i=1}^{n} e(\overline{\lambda} \otimes \pi^*\lambda_i) \, .$$

Write t for $e(\lambda) \in H^2(\mathbb{P}(\xi))$, so that $e(\overline{\lambda}) = -t$. Also write $t_i = e(\lambda_i)$, so that

$$e(\overline{\lambda} \otimes \pi^*\lambda_i) = \pi^*t_i - t$$

and

$$e(\overline{\lambda} \otimes \pi^*\xi) = \prod_{i=1}^{n}(\pi^*t_i - t) = (-1)^n \sum_{j=0}^{n}(\pi^*c_j(\xi))t^{n-j} \, .$$

Since $e(\overline{\lambda} \otimes \pi^*\xi) = 0$, this shows that our new Chern classes satisfy the identity Grothendieck used to define them. Since these coefficients were unique, this identifies Grothendieck's definition with the others we have introduced.

Stiefel-Whitney classes

Same story! Well, almost. We don't have the even/odd argument working for us anymore. We want to know that the Euler class is a non-zero-divisor. We do have the splitting principle, which assures us that

$$f^* : H*(BO(n); \mathbb{F}_2) \hookrightarrow H^*(BC_2^n; \mathbb{F}_2)^{\Sigma_n}.$$

By multiplicativity of the Euler class, it maps to $t_1 \cdots t_n \in H^n(BC_2^n; \mathbb{F}_2)$, which is nonzero in this integral domain and so is a non-zero-divisor. The result:

Proposition 74.1. $H^*(BO(n); \mathbb{F}_2) = \mathbb{F}_2[w_1, \ldots, w_n]$.

While we are talking about Stiefel-Whitney classes, let me point out that $w_1 \in H^1(B; \mathbb{F}_2)$ is precisely the obstruction to orientability of $\xi : E \downarrow B$. If B is path-connected, it can be identified with the homomorphism $\pi_1(B) \to C_2$ that takes on the value -1 on σ if the orientation of the fiber is reversed under the homotopy endomorphism of the fiber given by σ. You can check this in the universal case: The class $w_1 \in H^1(BO(n); \mathbb{F}_2)$ is represented by a map $BO(n) \to K(\mathbb{F}_2, 1)$. This map is the bottom Postnikov stage of $BO(n)$, and its homotopy fiber is the simply connected Whitehead cover of $BO(n)$. We know what that is, since $SO(n) \hookrightarrow O(n)$ is the connected component of the identity (and is the kernel of $\det : O(n) \to C_2$).

The map $BSO(n) \to BO(n)$ is (at least homotopy theoretically) a double cover; the fiber is S^0, so we are entitled to a Gysin sequence. The Euler class of this spherical fibration is exactly w_1, a non-zero-divisor, so we discover the short exact sequence

$$0 \to H^*(BO(n); \mathbb{F}_2) \xrightarrow{e \cdot} H^*(BO(n); \mathbb{F}_2) \to H^*(BSO(n); \mathbb{F}_2) \to 0.$$

This shows that $H^*(BSO(n); \mathbb{F}_2)$ is the polynomial algebra on the images of $w_2, \ldots, w_n$:

$$H^*(BSO(n); \mathbb{F}_2) = \mathbb{F}_2[w_2, \ldots, w_n].$$

It often happens that one cares about only the "stable" equivalence class of a vector bundle. This leads one to consider the direct limit or union

$$BO = \lim_{n \to \infty} BO(n).$$

Its cohomology is given by

$$H^*(BO) = \mathbb{F}_2[w_1, w_2, \ldots].$$

Of course the limit of the $BSO(n)$'s is written BSO. It is the simply-connected cover of BO. It's interesting to contemplate the rest of the Whitehead tower of BO. For a while the spaces involved have names:

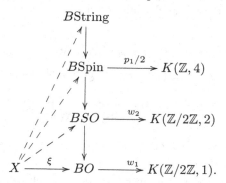

Pontryagin classes

Real vector bundles have integral characteristic classes too! They were studied by Lev Pontryagin (1908–1988, Steklov Institute, blinded in a stove accident at age 14). The idea is to use Chern classes to define such things. Given a real vector bundle ξ we can tensor up to the complex vector bundle $\mathbb{C} \otimes_{\mathbb{R}} \xi$, and study its Chern classes.

Complex vector bundles arising in this way have some additional structure. Any complex vector bundle $\zeta : E \downarrow B$ has a "complex conjugate" vector bundle $\overline{\zeta}$ with the same underlying real vector bundle but with complex structure defined by letting $z \in \mathbb{C}$ act on $\overline{\zeta}$ the way $\overline{z}$ acted on ζ. We've already seen this construction for line bundles, when $\lambda \otimes \overline{\lambda} = \epsilon$.

The complexification $\mathbb{C} \otimes_{\mathbb{R}} \xi$ of a real vector bundle comes equipped with an isomorphism

$$\mathbb{C} \otimes_{\mathbb{R}} \xi \cong \overline{\mathbb{C} \otimes_{\mathbb{R}} \xi}$$

given by $z \otimes v \mapsto \overline{z} \otimes v$. We discover that

$$c_i(\mathbb{C} \otimes_{\mathbb{R}} \xi) = c_i(\overline{\mathbb{C} \otimes_{\mathbb{R}} \xi}),$$

so we should ask: What are the Chern classes of the complex conjugate of a complex vector bundle?

Lemma 74.2. $c_i(\overline{\xi}) = (-1)^i c_i(\xi)$.

Proof. Exercise; use any one of the perspectives on Chern classes that we have developed. $\qquad\square$

This puts no restriction on $c_i(\mathbb{C} \otimes_\mathbb{R} \xi)$ for i even, but forces $2c_i(\mathbb{C} \otimes_\mathbb{R} \xi) = 0$ for i odd. The 2-torsion will get in the way, so let's work with coefficients in a ring R in which 2 is invertible — a $\mathbb{Z}[1/2]$-algebra, such as $\mathbb{Q}$ or $\mathbb{F}_p$ for $p \neq 2$. We already have Stiefel-Whitney classes with mod 2 coefficients, so this is not so bad.

Definition 74.3. The kth *Pontryagin class* of a real vector bundle ξ is

$$p_k(\xi) = (-1)^k c_{2k}(\mathbb{C} \otimes_\mathbb{R} \xi) \in H^{4k}(X; R)\dot{.}$$

Of course $p_k(\xi) = 0$ if $k > n/2$, since $\xi \otimes \mathbb{C}$ is of complex dimension n. The strange sign does not interfere with the Whitney sum formula:

$$p_k(\xi \oplus \eta) = (-1)^k \sum_{i+j=k} c_{2i}(\mathbb{C} \otimes_\mathbb{R} \xi)c_{2j}(\mathbb{C} \otimes_\mathbb{R} \eta) = \sum_{i+j=k} p_i(\xi)p_j(\eta)$$

since the odd terms contribute only 2-torsion, which we have eliminated by working over a $\mathbb{Z}[1/2]$-algebra.

The Pontryagin classes are defined for vector bundles, orientable or not. They are independent of the orientation if there is one. But an oriented $2k$-plane bundle over B has an Euler class $e(\xi) \in H^{2k}(B)$ as well, and we might ask how it is related to the Pontryagin classes. The sign is there in the definition of the Pontryagin classes so that the following important relation is satisfied.

Lemma 74.4. *For any oriented $2k$-plane bundle, $p_k(\xi) = e(\xi)^2$.*

Proof. We need to be careful about orientations. We have the isomorphism of *real* vector bundles

$$\xi \oplus \xi \xrightarrow{\cong} \mathbb{C} \otimes_\mathbb{R} \xi,$$

defined $(v, w) \mapsto v + iw$. We have establishes an orientation on $\mathbb{C} \otimes_\mathbb{R} \xi$. But suppose that ξ itself came equipped with an orientation. This puts an orientation on the direct sum. How are the two orientations related to each other? If $e_1, \ldots, e_n$ is a positive basis for an oriented vector space V, then we are comparing the ordered bases

$$e_1, e_2, \ldots, e_n, ie_1, ie_2, \ldots, ie_n \quad \text{for} \quad V \oplus V \quad \text{and}$$
$$e_1, ie_1, e_2, ie_2, \ldots, e_n, ie_n \quad \text{for} \quad \mathbb{C} \otimes_\mathbb{R} V.$$

Relating them requires

$$(n-1) + (n-2) + \cdots + 1 = \frac{n(n-1)}{2}$$

transpositions, so they give the same orientation if this number is even and opposite orientations if it is odd.

Now we can compute:

$$p_k(\xi) = (-1)^k c_{2k}(\mathbb{C} \otimes_{\mathbb{R}} \xi) = (-1)^k e(\mathbb{C} \otimes_{\mathbb{R}} \xi)$$
$$= (-1)^k (-1)^{2k(2k-1)/2} e(\xi \oplus \xi) = e(\xi)^2$$

since $2k(2k-1)/2 \equiv k \mod 2$. □

We can now systematically compute the cohomology of $BSO(n)$ away from 2 by induction on n using the Gysin sequence. Here's the result.

Theorem 74.5. *With coefficients in any $\mathbb{Z}[1/2]$-algebra, the cohomology of $BSO(n)$ is polynomial for all n. When $n = 2k + 1$, the generators are $p_1, \ldots, p_k$. When $n = 2k$, the generators are $p_1, \ldots, p_{k-1}, e_n$. The maps $H^*(BSO(n)) \to H^*(BSO(n-1))$ kill the Euler class and take Pontryagin classes to themselves except that $H^{4k}(BSO(2k+1)) \to H^{4k}(BSO(2k))$ sends p_k to e_{2k}^2.*

Here's a table of the algebra generators, with the squares of the Euler classes added in to indicate how p_k restricts.

	2	4	6	8	10	12
$H^*(BSO(2))$	e_2	(e_2^2)				
$H^*(BSO(3))$		p_1				
$H^*(BSO(4))$		p_1, e_4		(e_4^2)		
$H^*(BSO(5))$		p_1		p_2		
$H^*(BSO(6))$		p_1	e_6	p_2		(e_6^2)
$H^*(BSO(7))$		p_1		p_2		p_3

We can then compute $H^*(BO(n); R)$ for R a $\mathbb{Z}[1/2]$-algebra by using the fiber sequence

$$BSO(n) \to BO(n) \to \mathbb{RP}^\infty .$$

The spectral sequence has $E_2^{s,t} = H^s(\mathbb{RP}^\infty; H^t(BSO(n)))$. There are local coefficients here, but with any local coefficients the higher cohomology of $\mathbb{RP}^\infty$ is killed by 2 and so vanishes for us. As a result the edge homomorphism

$$H^*(BO(n); R) \to H^*(BSO(n); R)^{C_2}$$

is an isomorphism. The generator of $\pi_1(\mathbb{RP}^\infty)$ tracks the effect of reversing orientations: it fixes the Pontryagin classes and negates the Euler classes. The result is that

$$H^*(BO(2k); R) \xleftarrow{\cong} H^*(BO(2k + 1); R) \xrightarrow{\cong} H^*(BSO(2k + 1); R)$$

and all are given by

$$R[p_1, \ldots, p_k].$$

Exercises

Exercise 74.6. Determine the Stiefel-Whitney classes of the real bundle underlying a complex vector bundle, in terms of the mod 2 reductions of the Chern classes.

Exercise 74.7. (a) Embed C_2 into $O(n)$ as the scalar multiples of the identity matrix. This is a central subgroup, so the translation action $C_2 \times O(n) \to O(n)$ is a group homomorphism. Compute the effect of the induced map $BC_2 \times BO(n) \to BO(n)$ in mod 2 cohomology.

(b) Show that this action map restricts to an isomorphism $C_2 \times SO(n) \to O(n)$ if n is odd. Describe the inverse homomorphism. Describe the resulting isomorphism in mod 2 cohomology.

Exercise 74.8. (a) Exhibit a real 2-plane bundle that is nontrivial but whose Stiefel-Whitney classes vanish.

(b) Exhibit a complex 2-plane bundle that is nontrivial but whose Chern classes vanish.

Exercise 74.9. Express the mod 2 reduction of $c_k(\mathbb{C} \otimes \xi)$ as a polynomial in the Stiefel-Whitney classes of the real vector bundle ξ.

Exercise 74.10. Observe that if V is a complex vector space then $\mathbb{C} \otimes V = V \oplus \overline{V}$. If M is an "almost complex manifold" — endowed with a complex structure on its tangent bundle — give an expression for the Pontryagin classes of M in terms of its Chern roots (Theorem 72.4). In particular, determine the Pontryagin classes of $\mathbb{CP}^n$.

Exercise 74.11. Show that

$$H^*(BO(2); \mathbb{Z}) = \mathbb{Z}[x, y, p_1]/(2x, 2y, y^2 = xp_1), \quad |x| = 2, |y| = 3, |p_1| = 4.$$

Hint: Compare the spectral sequences for the fibration sequence $BSO(2) \to BO(2) \to BC_2$, with integer coefficients and with mod 2 coefficients. This will require the use of twisted coefficients.

Exercise 74.12. Follow Armand Borel's computation [8] of the mod 2 cohomology of flag varieties. Let $p_1 + \cdots + p_k = n$ be an ordered partition of the positive integer n, and embed $O(p_1) \times \cdots \times O(p_k)$ into $O(n)$ as a block diagonal subgroup. Write $F(p_1, \ldots, p_n)$ for the homogeneous space; this is the space of flags of type $(p_1, \ldots, p_k)$. For example $\mathbb{RP}^{n-1} = F(n-1,1)$, and $\mathrm{Gr}_p(\mathbb{R}^n) = F(p,q)$ with $p + q = n$. Write F_n for $F(1, \ldots, 1) = O(n)/Q(n)$, the quotient of $O(n)$ by its diagonal subgroup $Q(n) = O(1)^n = C_2^n$. This is the space of ordered sets of n mutually orthogonal lines in $\mathbb{R}^n$.

(a) By studying the fibration sequence $F(n) \to BQ(n) \to BO(n)$, show that $\dim H^1(F(n)) \geq n - 1$.

(b) Construct fibration sequences $F_{n-1} \to F_n \to \mathbb{RP}^{n-1}$, and use them to prove, inductively, that $\dim H^1(F_n) = n-1$ and that $H^*(F_n)$ is generated as an algebra by its 1-dimensional classes. Conclude that the spectral sequence from **(a)** collapses, and $H^*(BQ(n)) \to H^*(F_n)$ is surjective; and in fact that $H^*(F_n) = \mathbb{F}_2[t_1, \ldots, t_n]/(\sigma_1, \ldots, \sigma_n)$.

(c) Construct a fibration sequence (in which $p + q = n$) $F_p \times F_q \to F_n \to F(p,q)$ and use it to compute the Poincaré series $\sum_i \dim H^i(F(p,q))t^i$.

(d) Use the result of **(c)** to show that the spectral sequence associated to the fibration sequence $F(p,q) \to BO(p) \times BO(q) \to BO(n)$ collapses at E_2. Use this to determine the algebra structure for $H^*(F(p,q))$.

(e) Generalize this result to $F(p_1, \ldots, p_k)$.

Armand Borel (1923–2003) was a Swiss mathematician, student of Leray, working mainly at the Institute for Advanced Study, principally on algebraic groups. His early work, along with that of his contemporary Jean-Pierre Serre, served to introduce spectral sequences to a broad audience.

75 Steenrod operations

We worked hard to show that mod 2 cohomology takes values not just in graded $\mathbb{F}_2$-vector spaces, but actually in graded commutative $\mathbb{F}_2$-algebras. This additional structure has proven extremely useful. What other natural structure is there on mod 2 cohomology? Both the sum and the cup product are natural operations on two variables. The identity element $1 \in H^0$ is in a sense a natural operation on zero variables (and is the only nonzero natural element in mod 2 cohomology). This invites the question: are there nontrivial natural operations in one variable? Some of course are generated from the product: $x \mapsto x^r$, for example. When r is a power of 2, this is an *additive* operation. We know one other additive operation as well: the *Bockstein*,

$$\beta : H^n(X) \to H^{n+1}(X).$$

(All our coefficients will be in $\mathbb{F}_2$ in this lecture.) This is obtained as the boundary map in the long exact sequence associated to the short exact sequence $0 \to C_2 \to C_4 \to C_2 \to 0$.

Our goal in this lecture is to establish the following theorem, due to Norman Steenrod.

Theorem 75.1. *There is a unique family of additive natural transformations*

$$\mathrm{Sq}^k : H^n \to H^{n+k}, \quad n, k \geq 0,$$

such that

$$\mathrm{Sq}^0 x = x, \quad \mathrm{Sq}^k(x) = x^2 \text{ if } k = |x|, \quad \mathrm{Sq}^k x = 0 \text{ if } k > |x|,$$

and the "Cartan formula"

$$\mathrm{Sq}^k(xy) = \sum_{i+j=k} (\mathrm{Sq}^i x)(\mathrm{Sq}^j y)$$

is satisfied.

It will transpire that $\mathrm{Sq}^1 = \beta$.

By the Yoneda lemma, natural transformations $H^n \to H^{n+k}$ are classified by $H^{n+k}(K_n)$, where we write

$$K_n = K(\mathbb{F}_2, n).$$

We won't try to compute the whole of $H^*(K_n)$, at least not right away, though eventually it will turn out that the entire cohomology of a mod 2 Eilenberg Mac Lane space is generated as an algebra by iterates of the operations we will construct. But at least we can notice right off that

$$H^i(K_n) = 0 \quad \text{for} \quad 0 < i < n$$

and

$$H^n(K_n) = \mathbb{F}_2 \quad \text{for} \quad n > 0$$

by the Hurewicz theorem, so the only nonzero operation on n-dimensional classes that lowers degrees is the one sending every x to $1 \in H^0$.

The starting point is the failure of the Alexander-Whitney map (Lecture 28)

$$S_*(X \times Y) \to S_*(X) \otimes S_*(Y)$$

— or *any* natural chain map lifting the natural map $H_0(X \times Y) \to H_0(X) \otimes H_0(\times Y)$ — to be commutative, even with mod 2 coefficients. This failure reflects itself geometrically using the following construction.

Definition 75.2. The *extended square* of a space X is the balanced product

$$S^\infty \times_{C_2} X^2.$$

Here C_2 acts antipodally on S^∞, and swaps the factors in X^2.

This is the total space of the bundle with fiber X^2 associated to the universal principal C_2 bundle $S^\infty \downarrow \mathbb{RP}^\infty$. We will study it by means of the Serre spectral sequence.

Actually, it will be important to consider a pointed refinement of this. So suppose given a basepoint $* \in X$. It determines the subset

$$X \vee X \subseteq X \times X$$

consisting of the "axes" in the product. The *pair* $(X^2, X \vee X)$ is equivariant, and determines a bundle pair

$$S^\infty \times_{C_2} (X^2, X \vee X) \downarrow \mathbb{RP}^\infty.$$

A point in S^∞ determines a fiber inclusion

$$i : (X^2, X \vee X) \to S^\infty \times_{C_2} (X^2, X \vee X).$$

We'll be working with the cohomology Künneth theorem, so let's restrict ourselves to spaces whose mod 2 cohomology is of finite type. (We'll also suppose that the spaces are well-pointed.) Serre's mod $\mathcal{C}$ theory guarantees that K_n is in this category, and the Künneth theorem guarantees that the category is closed under products.

Proposition 75.3. *There is a unique natural transformation*

$$P : \overline{H}^n(X) \to H^{2n}(S^\infty \times_{C_2} (X^2, X \vee X))$$

such that

$$i^* P(x) = x^{\otimes 2} \in H^{2n}(X^2, X \vee X).$$

Proof. We'll study the associated Serre spectral sequence,

$$H^s(\mathbb{RP}^\infty; H^t(X^2, X \vee X)) \underset{s}{\Longrightarrow} H^{s+t}(S^\infty \times_{C_2} (X^2, X \vee X)).$$

While the chain-level cross product isn't equivariant, the cohomology cross product is: The cross relative product map (Exercise 33.4)

$$\overline{H}^*(X) \otimes \overline{H}^*(X) \to H^*(X^2, X \vee X)$$

is equivariant, if we let C_2 act by exchanging factors on the left and on the right. This map is an isomorphism if $H^*(X)$ is of finite type, and then the $\mathbb{F}_2[C_2]$-module featuring as coefficients in the spectral sequence can be written as $\overline{H}^*(X)^{\otimes 2}$. It's interesting and not hard to analyze this representation of C_2, but we do not need to know about that to construct Steenrod operations. All we need to know is that any $x \in \overline{H}^n(X)$ determines an invariant class $x \otimes x \in \overline{H}^n(X)^{\otimes 2}$.

Now comes the trick: It suffices to consider the universal example, $\iota_n \in \overline{H}^n(K_n)$. Since $\overline{H}^i(K_n) = 0$ for $i < n$, the entire E_2 term of

$$H^s(\mathbb{RP}^\infty; H^t(K_n^2, K_n \vee K_n)) \Longrightarrow H^*(S^\infty \times_{C_2} (K_n^2, K_n \vee K_n))$$

lies in vertical dimensions $t \geq 2n$.

So the group

$$E_2^{0,2n} = H^{2n}(K_n^2, K_n \vee K_n) = \langle \iota_n \otimes \iota_n \rangle$$

survives to $E_\infty^{0,2n}$. The element $\iota_n \otimes \iota_n$ lifts to an element of $H^{2n}(S^\infty \times_{C_2} (K_n^2, K_n \vee K_n))$, and this lift is unique because all the lower filtration degrees vanish. This lifted class is $P\iota_n$. By definition (and the edge homomorphism story) it restricts on $(K_n^2, K_n \vee K_n)$ to $\iota_n \otimes \iota_n$. $\qquad\square$

The resulting natural transformation $P : H^n(X) \to H^n(S^\infty \times_{C_2} (X^2, X \vee X))$ is the "total square." It's a prime example of a "power operation."

Now we "internalize," by pulling back under the diagonal map. The commutativity of the diagonal map becomes important:

$$\Delta : X \to X \times X$$

is equivariant, where C_2 acts trivially on X and by swapping the factors in $X \times X$. It induces a map

$$S^\infty \times_{C_2} (X, *) \to S^\infty \times_{C_2} (X^2, X \vee X).$$

But

$$S^\infty \times_{C_2} (X, *) = \mathbb{RP}^\infty \times (X, *)$$

so we have

$$\delta : \mathbb{RP}^\infty \times (X, *) \to S^\infty \times_{C_2} (X^2, X \vee X).$$

Pick $x \in \overline{H}^n(X)$ and consider the pullback $\delta^* P(x)$. By the Künneth theorem,

$$H^*(\mathbb{RP}^\infty \times (X, *)) = H^*(\mathbb{RP}^\infty) \otimes \overline{H}^*(X)$$

so $\delta^* P(x)$ has an expression as a polynomial in the generator $t \in H^1(\mathbb{RP}^\infty)$. The coefficients are the *Steenrod squares*:

$$\delta^* P(x) = (\text{Sq}^n x) + (\text{Sq}^{n-1} x)t + \cdots + (\text{Sq}^0 x)t^n , \quad \text{Sq}^i x \in \overline{H}^{n+i}(X) .$$

Since $\overline{H}^i(K_n) = 0$ for $i < n$, there are no natural transformations that decrease degree: so there are no negatively indexed squares; the sum terminates as indicated.

Any operation on $\overline{H}^*$ induces one on H^* by using the isomorphism

$$H^*(X) = \overline{H}^*(X_+) .$$

Note that $(X_+)^2 = X^2 \sqcup (X_+ \vee X_+)$ so the total square specializes to a natural transformation

$$P : H^n(X) \to H^{2n}(S^\infty \times_{C_2} X^2) .$$

Proposition 75.4. $\text{Sq}^n : H^n \to H^{2n}$ *is the squaring map* $x \mapsto x^2$.

Proof. This is the coefficient of $1 \in H^0(\mathbb{RP}^\infty)$, so we should pick a basepoint for $\mathbb{RP}^\infty$, and watch the evolution of the class Px in the cohomology of the commutative diagram

$$
\begin{array}{ccc}
* \times X & \xrightarrow{\Delta} & S^0 \times_{C_2} X^2 = X^2 \\
\downarrow & & \downarrow \\
\mathbb{RP}^\infty \times X & \xrightarrow{\delta} & S^\infty \times_{C_2} X^2
\end{array}
\qquad
\begin{array}{ccc}
1 \otimes \text{Sq}^n x = x^2 & \longleftarrow & x \otimes x \\
\uparrow & & \uparrow \\
1 \otimes \text{Sq}^n x + \cdots & \longleftarrow & Px .
\end{array}
$$

$\square$

Proposition 75.5. $\text{Sq}^1 = \beta$.

Proof. Acting on H^q for $q \geq 1$, both Sq^1 and β are nonzero. (Exercise: Provide examples.) We claim that $\dim H^{n+1}(K_n) = 1$ for $n \geq 1$, so the two must coincide. Since $K_1 = \mathbb{RP}^\infty$, we know that case. For the inductive step, use the Serre exact sequence on the fibration sequence

$$K_{n-1} \to PK_n \to K_n .$$

$\square$

How about Sq^0? Since $\overline{H}^n(K_n) = \mathbb{F}_2$, there are only two natural transformations $\overline{H}^n \to \overline{H}^n$: the identity and the zero map. The Steenrod operation Sq^0 is one or the other; which is it? In a sense the operations Sq^k get more sophisticated as k decreases; identifying Sq^0 is tricky. In fact there are many other contexts in which Steenrod operations can be defined, and in a sense the topological context is characterized by $\text{Sq}^0 = 1$. We'll study the simplest case first.

Proposition 75.6. $\mathrm{Sq}^0 = 1$ *on* $\overline{H}^1$.

Proof. It suffices to come up with a single example of a space with a nonzero class $x \in \overline{H}^1(X)$ such that $\mathrm{Sq}^0 x = x$. Our example will be S^1 with the generator $x \in \overline{H}^1(S^1)$.

It suffices to look at the subspace of the extended square in which S^∞ is replaced by S^1. Passing to the quotient space of the pair $S^1 \times_{C_2} (S^1 \times S^1, S^1 \vee S^1)$, we arrive at the pointed space

$$\frac{S^1 \times_{C_2} (S^1 \wedge S^1)}{S^1 \times_{C_2} *}$$

in which C_2 exchanges the two factors of S^1. The smash product may be identified with the one-point compactification of $\mathbb{R}^2$, with C_2 acting linearly by permuting the two basis vectors. This representation of C_2 is just $1 \oplus \sigma$, the sum of the trivial 1-dimensional representation with the sign representation.

We have the double cover $S^1 \downarrow \mathbb{RP}^1$. This is a principal C_2-bundle, and the space we are looking at is exactly the Thom space of the vector bundle over $\mathbb{RP}^1$ associated to this principal C_2 bundle and the representation $1 \oplus \sigma$: it is $\mathrm{Th}(\epsilon \oplus \lambda)$ where λ is the tautological line bundle over $\mathbb{RP}^1$. Thus we arrive at

$$\frac{S^1 \times_{C_2} (S^1 \wedge S^1)}{S^1 \times_{C_2} *} = \Sigma \mathbb{RP}^2 .$$

The fiber inclusion into the extended square corresponds under this identification with the fiber inclusion in the Thom space. So the nontrivial class in $H^2(\Sigma \mathbb{RP}^2)$ is the Thom class; it restricts to $x \otimes x$ in the fiber, and hence the Thom class is the total square Px.

The diagonal inclusion

$$\frac{S^1 \times_{C_2} S^1}{S^1 \times_{C_2} *} \to \frac{S^1 \times_{C_2} (S^1 \wedge S^1)}{S^1 \times_{C_2} *}$$

corresponds to including the fixed point subspace into the representation $1 \oplus \sigma$. This produces a bundle map $\epsilon \to \epsilon \oplus \lambda$ covering the inclusion $\mathbb{RP}^1 \hookrightarrow \mathbb{RP}^2$. We obtain a map of Thom spaces

$$\Sigma \overline{\mathbb{RP}}^1_+ \to \Sigma \mathbb{RP}^2$$

that (by naturality of the Thom isomorphism) is an isomorphism in dimension 2. This is generated by the class $t \otimes x$, and we conclude that $\mathrm{Sq}^0 x = x$. $\qquad\square$

The Cartan formula is quite easy to verify as well, but we won't carry that out here. Notice though that it has an important corollary.

Proposition 75.7. *The Steenrod operations are* stable: *For all n and q the diagram*

$$
\begin{array}{ccc}
\overline{H}^q(X) & \xrightarrow{\;\mathrm{Sq}^n\;} & \overline{H}^{q+n}(X) \\
\downarrow{\scriptstyle\sigma} & & \downarrow{\scriptstyle\sigma} \\
\overline{H}^{q+1}(\Sigma X) & \xrightarrow{\;\mathrm{Sq}^n\;} & \overline{H}^{q+n+1}(\Sigma X)
\end{array}
$$

commutes.

Proof. The suspension isomorphism is induced by the relative cross product

$$
\wedge : \overline{H}^1(S^1) \otimes \overline{H}^q(X) \to \overline{H}^{q+1}(\Sigma X) .
$$

The Cartan formula together with the fact that $\mathrm{Sq}^0 = 1$ on $\overline{H}^1$ gives the result. □

Corollary 75.8. Sq^0 *is the identity on* $\overline{H}^q$ *for any q.*

Proof. We just check this on $\iota_q \in H^q(K_q)$. The map $K_1 \times K_{q-1} \to K_q$ representing the cup product sends $\iota_1 \otimes \iota_{q-1}$ to ι_q, and the result then follows by induction and the Cartan formula. □

Corollary 75.9. $\mathrm{Sq}^n : \overline{H}^q \to \overline{H}^{q+n}$ *is additive.*

This is surprising, since the total power operation P is not additive.

Proof. Any stable operation $K_q \to K_{q+n}$ is additive: Being stable means that

$$
\begin{array}{ccc}
K_q & \xrightarrow{\;\mathrm{Sq}^n\;} & K_{q+n} \\
\downarrow{\scriptstyle\simeq} & & \downarrow{\scriptstyle\simeq} \\
\Omega^k K_{q+k} & \xrightarrow{\;\Omega^k \mathrm{Sq}^n\;} & K_{q+k+n}
\end{array}
$$

commutes up to homotopy. The H-space structure of K_q as a loop space is the structure representing the sum in H^q, so $\mathrm{Sq}^n : K_q \to K_{q+n}$ induces a homomorphism in $[X, -]$. □

The *Steenrod algebra A^** is the algebra of cohomology operations generated by the Steenrod operations. This is a noncommutative graded $\mathbb{F}_2$-algebra. It is not a free algebra: the Steenrod operations satisfy relations, starting with $\mathrm{Sq}^1\mathrm{Sq}^1 = 0$. In fact, all relations among them are determined by two facts:

- $\mathrm{Sq}^{2n-1}\mathrm{Sq}^n = 0$ and
- The assignment $\mathrm{Sq}^n \mapsto \mathrm{Sq}^{n-1}$ extends to a derivation on A^*.

An explicit generating family of relations is given by the *Adem relations*

$$\mathrm{Sq}^i\mathrm{Sq}^j = \sum_k \binom{j-k-1}{i-2k}\mathrm{Sq}^{i+j-k}\mathrm{Sq}^k, \quad i < 2j.$$

(José Adem, 1921–1991, was a student of Steenrod and a founding father of algebraic topology in Mexico.) This relation looks quadratic, and almost is, but fails to be whenever the binomial coefficient with $k = 0$ in the summation is nonzero. If n is not a power of 2, let j be the largest power of 2 less than n and let $i = n - j$. Then the binomial coefficient $\binom{j-1}{i}$ is nonzero, so the Adem relation shows that Sq^n is *decomposable*: a sum of products of positive-dimensional elements. From this we learn:

Proposition 75.10 (Adem). *A^* is generated as an algebra by Sq^1, Sq^2, Sq^4, Sq^8, ...*

This leads to information about the "Hopf invariant." Among its many interpretations, the Hopf invariant asks how far the sequence of 3-cell complexes $\mathbb{RP}^2$, $\mathbb{CP}^2$, $\mathbb{HP}^2$, can be extended. The "octonions" $\mathbb{O}$ provide us with one more, $\mathbb{OP}^2$. Adem's theorem puts a first restriction on such spaces:

Corollary 75.11. *Suppose there is a space X such that $H^*(X) = \mathbb{F}_2[x]/x^3$. Then $|x|$ is a power of 2.*

Proof. Let $n = |x|$. Then $\mathrm{Sq}^n x = x^2 \neq 0$. But if n is not a power of 2, this operation factors through groups between dimension n and $2n$. $\qquad\square$

This theorem was improved by Frank Adams to: $|x| = 1, 2, 4$ or 8; there are no examples beyond the classical ones. (John Frank Adams (1930–1989) was a key figure in the development of twentieth century homotopy theory, Lowndean Professor at Cambridge University.)

Exercises

Exercise 75.12. Use the splitting principal to prove Wu's formula for the action of Steenrod operations on Stiefel-Whitney classes:

$$\mathrm{Sq}^i w_j = \sum_k \binom{j + k - i - 1}{k} w_{i-k} w_{j+k}$$

(where $w_0 = 1$).

76 Cobordism

René Thom (1923–2002, IHES) discovered [71] how to use all this machinery to give a classification of closed smooth manifolds, which, while crude, is valid in all dimensions. His equivalence relation was *cobordism* (or "bordism" — opinions vary):

Definition 76.1. Let M and N be two closed smooth n-manifolds. A *cobordism* between them is an $(n + 1)$-manifold-with-boundary W together with a diffeomorphism

$$\partial W \cong M \sqcup N.$$

If there is a cobordism, M and N are said to be "cobordant."

If M and N are diffeomorphic, we may use $W = M \times I$ along with the diffeomorphism at one end to see that they are cobordant. Cobordism is an equivalence relation on the class of closed n-manifolds. Disjoint union endows the set (why "set"?)

$$\mathcal{N}_n = \Omega_n^O$$

of cobordism classes of n-manifolds with the structure of a commutative monoid. In fact it is a vector space over $\mathbb{F}_2$, since the same cylinder can be regarded as a null-bordism of $M \sqcup M$. The product of manifold actually renders the collection of bordism groups a graded commutative algebra. Thom proved:

Theorem 76.2 (Thom). *The unoriented bordism ring is given by*

$$\mathcal{N}_* = \mathbb{F}_2[x_i : i + 1 \text{ is positive and not a power of } 2],$$

where $|x_i| = i$.

We will sketch his proof of this amazing classification theorem over the next few lectures. (Bob Stong's notes [64] provide an excellent secondary source.)

Thom also addressed a question formulated by Norman Steenrod — but this question must have been in Poincaré's mind much earlier. There are two competing notions of an n-cycle: the singular one we have been using (or the equivalent but even more combinatorial version involving simplicial complexes), and the notion of the fundamental cycle of a closed n-manifold. Are they equivalent? Here's Steenrod's formulation of this question. Given an n-dimensional mod 2 homology class x in a space X, is there a closed n-manifold M and a continuous map $f : M \to X$ such that $f_*[M] = x$?

This question has an obvious integral variant as well, in which we demand that the manifold M is oriented.

Theorem 76.3 (Thom [71]). *The answers to these questions are: "Yes" in the unoriented case and "No" in the oriented case.*

The Pontryagin-Thom collapse

A smooth map $f : M \to N$ of manifolds is an *immersion* if it induces a monomorphism on all tangent spaces. One then has an embedding of vector bundles over M, $df : \tau_M \hookrightarrow f^*\tau_N$. The quotient bundle is the *normal bundle* of f, ν_f. If we equip τ_N with a metric, we receive a metric on $f^*\tau_N$ and can identify ν_f with the orthogonal complement of τ_M in $f^*\tau_N$:

$$\tau_M \oplus \nu_f \cong f^*\tau_N \, .$$

Suppose that M is compact. An *embedding* $f : M \to N$ is an injective immersion: an immersion without double points. In that case, the *tubular neighborhood theorem* (see [10, p. 93], for example) asserts that the subspace $f(M) \subseteq N$ admits a "regular" neighborhood that is equipped with a diffeomorphism rel M to the normal bundle ν_f. This regular neighborhood is moreover unique up to diffeomorphism rel M. In view of this identification we will denote the regular neighborhood by $E(\nu)$.

This observation provides a contravariant relationship between M and N: collapse the complement of $E(\nu)$ to a point. This provides a map

$$c : N_+ \to \mathrm{Th}(\nu)$$

from the one-point compactification of N to the Thom space of the normal bundle. This is the *Pontryagin-Thom collapse*. It's a special case of the fact that one-point compactification provides a *contravariant* functor on the category of locally compact Hausdorff spaces and open inclusions.

When $N = \mathbb{R}^{n+k}$, this construction associates to an embedded n-manifold $j : M \hookrightarrow \mathbb{R}^{n+k}$ a map $S^{n+k} \to \mathrm{Th}(\nu_j)$. If we vary the embedding through an isotopy (a smooth homotopy through embeddings) and vary the tubular neighborhood, the resulting maps vary through a homotopy.

Now comes Thom's observation: the normal bundle is classified by a map $M \to BO(k)$, which induces a map on the level of Thom spaces. By composing, we get a map

$$S^{n+k} \to \mathrm{Th}(\nu_j) \to \mathrm{Th}(\xi_k) = MO(k).$$

This provides a map from the set of isotopy classes of embeddings of n-manifolds into $\mathbb{R}^{n+k}$ to the homotopy group $\pi_{n+k}(MO(k))$. Disjoint unions get sent to the sum in the homotopy group. The empty manifold gets sent to zero.

But homotopy corresponds to a still broader equivalence relation on embedded n-manifolds. Given M_0 and M_1, both embedded in $\mathbb{R}^{n+k}$, an *ambient cobordism* between them is a manifold with boundary, W, embedded in $\mathbb{R}^{n+k} \times I$, meeting $\mathbb{R}^{n+k} \times 0$ and $\mathbb{R}^{n+k} \times 1$ transversely in M_0 (along $\mathbb{R}^{n+k} \times 0$) and M_1 (along $\mathbb{R}^{n+k} \times 1$). Isotopies provide cobordisms, but the cobordism could have some more complicated topology as well, and the ends of a cobordism do not have to be even homotopy equivalent. It's not hard to see that cobordisms produce homotopies. Here's the geometric content of Thom's work.

Theorem 76.4 (Thom). *The Pontryagin-Thom collapse map from the set of ambient cobordism classes of closed n-manifolds in $\mathbb{R}^{n+k}$ to the corresponding homotopy class in $\pi_{n+k}(MO(k))$ is bijective.*

For example, $MO(1) = \mathbb{R}P^\infty$, so $\pi_2(MO(1)) = 0$: a union of i circles embedded in $\mathbb{R}^2$ can be written as the boundary of a 2-sphere in $\mathbb{R}^2 \times I$ with i discs removed.

The inverse map is just as interesting. Start with a map

$$f : S^{n+k} \to MO(k).$$

Compress it through an approximation,

$$g : S^{n+k} \to \mathrm{Th}(\xi_{q,k} \downarrow \mathrm{Gr}_k(\mathbb{R}^q)).$$

Approximate this by a nearby (and hence homotopic) map that is smooth on the pre-image of $E(\xi_{q,k})$, and deform it further so that it meets the image

Z of the zero section transversely. Then the implicit function theorem guarantees that the preimage $g^{-1}(Z)$ is a submanifold $M \hookrightarrow S^{n+k}$. The zero section has codimension k in $E(\xi_{q,k})$, so M is an n-manifold.

This construction is pretty clearly inverse to the Pontryagin-Thom collapse. The whole story generalizes to allow structure on the normal bundle: for example an orientation or a complex structure or a trivialization. The key observation is that the normal bundle of the zero section in the Thom space of an appropriate manifold approximation of the relevant universal bundle can be identified with the restriction of the universal bundle and so inherits the same structure. The relevant homotopy groups are then $\pi_{n+k}(MSO(k))$ or $\pi_{n+k}(MU(k/2))$ in the first two cases. Giving a trivialization of a vector bundle is the same thing as giving an isomorphism with the pullback of a bundle over a point, so we can take a point as the corresponding classifying space. The Thom space is a sphere; so in that case the relevant homotopy group is $\pi_{n+k}(S^k)$. This gives a spectacular interpretation of the homotopy groups of spheres. It is the case Pontryagin considered.

Umkher maps

The Pontryagin-Thom collapse gives us a topological way to construct umkehr maps, studied earlier in Lecture 67. Let M and N be smooth manifolds, of dimension m and n, and $f : M \to N$ a smooth map. Embed M into $\mathbb{R}^{m+p}$ for some p, and consequently form an embedding $M \hookrightarrow N \times \mathbb{R}^{m+p}$ of M into a trivial vector bundle over N. Write ν for the normal bundle of this embedding; it's an $(n+p)$-plane bundle over M. The Pontryagin-Thom collapse gives us a map

$$c : \Sigma^{m+p} N_+ \to \mathrm{Th}(\nu).$$

Now an R-orientation of ν determines the Thom isomorphism in the following composite:

$$H^q(M) \cong \overline{H}^{n+p+q}(\mathrm{Th}(\nu)) \xrightarrow{c^*} \overline{H}^{n+p+q}(\Sigma^{m+p} N_+) \cong H^{q+n-m}(N).$$

If $f : M \to N$ is a fiber bundle, this map coincides with the one arising from the Serre spectral sequence.

Stabilization

Now it is definitely interesting to consider embedded manifolds, but perhaps abstract manifolds, without a chosen embedding, are even more interesting, or at least simpler. Whitney proved that any closed manifold embeds in Euclidean space of twice its dimension, and if you allow the ambient space to be of even higher dimension you find that any two embeddings are isotopic. Similarly, in high codimension the cobordisms become unconstrained.

Passing from an embedding in $\mathbb{R}^{n+k}$ to an embedding in $\mathbb{R}^{n+k+1}$ replaces the normal bundle ν with $\nu \oplus \epsilon$. Correspondingly, the map $BO(k) \to BO(k+1)$ classifies $\xi_k \oplus \epsilon$. This gives us maps

$$\Sigma MO(k) \to MO(k+1)$$

for each $k \geq 1$, and hence maps

$$\pi_{n+k}(MO(k)) \to \pi_{n+k+1}(MO(k+1)) \to \pi_{n+k+2}(MO(k+2)) \to \cdots$$

that correspond to considering manifolds embedded in higher and higher dimension. We also get maps in homology,

$$\overline{H}_{n+k}(MO(k)) \to \overline{H}_{n+k+1}(MO(k+1)) \to \overline{H}_{n+k+2}(MO(k+2)) \to \cdots.$$

This is a beautiful and motivating example of a (topological!) *spectrum*: A sequence of pointed spaces E_k together with maps $\Sigma E_k \to E_{k+1}$. The direct limit

$$\pi_n(E) = \lim_{k \to \infty} \pi_{n+k}(E_k)$$

is the *nth homotopy group* of the spectrum E. Similarly we can define the homology of the spectrum E as

$$H_i(E) = \lim_{k \to \infty} \overline{H}_{n+k}(E_k).$$

Spectra are by default "pointed"; there's no "unreduced" homology of a spectrum.

We have already seen a number of other spectra! For example, the *Eilenberg Mac Lane spectrum* HA for the abelian group A has $K(A, n)$ as its nth space, and the map $\Sigma K(A, n) \to K(A, n+1)$ that classifies the suspension of the fundamental class — the adjoint of the equivalence $K(A, n) \to \Omega K(A, n+1)$.

Spectra are the central objects of study in stable homotopy theory. Here's a tiny part of that theory. It is a consequence of the definition of homotopy equivalence for spectra that the following two proposed definitions of the suspension of a spectrum E are equivalent.

- $(\Sigma E)_n = \Sigma E_n$, and the bonding maps are the suspensions of the bonding maps in E;
- $(\Sigma E)_n = E_{n+1}$, and the bonding maps are the same.

So for example $\Sigma H A$ is equivalently given by

$$\Sigma K(A, 0), \Sigma K(A, 1), \ldots \quad \text{and} \quad K(A, 1), K(A, 2), \cdots.$$

The second definition of suspension is clearly a categorical equivalence on the category of spectra.

The spectrum built from Thom spaces as above is the *unoriented Thom spectrum*, and is denoted simply MO. The space $MO(k)$ is $(k-1)$-connected, so the Freudenthal suspension theorem assures us that the direct limit defining $\pi_n(MO)$ is achieved. We also have Thom spectra MSO and MU; the Thom spectrum corresponding to framed manifolds is the *sphere spectrum* S, with nth space S^n.

The ambient cobordism theorem stabilizes to give:

Theorem 76.5 (Thom). *The Pontryagin-Thom construction gives an isomorphism from the group of cobordism classes of closed n-manifolds to $\pi_n(MO)$:*

$$\mathcal{N}_n \xrightarrow{\cong} \pi_n(MO).$$

So Thom's classification theorem amounts to computing the homotopy groups of the Thom spectrum MO.

Characteristic numbers

To compute these homotopy groups we need a way to distinguish cobordism classes from each other: We need a supply of "cobordism invariants." Characteristic classes afford such invariants.

Let M be an n-manifold. Embed it in some Euclidean space, $M \hookrightarrow \mathbb{R}^{n+k}$, and denote the normal bundle of the embedding by ν. Its mod 2 characteristic classes are polynomials in the Stiefel-Whitney classes; there are lots of them. The ones that happen to lie in $H^n(M)$ can be paired against the fundamental class $[M]$. The resulting elements of $\mathbb{F}_2$ are *characteristic numbers*.

Lemma 76.6. *Characteristic numbers are cobordism invariants.*

Proof. We have to show that if $M = \partial N$ then

$$\langle w(\nu), [M] \rangle = 0$$

for any $w \in H^n(BO)$. The class $[M]$ is the boundary of the relative fundamental class $[N, M] \in H^{n+1}(N, M)$, so using the adjointness of the boundary and coboundary maps

$$\langle w(\nu), [M] \rangle = \langle \delta w(\nu), [N, M] \rangle.$$

We claim that $\delta w(\nu) = 0$, and we will show that by exhibiting a class in $H^n(N)$ that restricts to $w(\nu)$. By increasing the codimension if necessary, we can assume that the bounding manifold W embeds in $\mathbb{R}^{n+k} \times [0, \infty)$, meeting $\mathbb{R}^{n+k} \times 0$ transversely in M. So the normal bundle ν extends the normal bundle ν_N of $N \hookrightarrow \mathbb{R}^{n+k} \times [0, \infty)$, and $w(\nu) = w(i^*\nu_N) = i^*w(\nu_N)$ (where $i : M \hookrightarrow N$ is the inclusion of the boundary). $\square$

Putting all the characteristic numbers in play at once, we get the "characteristic number map"

$$\mathcal{N}_n \to \operatorname{Hom}(H^n(BO), \mathbb{F}_2) = H_n(BO).$$

We'll reinterpret this map in terms of the Thom spectrum MO.

Let ξ be a real n-plane bundle over a space B. The cohomology Thom isomorphism relied on the pairing

$$\operatorname{Th}(\xi) \to B_+ \wedge \operatorname{Th}(\xi),$$

and was given by pairing with the Thom class $U \in H^n(\operatorname{Th}(\xi))$. In homology, this pairing produces the top row in

$$\overline{H}_{*+n}(\operatorname{Th}(\xi)) \longrightarrow H_*(B) \otimes \overline{H}_n(\operatorname{Th}(\xi))$$

with diagonal map $\cong$ to $H_*(B)$ and vertical map $1 \otimes <U,->$ to $H_*(B)$.

The vertical map is defined using the Kronecker pairing with the Thom class. The diagonal map is the *homology Thom isomorphism*.

In the universal case we have isomorphisms

$$\overline{H}_{*+n}(MO(n)) \xrightarrow{\cong} H_*(BO(n)).$$

These maps are compatible with stabilization and give the stable Thom isomorphism

$$\Phi : H_*(MO) \xrightarrow{\cong} H_*(BO).$$

These constructions fit together in the commutative diagram:

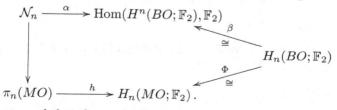

Thom proved that the mod 2 Hurewicz homomorphism h is a monomorphism. As a corollary:

Corollary 76.7. *If the closed n-manifolds M and N have the same Stiefel-Whitney numbers, then they are cobordant.*

This uses algebraic topology to guarantee a very geometric outcome! For example, if all the Stiefel-Whitney numbers vanish then the manifold is *null-bordant*: it is the boundary of some $(n+1)$-manifold-with-boundary.

Thom's basic homotopy-theoretic theorem is this:

Theorem 76.8 (Thom). *The spectrum MO is a product of suspensions of the mod 2 Eilenberg Mac Lane spectrum.*

This implies a positive solution to Steenrod's question. A convenient way to explain this is via an observation of Michael Atiyah [6]. Let X be any space (a "background," in physics parlance), and consider the set of continuous maps from closed n-manifolds into X, modulo the equivalence relation given by cobordism of manifolds together with extension of the maps. This is an abelian group depending covariantly on X,

$$X \mapsto \Omega_n^O(X).$$

Atiyah showed that it is a generalized homology theory. Its "coefficients" are

$$\Omega_n^O(*) = \mathcal{N}_n.$$

There is a natural map, the "Thom reduction,"

$$\Omega_n^O(X) \to H_n(X; \mathbb{F}_2)$$

given by sending $f : M \to X$ to $f_*([M]) \in H_n(X; \mathbb{F}_2)$. Steenrod's question asks whether this map is surjective.

Generalized homology theories are "represented" by spectra. Given a spectrum E and a pointed space Y, one can form the "smash product" spectrum $E \wedge Y$ with

$$(E \wedge Y)_n = E_n \wedge Y$$

and the obvious bonding maps.

Theorem 76.9 (George Whitehead and Edgar Brown, [12,79]; see also [3]). *Given any spectrum E, the functors*

$$E_* : X \mapsto \pi_n(E \wedge X)$$

constitute a reduced homology theory, and any homology theory admits such a representation.

In particular

$$\Omega_n^O(X) = \pi_n(MO \wedge X) \quad \text{and} \quad H_n(X; \mathbb{F}_2) = \pi_n(H\mathbb{F}_2 \wedge X)$$

so the fact that there is a section of the Thom class $U : MO \to H\mathbb{F}_2$ (given by including the bottom factor into the product) implies a positive answer to Steenrod's question.

Exercise

Exercise 76.10. Suppose that a compact smooth R-oriented manifold M embeds into $\mathbb{R}^k$. Show that the Euler class of the normal bundle vanishes.

77 Hopf algebras

Product structure

There is more structure to exploit in our study of the bordism groups. The product of a closed m-manifold M and a closed n-manifold N is a closed $(m + n)$-manifold. This is what gives $\Omega_*^O = \Omega_*^O(*)$ its structure as a commutative graded ring. To pass this through the Pontryagin-Thom collapse, notice that $M \times N$ embeds into the product of ambient Euclidean spaces, and the resulting normal bundle is the product of the two normal bundles. The universal case of a product of m-plane and n-plane bundles is represented by a map

$$BO(m) \times BO(n) \to BO(m + n)$$

which is covered by the bundle map $\xi_m \times \xi_n \to \xi_{m+n}$ and hence induces a map on the level of Thom spaces:

$$MO(m) \wedge MO(n) \to MO(m + n).$$

These maps render MO a "ring spectrum," making $\pi_*(MO)$ a graded ring, and the map

$$\Omega_*^O \to \pi_*(MO)$$

is a ring isomorphism. Equally, $H_*(MO)$ is a graded ring and the Hurewicz map is a ring homomorphism. The homology Thom isomorphism is also multiplicative: The space BO has a commutative H-space structure derived from Whitney sum, and the map $\Phi : H_*(BO) \to H_*(MO)$ is an isomorphism of graded rings.

Hopf algebras

With a field for coefficients, the Künneth theorem delivers for any space X a map

$$\Delta : H_*(X) \to H_*(X \times X) \xleftarrow{\cong} H_*(X) \otimes H_*(X)$$

(all tensors over the coefficient field k) variously termed a "coproduct," "comultiplication," or "diagonal." The unique map $X \to *$ gives us a "counit" $H_*(X) \to k$. This renders $H_*(X)$ a graded coalgebra, as observed in Definition 26.1. We'll say that a graded k-coalgebra is *connected* if it vanishes in negative dimensions and the counit is an isomorphism in dimension 0; so $H_*(X)$ is a connected coalgebra just when X is path connected.

The diagonal in $H_*(X)$ is dual to the cup product: the universal coefficient isomorphism

$$\mathrm{Hom}(H_*(X), k) \cong H^*(X)$$

sends the diagonal to the cup product and ε to the unit map $k \to H^*(X; k)$.

Now, if X is an H-space, the product induces the "Pontryagin product" $\mu : H_*(X) \otimes H_*(X) \to H_*(X)$. Since the product and the basepoint inclusion $* \to X$ are maps of spaces, they are maps of coalgebras. We have to say what the coalgebra structure is on a tensor product of coalgebras, say A and B: define

$$\Delta_{A \otimes B} : A \otimes B \xrightarrow{\Delta \otimes \Delta} (A \otimes A) \otimes (B \otimes B) \xrightarrow{1 \otimes T \otimes 1} (A \otimes B) \otimes (A \otimes B)$$

and

$$\varepsilon_{A \otimes B} : A \otimes B \xrightarrow{\varepsilon \otimes \varepsilon} k \otimes k = k.$$

Then the dual of Proposition 29.3 asserts that

$$H_*(X) \otimes H_*(Y) \to H_*(X \times Y)$$

is a map of coalgebras.

We have described the structure of a *bialgebra*: an associative multiplication with unit and an associative comultiplication with counit on the same (possibly graded) vector space, that are compatible in the sense that the unit and multiplication are coalgebra maps, or, equivalently, that the counit and comultiplication are algebra maps.

If the H-space X has an "inverse" — a map $x \mapsto x^{-1}$ making it into a group in the homotopy category — then $A = H_*(X)$ becomes a *Hopf algebra*: it is a bialgebra, and there is a map $\chi : A \to A$ such that

$$A \xrightarrow{\Delta} A \otimes A \xrightarrow{1 \otimes \chi} H \otimes A \xrightarrow{\mu} A$$

with diagonal maps ε from A to A and η to A.

commutes. This "canonical anti-automorphism" χ exists uniquely if A is a connected graded bialgebra.

The authoritative reference for the theory of Hopf algebras is the paper [45] by Milnor and Moore.

An important and motivating example of an ungraded Hopf algebra is given by the group algebra of a group G: $k[G]$ admits the diagonal determined by $\Delta g = g \otimes g$ for $g \in G$. The anti-automorphism is induced by the map $g \mapsto g^{-1}$. Indeed, a Hopf algebra with commutative diagonal is just a group object in the category of commutative coalgebras.

The k-linear dual of a k-coalgebra is a k-algebra. If a Hopf algebra is of finite type, its dual is again a Hopf algebra. So if X is an H-space of finite type then $H^*(X)$ is also a Hopf algebra; the coproduct comes from the multiplication in X. It's a good exercise to go through our list of H-spaces and understand the Hopf algebra structure on their homology and cohomology. Here's an example, with coefficients in $\mathbb{F}_2$.

Proposition 77.1 (e.g. [69, Theorem 16.17]). *Whitney sum renders BO a commutative H-space, and the map $\mathbb{RP}^\infty = BO(1) \to BO$ sends the vector space generators of $\overline{H}_*(BO(1))$ to polynomial generators a_i:*

$$H_*(BO) = \mathbb{F}_2[a_1, a_2, \ldots].$$

Thus $H_*(BO)$ is "bipolynomial": both homology and cohomology are polynomial algebras.

The diagonal puts strong restrictions on the algebra structure of a Hopf algebra.

Proposition 77.2 (Hopf and Leray; e.g. [45, Theorem 7.5]). *Let A be a connected commutative graded algebra over a field of characteristic zero that admits the structure of a Hopf algebra. Then A is a free commutative graded algebra.*

This means that A is a tensor product of a polynomial algebra on even generators and an exterior algebra on odd generators.

Corollary 77.3 (Hopf). *The rational cohomology of any connected Lie group is an exterior algebra on odd generators.*

Here's an analogue in finite characteristic.

Proposition 77.4 (Borel; e.g. [45, Theorem 7.11]). *Let A be a connected commutative graded algebra of finite type over a perfect field of characteristic p that admits the structure of a Hopf algebra. If p is odd, A is an* *exterior algebra on odd generators tensored with a polynomial algebra on even generators modulo the ideal generated by p^k th powers of some of those generators. If $p = 2$, it is a polynomial algebra modulo 2^k th powers of some generators.*

The Steenrod algebra and its dual

Given two modules M and N over a Hopf algebra A, their tensor product over k has a canonical structure of module over A again:

$$A \otimes M \otimes N \xrightarrow{\Delta \otimes 1 \otimes 1} A \otimes A \otimes (M \otimes N) \xrightarrow{1 \otimes T \otimes 1} (A \otimes M) \otimes (A \otimes N) \xrightarrow{\varphi \otimes \varphi} M \otimes N.$$

When $A = k[G]$, this is the familiar diagonal tensor product of representations.

John Milnor [42] made the observation that the Cartan formula may be formulated in terms of a Hopf algebra structure on the Steenrod algebra itself:

Proposition 77.5. *The association*

$$\Delta : \mathrm{Sq}^k \to \sum_{i+j=k} \mathrm{Sq}^i \otimes \mathrm{Sq}^j$$

extends to an algebra map, and provides the (commutative!) coproduct in a Hopf algebra structure on the Steenrod algebra A^*.

The Cartan formula then merely asserts that the cup product $H^*(X) \otimes H^*(X) \to H^*(X)$ is a map of A^*-modules.

This is pleasant, but much more striking is the insight this gives you into the structure of the Steenrod algebra. Write A_* for the Hopf algebra dual to A^*.

Proposition 77.6 ([42]). *There exist elements $\zeta_i \in A_{2^i-1}$ such that*

$$A_* = \mathbb{F}_2[\zeta_1, \zeta_2, \ldots]$$

and (with $\zeta_0 = 1$)

$$\Delta \zeta_k = \sum_{i+j=k} \zeta_i^{2^j} \otimes \zeta_j.$$

This is equivalent to the Adem relations, but it's much easier to remember!

Lagrange and Thom

Theorem 77.7 ([71]). $H^*(MO)$ *is free as module over the Steenrod algebra* A^*.

Thom gave a fairly elaborate combinatorial proof of this theorem, writing down a basis. It turns out that a little bit of Hopf algebra technology makes this a lot simpler (or at least more believable).

Lemma 77.8 ("Lagrange"; see e.g. [64], p. 94). *Let* A *be a connected Hopf algebra and* C *a connected coalgebra with compatible* A-*module structure (so that the counit and diagonal are* A-*module maps). Let* $u \in C^0$ *be such that* $\varepsilon u = 1$. *If* Au *is free, then* C *is free as* A-*module.*

The reference to Lagrange is this: A common application of this lemma is to take C to be a Hopf algebra containing A as a subalgebra. The result is that C is automatically free as an A-module. This is analogous to an observation attributed to Lagrange: If G is a group and $H < G$ a subgroup then the translation action of H on G is free.

We will apply it with $A = A^*$ and $C = H^*(MO)$. Then $H^0(MO)$ is generated by the Thom class U, so what we have to do is to check that A^* acts freely on the Thom class.

This is proved using the following amazing observation of Thom's:

Proposition 77.9 ([70]). *Let* ξ *be a real* n-*plane bundle over* B, *with Thom space* $\text{Th}(\xi)$. *Then*

$$\text{Sq}^i U = w_i \cup U \in H^{n+i}(\text{Th}(\xi)).$$

This provides a definition of the Stiefel-Whitney classes that only uses the spherical fibration determined by the vector bundle, and indeed one that makes sense for any spherical fibration. It's quite easy to prove that these classes satisfy the axioms.

Conclusion

Stably, cohomology is represented by the Eilenberg Mac Lane spectrum. Pick a basis B for $H^*(MO)$ as an A^*-module. Each element $b \in B$ determines a homotopy class $MO \to \Sigma^{|b|} H\mathbb{F}_2$. Assembling them gives a map

$$MO \to \prod_{b \in B} \Sigma^{|b|} H\mathbb{F}_2$$

that is an isomorphism in mod 2 cohomology. Since the homotopy of MO is all 2-torsion, this map is actually weak equivalence.

The Eilenberg Mac Lane spectrum $H\mathbb{F}_2$ is a commutative ring spectrum as well; the ring structure represents the cup product in cohomology. Its homology is thus a graded commutative algebra, namely the dual of the Steenrod algebra (which is the *cohomology* of $H\mathbb{F}_2$!). We can now estimate the size of $\pi_*(MO)$: Each basis element produces a suspended copy of A_* in $H_*(MO) = \mathbb{F}_2[a_1, a_2, \ldots]$. It looks like the Milnor generators, $\zeta_i \in A_{2^i-1}$ account for some of the a_i's. The rest must come from the homotopy. Some further argumentation leads to the conclusion that

$$\mathcal{N}_* = \pi_*(MO) = \mathbb{F}_2[x_i : i + 1 \text{ is not a power of } 2].$$

Exercises

Exercise 77.10. Prove Lemma 77.8.

Exercise 77.11. Let M be a closed smooth n-manifold. By Poincaré duality, there is for each k a unique class $v_k \in H^k(M)$ such that $\langle v_k x, [M] \rangle = \langle \mathrm{Sq}^k x, [M] \rangle$ for all $x \in H^{n-k}(M)$. These are the *Wu classes* of the manifold. Show that

$$w_k(\tau_M) = \sum_{i+j=k} \mathrm{Sq}^i v_j.$$

The tangential Stiefel-Whitney classes are therefore homotopy invariants of the manifold. Show that the normal Stiefel-Whitney classes are as well, and conclude that if two closed manifolds are homotopy equivalent then they are cobordant.

Exercise 77.12. Verify that the classes defined in Proposition 77.9 satisfy the axioms for Stiefel-Whitney classes.

78 Applications of cobordism

Oriented cobordism

The Pontryagin-Thom collapse/transversality story is very general, and provides for example an isomorphism

$$\Omega_*^{SO} \cong \pi_*(MSO).$$

The oriented bordism groups were computed completely by C.T.C. Wall [74]. All torsion is killed by 2. The first few groups are

n	0	1	2	3	4	5	6	7
Ω_n^{SO}	$\mathbb{Z}$	0	0	0	$\mathbb{Z}$	$\mathbb{Z}/2\mathbb{Z}$	0	0

Wall's computation is involved, but Thom computed $\pi_*(MSO) \otimes \mathbb{Q}$. This is quite easy, by virtue of a general observation.

Proposition 78.1. *For any spectrum E, the rational Hurewicz map*

$$\pi_*(E) \otimes \mathbb{Q} \to H_*(E; \mathbb{Q})$$

is an isomorphism.

There are many ways to see this. For example, up to weak equivalence we may build up a spectrum by attaching cells. Both π_*^s and H_* are generalized homology theories; they send cofiber sequences to long exact sequence. So it's enough to show that the map is an isomorphism for the case of the sphere spectrum, where it follows from Serre's calculation of the rational homotopy of spheres.

So we have the commutative diagram of algebra isomorphisms

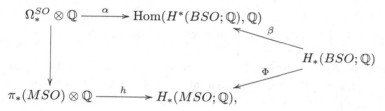

where the top arrow is the characteristic number map sending $[M]$ to $(p \mapsto \langle p(\nu), [M]\rangle)$. This already says something important: The rational Pontryagin numbers of a manifold determine is position in the rational oriented bordism ring. If they all vanish on a manifold M, some multiple of M bounds an oriented manifold-with-boundary.

Again, BSO is a commutative H-space, so $H_*(BSO; \mathbb{Q})$ is a $\mathbb{Q}$-Hopf algebra, and so by the Hopf-Leray theorem (Proposition 77.2) it is a polynomial algebra. Since $H^*(BSO; \mathbb{Q}) = \mathbb{Q}[p_1, p_2, \ldots]$, we find that the homology is also a polynomial algebra on generators of dimension $4k$. An analysis of the characteristic numbers of projective spaces shows that we may take the classes of the even complex projective spaces as the polynomial generators:

$$\Omega_*^{SO} \otimes \mathbb{Q} = \mathbb{Q}[[\mathbb{CP}^2], [\mathbb{CP}^4], \ldots].$$

Steenrod operations on the Thom class

When Thom tried to move beyond this rational calculation, and follow his analysis of the homotopy type of MO, he ran into trouble at odd primes.

There are odd primary Steenrod operations, constructed in the same way as the squares were. (A nice reference for this is [24].) They take the form

$$P^i : H^n(X; \mathbb{F}_p) \to H^{n+2(p-1)i}(X; \mathbb{F}_p).$$

Now $P^0 x = x$, $P^n x = x^p$ if $|x| = 2n$, $P^n x = 0$ if $|x| < 2n$. There is also the Bockstein operation $\beta : H^n(X; \mathbb{F}_p) \to H^{n+1}(X; \mathbb{F}_p)$. These operations generate all the additive operations on mod p cohomology. The dual of A^*, for p odd, has the form [42]

$$A_* = E[\tau_0, \tau_1, \ldots] \otimes \mathbb{F}_p[\xi_1, \xi_2, \ldots], \quad |\tau_i| = 2p^i - 1, \quad |\xi_i| = 2p^i - 2.$$

Now $H^1(BSO) = 0$ (we've killed w_1!), so $H^1(MSO) = 0$ as well; the Thom class $U \in H^0(MSO)$ is killed by the Bockstein. It turns out that at $p = 2$, $\beta = \mathrm{Sq}^1$ generates the annihilator ideal of U. This isn't so bad, since in fact

$$H^*(H\mathbb{Z}; \mathbb{F}_2) = A^*/A^*\mathrm{Sq}^1$$

and indeed $MSO_{(2)}$ splits as a product of Eilenberg Mac Lane spectra (but now not just $H\mathbb{F}_2$'s but also $H\mathbb{Z}_{(2)}$'s).

But at an odd prime the situation is worse; the annihilator of $U \in H^0(MSO; \mathbb{F}_p)$ is the left ideal generated by βP^i for all i. This implies, for example, that βP^1 kills the Thom class of the normal bundle for any embedding of an oriented manifold into Euclidean space. The Thom spectrum MSO does not split as a product of Eilenberg Mac Lane spectra at an odd prime.

Duality

To see how this behavior of Steenrod operations on the Thom class leads to Thom's counterexample to the oriented form of Steenrod's question, we have to explain something about duality in homotopy theory. One of the motivations for the development of the stable homotopy category was a desire to make this story smooth. We will be brief, however.

Any finite complex K may be embedded into some finite dimensional Euclidean space $\mathbb{R}^m$. It can be arranged that the complement has a finite subcomplex L as a deformation retract. Alexander duality (Theorem 38.4) then gives us an isomorphism

$$\alpha : H_q(K) \cong \widetilde{H}^{m-q-1}(L)$$

for any q.

A homotopy-theoretic duality underlies this homological duality: L (or an appropriate desuspension of it in the stable homotopy category) is the "Spanier-Whitehead dual" of K_+. This geometry implies that with mod p coefficients this isomorphism commutes with the action of Steenrod operations. To make sense of this, use the universal coefficient theorem to reexpress homology as the linear dual of cohomology:

$$H_q(K) = H^q(K)^\vee.$$

This imposes a "contragredient" *right* action of A^* on homology, with $\theta \in A^r$ acting in such a way that

$$\langle x, c\theta \rangle = \langle \theta x, c \rangle, \quad x \in H^{q-r}(K), \; c \in H_q(K), \; \theta \in A^r.$$

The isomorphism α demands a *left* action of A^*, which is achieved by acting in homology by $\bar{\theta}$ where $\theta \mapsto \bar{\theta}$ is the Hopf anti-automorphism. The duality isomorphism is compatible with this action; that is, for $c \in H_q(K)$,

$$\theta(\alpha c) = \alpha(c\bar{\theta}).$$

Now suppose that $M \hookrightarrow \mathbb{R}^{n+k}$ is an embedding of a closed smooth n-manifold, with normal bundle ν. Let N be the closure of a regular neighborhood of M; it may be identified with $\mathbb{D}(\nu)$.

The complement $\mathbb{R}^{n+k} - E(\nu)$ is our finite complex L. Here's an important point: we have equivalent cofiber sequences

$$
\begin{array}{ccccc}
\mathbb{R}^{n+k} - E(\nu) & \longrightarrow & \mathbb{R}^{n+k} & \longrightarrow & \mathbb{R}^{n+k}/(\mathbb{R}^{n+k} - E(\nu)) \cong \mathbb{D}(\nu)/\mathbb{S}(\nu) = \mathrm{Th}(\nu) \\
\downarrow \simeq & & \downarrow \simeq & & \downarrow \simeq \\
L & \longrightarrow & CL & \longrightarrow & \Sigma L
\end{array}
$$

so

$$\mathrm{Th}(\nu) \simeq \Sigma L.$$

In short, the Thom space of the normal bundle is (up to suspension) the Spanier-Whitehead dual of M_+. This is *Milnor-Spanier duality*. Michael Atiyah [5] proved a version of this for manifolds-with-boundary and it is often called "Atiyah duality."

The duality isomorphism is thus

$$\alpha : H_q(M) \xrightarrow{\cong} \overline{H}^{n-q+k}(\mathrm{Th}(\nu)).$$

Combining this with the Thom isomorphism gives an isomorphism

$$H_q(M) \xrightarrow{\cong} H^{n-q}(M).$$

This is Poincaré duality! and indeed a proof of it can be given along these lines.

Thom's counterexample

The duality map sends the fundamental class $[M] \in H_n(M)$ to the Thom class $U \in H^k(\mathrm{Th}(\nu))$. Thus if $\theta \in A^r$ annihilates the Thom class, we find that

$$\alpha([M]\overline{\theta}) = \theta(\alpha[M]) = \theta U = 0,$$

so for any $x \in H^{n-r}(M)$

$$0 = \langle x, [M]\overline{\theta} \rangle = \langle \overline{\theta}x, [M] \rangle.$$

The image of $\overline{\theta}$ in $H^n(M)$ annihilates the fundamental class.

Let $f : M \to X$ be any map, and $x \in H^{n-r}(X)$, and compute

$$\langle \overline{\theta}x, f_*[M] \rangle = \langle f^*\overline{\theta}x, [M] \rangle = \langle \overline{\theta}(f^*x), [M] \rangle = 0.$$

So in order for a class in $H_n(X)$ to be carried by an oriented n-manifold the image of $\overline{\theta} : H^{n-r}(X) \to H^n(X)$ must annihilate it.

For a specific example, Thom looked at $K_1 = K(\mathbb{Z}/3\mathbb{Z}, 1)$. This is an infinite "lens space." The cohomology is

$$H^*(K_1; \mathbb{F}_3) = E[e] \otimes \mathbb{F}_3[x], \quad |e| = 1, \, |x| = 2.$$

The Steenrod action is determined by

$$\beta e = x, \quad P^1 x = x^3.$$

The anti-automorphism is easily seen to send both β and P^1 to their negatives, so

$$\overline{\beta P^1} = P^1 \beta.$$

The class $x^3 \in H^6(K_1; \mathbb{F}_3)$ is in the image of this class, so the dual homology class cannot be carried by an oriented closed manifold.

This is a counter-example to Steenrod's question in mod p homology; how about integrally? The Bocksteins tell us that $\overline{H}_*(K_1; \mathbb{Z})$ is unfortunately concentrated in odd degrees, while $P^1\beta H^*(K_1; \mathbb{F}_3) = 0$ in odd degrees. So Thom moves up a dimension to $K_2 = K(\mathbb{Z}/3\mathbb{Z}, 2)$. It's known, and not hard to verify by pulling back under the map $K_1 \times K_1 \to K_2$ classifying the cup product, that $\beta P^1 \beta \iota_2 \neq 0$. In homology, then, there is a class $c \in H_8(K_2; \mathbb{F}_3)$ such that $c\beta P^1\beta \neq 0$ in $H_2(K_2; \mathbb{F}_3)$. The class $c\beta \in H_7(K_2; \mathbb{F}_3)$ can't be carried by an oriented manifold since

$$\langle P^1\beta\iota, c\beta \rangle = \langle \beta P^1(\beta\iota), c \rangle \neq 0.$$

But the Bockstein factors as

$$H_8(K_2; \mathbb{F}_3) \xrightarrow{\partial} H_7(K_2; \mathbb{Z}) \xrightarrow{\rho} H_7(K_2, \mathbb{F}_3),$$

so $\partial c \in H_7(K_2; \mathbb{Z})$ can't be carried by a manifold since its reduction $\beta c \in H_7(K_2; \mathbb{F}_3)$ can't be.

The Postnikov system for MSO provides further obstructions.

The Brown-Peterson spectrum

The annihilator ideal of $U \in H^0(MSO)$ at an odd prime is the two-sided ideal generated by the Bockstein. The quotient by this ideal turns out to be the cohomology of a ring spectrum — not an Eilenberg Mac Lane spectrum, but rather a new gadget called the "Brown-Peterson spectrum" and denoted (without reference to the prime p) by BP. (Frank Peterson, 1930–2000, was an MIT faculty member and long-time treasurer of the AMS.) At odd primes, MSO splits into a product of suspensions of BP. The mod p Thom class restricts to a map $BP \to H\mathbb{F}_p$ that induces an embedding of $H_*(BP) \hookrightarrow A_*$ as the polynomial algebra on the ξ's.

The homotopy type of MU was studied by Milnor using the Adams spectral sequence. It turns out that

$$\pi_*(MU) = \mathbb{Z}[x_1, x_2, \ldots], \quad |x_i| = 2i,$$

and that MU localized at any prime p splits as a product of the p-local Brown-Peterson spectrum as well (even if $p = 2$). The homotopy of BP is also a polynomial algebra, but now much sparser:

$$\pi_*(BP) = \mathbb{Z}_{(p)}[v_1, v_2, \ldots], \quad |v_i| = 2p^i - 2.$$

Surgery

There is a simple way to modify a manifold to give a new manifold with different topology but related by a cobordism. The most classical example of surgery occurs in dimension 2. Start with an embedded loop L in a closed surface M. Assume that the normal bundle of L is framed (always the case if M is orientable), so that we have an embedding of $S^1 \times D^1$ into M. This kind of product is familiar! In general

$$\partial(D^p \times D^q) = S^{p-1} \times D^q \cup_{S^{p-1} \times S^{q-1}} D^p \times S^{q-1}.$$

In our case $p = 2$ and $q = 1$. We can remove the interior of $\partial D^2 \times D^1$ and replace it with the interior of $D^2 \times \partial D^1 = D^2 \times S^0$, to get a new manifold M'. If the regular neighborhood of the loop was a belt around a waste (or "handle"), this has the effect of removing the belt and capping off the two body parts. This process is called "surgery."

What's a little harder to see is that $D^p \times D^q$ can be used to construct a cobordism between M and M'.

A proof using Morse theory [44] shows that any two closed smooth n manifolds in the same bordism class can be connected by a bordism constructed by a series of surgeries.

The surgery operation, pioneered by Milnor and Wallace and later Browder, Novikov, and Wall, led to an enormous research program aimed at the classification of manifolds up to diffeomorphism.

Remark 78.2. The surgery process involves killing homology groups in a manifold. It requires establishing that (1) the class is spherical — in the image of the Hurewicz map; (2) the map from a sphere is a smooth embedding; and (3) the normal bundle of this embedded sphere is trivial.

Typically the first requirement is met using the Hurewicz theorem; we try to kill bottom dimensional homology. The second can be achieved by Whitney embedding theorem as long as we are below the middle dimension of the manifold. The third is much more problematic. One way to ensure that the process can continue above dimension one is to work with framed bordism. The Pontryagin-Thom theorem identifies this with stable homotopy, so there is considerable interest in this case. The surgery process then works to find a "highly connected" representative of a framed bordism class in which the homology is concentrated in the middle dimension. When n is odd, any class in Ω_n^{fr} has is represented by a homotopy sphere, since there is then no middle dimension. The same turns out to be true when $n = 4k$. When $n = 4k + 2$, there is a potential obstruction, the *Kervaire invariant*, with values in C_2. It's already visible in dimension 2, when the square of the nontrivially framed circle (which represents the stable homotopy class η of the Hopf map $S^3 \to S^2$) is not framed null-bordant (since in fact $\eta^2 \neq 0$). The higher dimensional Hopf fibrations give other examples in dimensions 6 and 14. Bill Browder proved that the invariant could be nonzero only in dimensions of the form $2^j - 2$, and identified the invariant in terms of the Adams spectral sequence. In the 1970's examples were constructed using homotopy theory in dimensions 30 and 62, and in 2015 work of Mike Hill, Mike Hopkins, and Doug Ravenel finally showed that the invariant is in fact trivial for dimensions larger than 126 (where it remains unknown today).

Signature

This ability to move around within a cobordism class suggests that there are very few bordism invariants that one an derive from cohomology. What homological features of a manifold are cobordism invariants?

When M is an oriented $4k$-manifold, $H^{2k}(M; \mathbb{Q})$ supports a symmetric bilinear form, the "intersection form"

$$x \cdot y = \langle xy, [M] \rangle$$

which is nondegenerate on account of Poincaré duality. A fact from linear algebra: Any symmetric bilinear form over $\mathbb{Q}$ is diagonal with respect to some basis. If it is nondegenerate then all the diagonal entries in the diagonalization are nonzero, and the difference between the number of positive entries and the number of negative entries is a independent of the diagonalizing basis. It is the *signature* of the bilinear form.

Lemma 78.3 (Thom). *The signature of the intersection form of an oriented 4k-manifold is a multiplicative oriented bordism invariant.*

This follows from Lefschetz duality 37.7 and the Künneth theorem. The result is a graded ring homomorphism

$$\sigma : \Omega_*^{SO} \to \mathbb{Z}[u] , \quad |u| = 4 .$$

Such a ring homomorphism is a *genus*. (This term entered mathematics from biology through Gauss's work on quadratic forms, and then spread to the genus of a surface, and then to other numerical invariants of manifolds.) Since the characteristic number map is a rational isomorphism, the value of a rational genus on a $4k$-manifold M is some Pontryagin number.

The even complex projective spaces generate Ω_*^{SO} rationally, so giving the value of a genus on them completely specifies the value of the genus on any oriented manifold. Since $\mathbb{CP}^{2k}$ obviously has signature 1 for any k, the signature is in a sense the simplest genus. For each k there is a polynomial

$$L_k \in H^{4k}(BSO; \mathbb{Q})$$

in the Pontryagin classes such that for any closed oriented $4k$-manifold M

$$\sigma(M) = \langle L_k(\tau_M), [M] \rangle .$$

This is the "Hirzebruch signature theorem." Identifying these polynomials is a beautiful story [25]. The results are for example that

$$L_1 = \frac{p_1}{3} , \quad L_2 = \frac{7p_2 - p_1^2}{45} , \quad L_3 = \frac{62p_3 - 13p_2p_1 + 2p_1^3}{945} , \dots .$$

These formulas put divisibility conditions on certain combinations of Pontryagin classes of the tangent bundle of an oriented closed smooth manifold: while the L-class has denominators, you get an integral class when you pair it against the fundamental class. The first tangential Pontryagin class of an orientable 4-manifold has to be divisible by 3, for example. This was observed even earlier by Rohlin.

The signature theorem in dimension 8 played a key role in Milnor's proof [40] that certain S^3-bundles over S^4 are not diffeomorphic to the standard 7-sphere despite being homeomorphic to it.

Exercises

Exercise 78.4. Show that any positive dimensional oriented bordism class contains a connected manifold. Show that any oriented cobordism class of dimension at least 2 contains a simply connected manifold. Display counterexamples to these to statements in lower dimensions.

Exercise 78.5. Identify representatives of the four elements of $\mathcal{N}_4$.

Exercise 78.6. Verify Thom's Lemma 78.3.

Exercise 78.7. Using Exercise 74.10, verify the signature theorem for $\mathbb{CP}^2$, $\mathbb{CP}^4$, and $\mathbb{CP}^6$.

Bibliography

[1] J. Frank Adams, On the non-existence of elements of Hopf invariant one, Annals of Mathematics **72** (1960) 20–104.

[2] J. Frank Adams, Vector fields on spheres, Annals of Mathematics **75** (1962) 603–632.

[3] J. Frank Adams, *Stable Homotopy and Generalized Homology*, Chicago Lectures in Mathematics, 1974.

[4] Martin Arkowicz, *Introduction to Homotopy Theory*, Springer, 2011.

[5] Michael Atiyah, Thom complexes, Proceedings of the London Philosophical Society **11** (1961) 291–310.

[6] Michael Atiyah, Bordism and cobordism, Proceedings of the Cambridge Philosophical Society **57** (1961) 200–208.

[7] Michael Barratt and John Milnor, An example of anomalous singular homology, Proceedings of the American Mathematical Society **13** (1962) 293–297.

[8] Armand Borel, La cohomologie mod 2 de certains espaces homogènes, Commentarii mathematici Helvetici **27** (1953) 165–197.

[9] Aldridge Knight Bousfield, The localization of spaces with respect to homology, Topology **14** (1975) 133–150.

[10] Glen Bredon, *Topology and Geometry*, Graduate Texts in Mathematics **139**, Springer, 1993.

[11] Glen Bredon, *Sheaf Theory*, Graduate Texts in Mathematics **170**, Springer, 1997.

[12] Edgar Brown, Cohomology Theories, Annals of Mathematics **75** (1962) 467–484.

[13] Henri Cartan and Samuel Eilenberg, *Homological Algebra*, Princeton University Press, 1956.

[14] James Davis and Paul Kirk, *Lecture Notes in Algebraic Topology*, American Mathematical Society, 2001.

[15] Albrecht Dold, *Lectures on Algebraic Topology*, Springer, 1995.

[16] Andreas Dress, Zur Spektralsequenz einer Faserung, Inventiones Mathematicae **3** (1967) 172–178.

[17] Björn Dundas, *A Short Course in Differential Topology*, Cambridge University Press, 2018.

[18] Samuel Eilenberg and Norman Steenrod, *Foundations of Algebraic Topology*, Princeton University Press, 1952.

[19] Samuel Eilenberg and John Moore, Homology and fibrations, I: Coalgebras, cotensor product and its derived functors, Commentarii Mathematici Helvitici **40** (1965) 199–236.

[20] Rudolph Fritsch and Renzo Piccinini, *Cellular Structures in Topology*, Cambridge University Press, 1990.

[21] Martin Frankland, Math 527 - Homotopy Theory: Additional notes, `http://uregina.ca/~franklam/Math527/Math527_0204.pdf`

[22] Roger Godement, *Théorie des faisceau*, Publications de l'Institute de Mathématique de l'Université de Strasbourg **XIII**, Hermann, 1964.

[23] Paul Goerss and Rick Jardine, *Simplicial Homotopy Theory*, Progress in Mathematics **174**, Springer, 1999.

[24] Alan Hatcher, *Algebraic Topology*, Cambridge University Press, 2002.

[25] Fritz Hirzebruch, *Topological Methods in Algebraic Geometry*, Springer, 1966 and reprints.

[26] John Hocking and Gail Young, *Topology*, Addison-Wesley, 1961.

[27] Dale Husemoller, *Fiber Bundles*, Graduate Texts in Mathematics **20**, Springer, 1993.

[28] Sören Illman, The equivariant triangulation theorem for actions of compact Lie groups. Mathematische Annalen **262** (1983) 487–501.

[29] Niles Johnson, Hopf fibration – fibers and base, `https://www.youtube.com/watch?v=AKotMPGFJYk`.

[30] Dan Kan, Adjoint functors, Transactions of the American Mathematical Society **87** (1958) 294–329.

[31] Anthony Knapp, *Lie Groups Beyond an Introduction*, Progress in Mathematics **140**, Birkaüser, 2002.

[32] Wolfgang Lück, Survey on classifying spaces for families of subgroups, *Infinite groups: geometric, combinatorial and dynamical aspects*, Progress in Mathematics **248** (2005) 269–322.

[33] Saunders Mac Lane, *Homology*, Springer, 1967.

[34] Saunders Mac Lane, *Categories for the Working Mathematician*, Graduate Texts in Mathematics **5**, Springer, 1971 and 1998.

[35] William Massey, *Algebraic Topology: An Introduction*, Graduate Texts in Mathematics **56**, Springer, 1977.

[36] Jon Peter May, *A Consise Course in Algebraic Topology*, University of Chicago Press, 1999.

[37] Haynes Miller, Leray in Oflag XVIIA: The origins of sheaf theory, sheaf cohomology, and spectral sequences, *Jean Leray (1906–1998)*, Gazette des Mathematiciens **84** suppl (2000) 17–34, `http://math.mit.edu/~hrm/papers/ss.pdf`

[38] Haynes Miller, George William Whitehead, Jr. (1918–2004), Biographical Memoirs of the National Academy of Sciences, 2015, `http://math.mit.edu/~hrm/papers/whitehead-note.pdf`

[39] Haynes Miller and Douglas Ravenel, Mark Mahowald's work on the homotopy groups of spheres, *Algebraic Topology, Oaxtepec 1991*, Contemporary Mathematics **146** (1993) 1–30.

[40] John Milnor, On manifolds homeomorphic to the 7-sphere, Annals of Mathematics **64** (1956) 399–405.

[41] John Milnor, The geometric realization of a semi-simplicial complex, Annals of Mathematics **65** (1957) 357–362.

[42] John Milnor, The Steenrod algebra and its dual, Annals of Mathematics **67** (1958) 150–171.

[43] John Milnor, On axiomatic homology theory, Pacific Journal of Mathematics **12** (1962) 337–341.

[44] John Milnor, A procedure for killing homotopy groups of differentiable manifolds, Proceedings of Symposia in Pure Mathematics, **III** (1961) 39–55.

[45] John Milnor and John Moore, Hopf Algebras, Annals of Mathematics **81** (1965) 211–264.

[46] John Milnor and Jim Stasheff, *Characteristic Classes*, Annals of Mathematics Studies **76**, 1974.

[47] Steve Mitchell, Notes on principal bundles and classifying spaces.

[48] Steve Mitchell, Notes on Serre Fibrations.

[49] John Moore, On the homotopy groups of spaces with a single non-vanishing homology group, Annals of Mathematics **59** (1954) 549–557.

[50] James Munkres, *Elements of Algebraic Topology*, Addison-Wesley, 1984.

[51] James Munkres, *Topology*, Prentice-Hall, 2000.

[52] nLab, *Relation between type theory and category theory*, https://ncatlab. org/nlab/show/relation+between+type+theory+and+category+theory.

[53] Maximilien Peyroux, The Serre spectral sequence, http://homepages.math. uic.edu/~mholmb2/serre.pdf.

[54] Daniel Quillen, *Homotopical Algebra*, Springer Lecture Notes in Mathematics **43**, 1967.

[55] Daniel Quillen, The spectrum of an equivariant cohomology ring, Annals of Mathematics **94** (1971) 549–572.

[56] Melvin Rothenberg and Norman Steenrod, The cohomology of classifying spaces of H-spaces, Bulletin of the American Mathematical Society **71** (1965) 872–875.

[57] Marian Schmitt, *Hommes de Science: 28 portraits*, Hermann, 1990.

[58] Graeme Segal, Classifying spaces and spectral sequences, Publications mathématiques de l'IHES **34** (1968) 105–112.

[59] Paul Selick, *Introduction to Homotopy Theory*, American Mathematical Society, 1997.

[60] Jean-Pierre Serre, Homologie singulière des espaces fibrés. Applications, Annals of Mathematics **54** (1951) 425–505.

[61] Jean-Pierre Serre, Cohomologie modulo 2 des complexes d'Eilenberg-Maclane, Commentarii Mathematici Helvetici **27** (1953) 198–232.

[62] William Singer, Steenrod squares in spectral sequences. I, II, Transactions of the American Mathematical Society **175** (1973) 327–336 and 337–353.

[63] Edwin Spanier, *Algebraic Topology*, McGraw Hill, 1966, and later reprints.

[64] Robert Stong, *Notes on Cobordism Theory*, Princeton University Press, 1968.

[65] Neil Strickland, The category of CGWH spaces, `http://www.neil-strickland.staff.shef.ac.uk/courses/homotopy/cgwh.pdf`.

[66] Neil Strickland, An abelian embedding for Moore spectra, `http://de.arxiv.org/pdf/1205.2247.pdf`.

[67] Arne Strøm, A note on cofibrations, Mathematica Scandinavica **19** (1966) 11–14.

[68] Arne Strøm, The homotopy category is a homotopy category, Archiv der Mathematik **23** (1972) 435–441.

[69] Robert Switzer, *Algebraic Topology – Homotopy and Homology*, Springer, 1975.

[70] René Thom, Espaces fibrés en sphères et carrés de Steenrod, Annales Scientifique de l'École Normale Supérieure **69** (1952) 109–182.

[71] René Thom, Quelques propriétés globales des variétés différentiables, Commentarii Mathematici Helvitici **28** (1954) 17–86.

[72] Tammo tom Dieck, *Algebraic Topology*, European Mathematical Society, 2008.

[73] Hiroshi Uehara, On a Hopf homotopy classification theorem, Nagoya Mathematics Journal **3** (1951) 49–54.

[74] Charles Terrence Clegg Wall, Determination of the cobordism ring, Annals of Mathematics **72** (1960) 292–311.

[75] Charles Terrence Clegg Wall, Finiteness conditions for CW-complexes, Annals of Mathematics **81** (1965) 56–69.

[76] Stefan Waner, Equivariant homotopy theory and Milnor's theorem, Transactions of the American Mathematical Society **258** (1980) 351–368.

[77] Charles Weibel, *An Introduction to Homological Algebra*, Cambridge Studies in Advanced Mathematics **38**, 1994.

[78] James West, Mapping Hilbert cube manifolds to ANR's: a solution of a conjecture of Borsuk, Annals of Mathematics **106** (1977) 1–18.

[79] George Whitehead, Cohomology Theories, Transactions of the American Mathematical Society **102** (1962) 227–283.

[80] George Whitehead, *Elements of Homotopy Theory*, Graduate Texts in Mathematics **61**, Springer, 1978.

Index

Printed in the United States
by Baker & Taylor Publisher Services

Preface

APRES 2015 was the second edition of the Asia-Pacific Requirements Engineering Symposium (APRES 2015) that serves as a highly interactive forum for in-depth discussion of all issues related to requirements engineering following the success of APRES 2014 in New Zealand. The symposium took place on the beautiful campus of Wuhan University, in the historical city of Wuhan, capital of Hubei Province, central China, with the Yangzi River running across the city landscape.

Requirements engineering (RE) as an established discipline of research and practice in software and systems development has established an RE community worldwide. The importance of developing and following effective RE practices has long been recognized by researchers and practitioners alike, especially when facing the latest advancements in socio-technological needs in various industrial sectors, such as e-commerce, manufacturing, health care, etc. This year, the discussions concentrated on "RE in the Big Data Era". According to the growing interest in RE research and practice, the APRES symposium series aims to develop and expand the RE research and practice community specifically in the Asia-Pacific region and to foster collaborations among local researchers and practitioners in Asia and Oceania.

This year the symposium received 18 submissions from active researchers from all over the world, among which nine full papers, one short paper, and three tool demos were accepted. All papers were carefully reviewed by at least three Program Committee members, and detailed constructive feedback was provided to the authors. We had participants and presentations from Asia, Oceania, Europe, and North America on both traditional RE topics, such as requirements acquisition, modeling and analysis, as well as new topics such as crowdsourcing, extreme RE, and creativity.

APRES 2015 had a focused program including two keynote speeches covering RE research and big data challenges. There were four paper presentation sessions, including RE tool demos and case studies. A plenary panel was arranged at the end of the conference to discuss future steps for RE improvement. APRES 2015 also hosted a dedicated industrial RE Day, which attracted a very good number of participants and enriched the overall offering of the conference venues for the discussion of problems and sharing of good practices in RE, especially so as to foster RE research–industrial collaborations in the Asia-Pacific region.

Our greatest thanks go to the authors and presenters, whose contributions made APRES 2015 a success. We are grateful to the Program Committee members for their thorough and timely reviews on the submissions. We thank the Steering Committee for their valuable guidance. We would like to thank Yong Xia for organizing and managing the industrial track for APRES 2015. We thank the sponsors of APRES 2015, Wuhan University and International Requirements Engineering Board (IREB). Our thanks also go to Springer, the publisher of the APRES proceedings, for the continuous support. Finally, thanks to EasyChair for making conference managements such a straightforward task.

We hope you all enjoy the APRES 2015 proceedings.

August 2015

Lin Liu
Mikio Aoyama

Organization

Steering Committee

Didar Zowghi	University of Technology, Sydney, Australia
Zhi Jin	Peking University, China
Jim Buchuan	Auckland University of Technology, New Zealand

General Chair

Rong Peng	Wuhan University, China

Program Chairs

Lin Liu	Tsinghua University, China
Mikio Aoyama	Nanzan University, Japan

Program Committee

Muneera Bano	University of Technology, Sydney, Australia
Tony Clear	Auckland University of Technology, New Zealand
Xiaohong Chen	Eastern China Normal University, China
Chad Coulin	Barrick Gold, Papua New Guinea
Zuohua Ding	Zhejiang University of Science and Technology, China
Smita Ghaisa	Tata Research Design and Development Center, India
Naveed Ikram	Riphah International University, Pakistan
Massila Kamalrudin	Universiti Teknikal Malaysia Melaka, Malaysia
Seok-Won Lee	Ajou University, South Korea
Tong Li	Yunnan University, China
Xiaohong Li	Tianjin University, China
Zhi Li	Guangxi Normal University, China
Peng Liang	Wuhan University, China
Huaxiao Liu	Jilin University, China
Xiaodong Liu	University of Edinburg Napier, UK
Stuart Marshall	Victoria University of Wellington, New Zealand
Xin Peng	Fudan University, China
Shahida Sulaiman	University Teknologi Malaysia, Malaysia
Hongji Yang	Bath SPA University, UK
Gang Yin	National University of Defence Technology, China
Yijian Wu	Fudan University, China

Requirements Engineering Tools

Contents

Urban Computing: Using Big Data to Solve Urban Challenges

Yu Zheng

Microsoft Research, China

Abstract. Urban computing is a process of acquisition, integration, and analysis of big and heterogeneous data generated by a diversity of sources in cities to tackle urban challenges, e.g. air pollution, energy consumption and traffic congestion. Urban computing connects unobtrusive and ubiquitous sensing technologies, advanced data management and analytics models, and novel visualization methods, to create win-win-win solutions that improve urban environment, human life quality, and city operation systems. Urban computing is an inter-disciplinary field where computer science meets urban planning, transportation, economy, the environment, sociology, and energy, etc., in the context of urban spaces. In this talk, I will present our recent progress in urban computing, introducing the key applications and technologies for integrating and deep mining big data. Examples include fine-grained air quality inference throughout a city, city-wide estimation of gas consumption and vehicle emissions, and diagnosing urban noises with big data. The research has been published at prestigious conferences (such as KDD and UbiComp) and deployed in the real world. More details can be found on http://research.microsoft.com/en-us/projects/urbancomputing/default.aspx

Requirements Engineering in the Age of Disruptive Digital Transformation

Eric Yu

University of Toronto, Canada

Abstract. Recent emerging technologies such as social, mobile, cloud and particularly data analytics are rapidly transforming how organizations operate, leading them to rethink their business models and strategic positioning. What requirements engineering techniques are needed in this age of rapid and frequently disruptive change? Goal modeling with strategic actor relationships, such as in i*, could be a starting point. In this talk, I will outline some tentative steps towards a modeling framework.

Keynotes